FESTKÖRPERPROBLEME

ADVANCES IN SOLID STATE PHYSICS 27

FESTKÖRPER PROBLEME
ADVANCES IN SOLID STATE PHYSICS 27

Plenary Lectures of the Divisions
"Semiconductor Physics"
"Thin Films"
"Dynamics and Statistical Mechanics"
"Magnetism"
"Metal Physics"
"Low Temperature Physics"
of the German Physical Society (DPG)

Münster, March 9 ... 13, 1987

Edited by
P. Grosse, Aachen

With 208 Figures

Springer Fachmedien Wiesbaden GmbH

ISSN   04030-3393

© Springer Fachmedien Wiesbaden 1987
Originally published by Friedr. Vieweg & Sohn Verlagsgesellschaft mbH, Braunschveig, in 1987.

Set by Vieweg, Braunschweig
Printed by Lengericher Handelsdruckerei, Lengerich
Bound by W. Langelüddecke, Braunschweig
Cover design: Barbara Seebohm, Braunschweig

ISBN 978-3-662-16087-9   ISBN 978-3-540-75356-8 (eBook)
DOI 10.1007/978-3-540-75356-8

Foreword

In 1987 the spring meeting of the „Arbeitskreis Festkörperphysik" of the Deutsche Physikalische Gesellschaft (DPG) was held in Münster. It was organized together with the Nederlande Natuurkundige Vereniging (NNV) and the Österreichische Physikalische Gesellschaft (ÖPG).

At the conference about 900 invited and contributed papers have been presented covering a wide range of topics from fundamental physics to technical application.

Volume 27 of the „Festkörperprobleme" contains a selection of invited talks hold at the conference. In choosing the authors, the editor has avoided double publication in case the authors in question recently published a similar review on the same subject. In some other cases, however, the editor has been interested to get a manuscript, but the invited authors have not been able to write it due to some other occupations. That's a pity! The lot of many editors!

The first contribution of B. Ewen and D. Richter deals with the analysis of the dynamics of macromolecules in solutions and melts. This work was honoured by the Walter-Schottky-Prize 1987. The results are a fascinating example of the power of the highly resolving spin-echo neutron spectroscopy.

The following paper reports on demonstrations of temporal and spatial structures in non-linear systems. H. G. Purwins gave pleasure to an audience of about 900 people with his excellent talk and the well organized experiments in Münster.

D. Vollhardt's contribution gives an introduction to the phenomenon of localization and the related theoretical models, and E. Wassermann reports on a modern approach to explain the properties of the Invar alloys. Two papers are dedicated to new experimental results obtained at the 2-dimensional electron gas observed in heterostructures (U. Merkt) and in superlattices (J. K. Maan). In two other papers, the structure and reactivity of metal surfaces (G. Ertl) and of silicon surfaces (M. Henzler) are considered.

The last five contributions concern topics more or less related to applications: P. Kocevar discusses, on the basis of theoretical models, the energy transfer from a highly laser excited electron gas to the phonon system leading to "hot phonons", and H. R. Zeller presents the actual knowledge of the electrical breakdown in dielectrics.

The contribution of W. Uelhoff deals with crystal growth by means of the Czochralski-method, one of the most important steps in microelectronics. E. Kasper and K. Kempter show the chances of unconventional semiconductor materials for the production of devices: silicon-germanium heterostructures and amorphous silicon.

The authors as well as the editor and the publisher hope that volume 27 of the „Festkörperprobleme" will again help interested physicists to make approaches to some actual problems of modern physics of condensed matter and to learn something about the state of the art.

Finally, I thank my secretary, my colleagues in Aachen and the chairmen of the „Fachausschüsse" for their support in preparing the scientific programme of the conference, and Mr W. Große-Nobis and the local organizing committee for their excellent work.

As the editor of this volume, I thank all the authors and the publisher for their efficient collaboration. In particular, I have the pleasure to thank Mr. E. Gerlach for his critical reading of the manuscripts during the conference and B. Gondesen from Vieweg-Verlag for his congenial cooperation. Mostly, however, I thank Mr. J. Brunn. As in previous years, he assisted me again as a lector and a very personal advisor during my editorship.

This is the last volume of a series of six volumes I have edited. It has been an interesting work, and I have enjoyed very much the interaction with my fellow-physicists and with the publisher. The next volume will be edited by U. Rößler, Regensburg, who is also my successor to the chair of the „Fachausschuß Halbleiterphysik". I wish him all the best for his future activities and the same cordial support of all the semiconductor physicists as I have found throughout the last years!

„Auf Wiedersehen!"

Aachen, May 1987   *Peter Grosse*

Contents

Neutron Spin Echo Studies on the Segmental Dynamics of Macromolecules

Bernd Ewen

Max-Planck-Institut für Polymerforschung, D-6500 Mainz, Federal Republic of Germany

Dieter Richter

Institut Laue-Langevin, F-38042 Grenoble, France

Summary: The ultrahigh resolution neutron spin echo (NSE) technique provides the unique opportunity to investigate internal dynamic relaxation processes of polymer molecules simultaneously in spcae and time. In particular, information on the single chain behaviour is not restricted to dilute solutions but may also be obtained, if an appropriate mixture of protonated and deuterated polymers is used, from concentrated systems, the melt included. This paper gives a review on recent NSE studies on liquid polydimethylsiloxane (PDMS) systems. Single chain relaxation as well as collective fluctuations are considered in the whole range of concentrations. Thereby especially the cross over phenomena between different dynamic regimes are emphasized. The results are discussed in terms of microscopic models and compared with the theoretical predictions concerning universal and critical behaviour.

1 Introduction

Macromolecules are chemical compounds build-up from a large number of covalent bonded repeat units (monomers) [1]. In general such chain or polymer molecules may assume an enormous array of spatial configurations, which is the consequence of the rotational freedom around the single bonds of the backbone atoms [2]. In solution as well as in the melt and the glassy solid state the overall shape of the individual polymer molecules is coil like. On length scales, where the details of the chemical structure of the monomer units become of secondary importance and the polymer molecules can be described by a sequence of freely jointed statistical segments, certain molecular properties exhibit universal behaviour [3]. For example, the mean square end to end distance $\langle R_0^2 \rangle$ and the mean square radius of gyration $\langle R_g^2 \rangle$ scale with $M^{2\nu}$, where M is the molar mass and ν the Flory exponent [2]. In the melt and under Θ-conditions in solution, where the repulsive forces between different segments are totally balanced by the attractive van der Waals forces and no excluded volume interactions occur, the numerical value of ν is $1/2$. In contrast, in solutions under good solvent conditions ν increases to $3/5$. Similarly the coil diffusion coefficient D scales with $D \sim M^{-\nu}$ in dilute solutions, whereas in melts this relation is altered to $D \sim M^{-2}$ [2, 3].

Correspondingly, universal behaviour is also predicted by the microsopic models [4 ... 9] which were introduced in order to describe the dynamics of polymers in solution and in the melt on a segmental scale.

Such universal behaviour became quite obvious, when the analogy between polymeric and magnetic systems was pointed out [3, 10]. In the framework of this concept it can be shown, for example, that the transition from Θ- to good solvent conditions in dilute solutions, which is driven by the temperature, or the transition from dilute to semidulute solution at increasing concentrations exhibit the features of critical phenomena with a smooth cross over between the different phases, well known in solid state physics [11].

In order to investigate these predictions of universal behaviour and to test the different microscopic dynamic models on appropriate length and time scales, microscopic experimental methods are required, which can provide information on the segmental motion in space and time. Up to the end of the seventies no such methods were available. The development of the neutron spin echo (NSE) technique [12 ... 14] which gives direct access to the pair correlation function in the space and time domain of segmental diffusion marked the breakthrough. Using labelling methods the information on the single chain behaviour is not restricted to dilute solutions, but may also be obtained for concentrated systems including the melt.

This paper will give a review on recent neutron spin echo studies on segmental diffusion of polydimethylsiloxane [PDMS:$-(Si(CH_3)_2-O)-_n$] systems [15 ... 23] in the full concentration range. After commenting on the principles of the method and addressing to some details of the microscopic models for the internal chain relaxation, segmental diffusion in dilute Θ-solutions is compared with the predictions of the Zimm model. This model is based on the assumption that hydrodynamic interactions between the different segments occur down to microscopic scales. In addition, the cross over from Θ- to good solvent conditions which is initiated by increasing the temperature is studied. In semidilute solutions, where the different polymer coils interpenetrate each other, the collective response is characterized by a cross over from Zimm dynamics at short scales to collective relaxation at larger scales, whereas the single chain relaxation is dominated by the screening of the hydrodynamic interactions. Finally, in polymer melts and melt-like concentrated solutions the microscopic data of single chain relaxation seem to favour the Rouse model (Zimm model without any hydrodynamic interactions), while the macroscopic data require topological constraints between the different chains to be taken into account.

2 The Spin Echo Method

The unique feature of neutron spin echo (NSE) is its ability to measure energy changes of neutrons which occur during a scattering process in a direct way [14]. This differentiates it from conventional scattering techniques, where the scattering experiment takes place in two steps: first the monochromatization of the incoming and thereafter the analysis of the scattered beam. The energy and momentum changes during scattering then are obtained from evaluating the appropriate difference from the two initial measurements. Consequently, conventional high energy resolution techniques require to cut out a very small primary energy interval from the comparatively weak spectrum of the neutron source and in general have to fight with low neutron intensities. Other than the conventional techniques, NSE measures the neutron velocities of the incoming and scattered neutrons utilizing the Larmor precessions of the neutron spin in an external guide field. This measurement is performed for each neutron individually, since the neutron spin vector acts like the hand of an "internal clock" attached to each neutron which stores the result of the velocity measurement on the neutron itself. Therefore, the incoming and outgoing velocities of one and the same neutron can be compared directly, and a velocity difference measurement becomes possible. Thus, energy resolution and monochromatization of the primary beam or the proportional neutron intensity are decoupled and an energy resolution in the order of 10^{-5} can be achieved with an incident neutron spectrum of 20% bandwidth. In the following we first explain basic neutron spin manipulations by means of external magnetic fields and thereafter discuss the NSE principles.

2.1 Neutron Spin Manipulations by Magnetic Fields

The motion of a neutron spin in a magnetic field $\vec{H}$ is described by

$$\frac{d}{dt} \hat{S}_\alpha = \frac{2|\gamma|\mu i}{\hbar} [\vec{H} \cdot \vec{\hat{S}}, \hat{S}_\alpha] \qquad (\alpha: x, y, z) \qquad (1)$$

where $\vec{\hat{S}} = (\hat{S}_x, \hat{S}_y, \hat{S}_z)$ is the spin operator; $\gamma = -1.91$, μ the nuclear magneton, and $i = \sqrt{-1}$; [ ] denotes the commutator. The neutron polarization P_α is obtained from the expectation values $\langle \hat{S}_\alpha \rangle \equiv s_\alpha$ for which the classical equations of motion are valid

$$\frac{d}{dt} s_\alpha = \frac{2|\gamma|\mu}{\hbar} (\vec{H}(t) \times \vec{s}(t))_\alpha, \qquad (2)$$

($\hbar$ Planck constant h divided by 2π).

Eq. (2) is the basis for the manipulation of the neutron polarization by external fields. We discuss two simple spin turn operations: Let us consider a neutron beam which propagates in z-direction with a spin polarization parallel to its momentum. A guide field $\vec{H}_0$ parallel to z stabilizes the polarization. First we explain a so-called π-coil. There, two components of the neutron spin are reversed. Its principle is shown in Fig. 1a and b. A flat long rectangular coil, a so-called Mezei coil is turned slightly out of the x-y plane. A field $\vec{H}_c$ is produced in such a way that the resulting field $\vec{H}_\pi = \vec{H}_0 + \vec{H}_c$ is pointing in x-direction and starts to precess around this axis. During a time $t = d/v$, where d is the coil thickness and v the neutron velocity, a

3

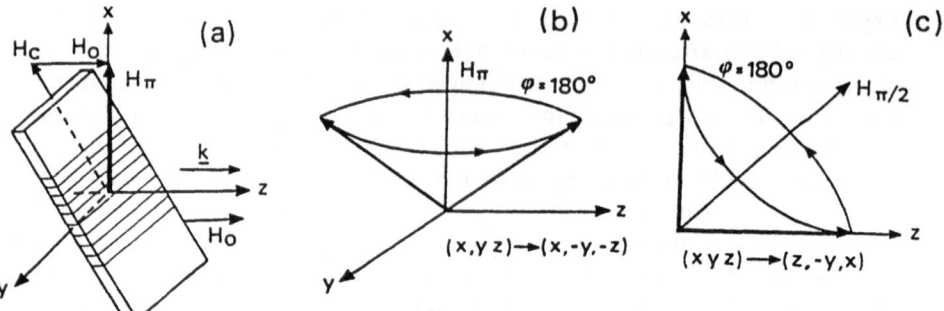

Fig 1 Basic spin turn operation relevant for NSE:
a) Mezei-coil set-up for a π-turn of the neutron spins
b) Motion of the neutron polarization on the action of the π-turn coil
c) Motion of the neutron polarization during a $\pi/2$ turn.

phase angle $\Phi = \omega_L t$ is covered. With the Larmor frequency $\omega_L = \frac{2|\gamma|\mu}{\hbar} H_\pi$ and $v = h/m\lambda$ we find

$$\Phi = \frac{2|\gamma|\mu m}{\hbar^2 2\pi} d \cdot \lambda \cdot H_\pi \tag{3}$$

where m is the neutron mass and λ the neutron wavelength. Thus, Φ is given by the path integral $\int H ds$ and is proportional to λ. E. g., for a coil thickness of $d = 1$ cm and $\lambda = 8\text{Å}$ a field H_π of 8.5 Oe is needed for a $180°$ turn. Such a spin turn is schematically shown in Fig. 1b. As can be seen it transforms the spin components (x, y, z) to $(x, -y, -z)$.

The second important spin turn operation is the $\pi/2$ turn. There, the polarization is turned from the z- into the x-direction or vice versa. A Mezei coil in the x-y plane is tuned such that the resulting field points into the direction of the bisector of the angle between x and z. Then, a $180°$ spin turn around this axis transforms the z-component into the x-direction. At the same time the sign of the y-component is inversed (Fig. 1c).

2.2 The Principles of Neutron Spin Echo

The basic experimental set up of a neutron spin echo spectrometer is shown in Fig. 2. A velocity selector in the primary neutron beam selects a wavelength interval of about 20 % full width half maximum. The spectrometer offers primary and secondary neutron flight paths, where guide fields $\vec{H}$ and $\vec{H}'$ can be applied. At the beginning of the first flight path a supermirror polarizer polarizes the neutrons in direction of propagation. A first $\pi/2$ coil turns the neutron spins into the x-direction perpendicular to the neutron momentum $\hbar\vec{k}$. Starting with this well defined initial

4

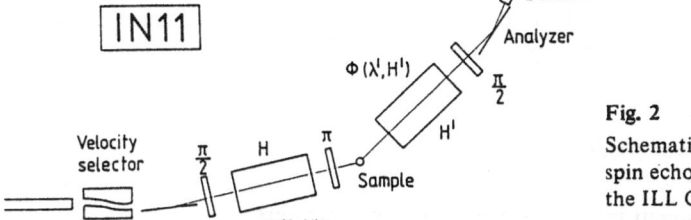

Fig. 2
Schematic scetch of the neutron
spin echo spectrometer IN11 at
the ILL Grenoble.

condition the neutrons commence to precess in the applied guide fields. Without the action of the π-coil each neutron performs a phase angle $\Phi \sim \lambda \int H ds$. Since the wavelength are distributed over a wide range, in front of the second $\pi/2$-coil, the phase angle will be different for each neutron and the beam will be completely depolarized. A π-coil positioned at the half value of the total field integral avoids this effect: On its way to the π-coil the neutron may pass an angle $\Phi_1 = 2\pi n + \Delta\Phi_1$. The action of the π-coil transforms the spin components $(x, y, z) \rightarrow (x, -y, -z)$ or the angle $\Delta\Phi_1$ to $-\Delta\Phi_1$. In a symmetric set up (both field integrals before and after the π-coil are identical) the neutron spin turns by another phase angle $\Phi_2 = \Phi_1 = 2n\pi + \Delta\Phi_1$. The spin transformation at the π-coil just compensates the residual angles $\Delta\Phi_1$ and in front of the second $\pi/2$-coil the neutron spin points again into x-direction independent of its velocity. Finally, the second $\pi/2$-coil projects the x-component of the polarization in the z-direction, and at the super-mirror analyzer the total polarization is recovered. The experimental set up is spin focussing: Similar to NMR spin echo methods in front of the second $\pi/2$ coil for each spin separately the phase is focussed to its initial value.

In the spin echo spectrometer IN11 realized at the Institut Laue-Langevin, Grenoble, the sample is positioned in front of the π-coil (Fig. 2). With exception of losses due to field inhomogeneities in the case of elastic scattering the polarization remains preserved. If the neutron energy is changed due to inelastic scattering at the sample, the neutron wavelength is modified from λ to $\lambda' = \lambda + \delta\lambda$. Then the phase angles Φ_1 and Φ_2 do not compensate each other and the second $\pi/2$-coil projects only the x-component of the polarization into the z-direction which passes afterwards through the analyzer. Apart from resolution corrections the final polarization P_f is then related to initial polarization P_i by

$$P_f = P_i \cdot \cos(\Phi_2 - \Phi_1). \tag{4}$$

Using Eq. (3) for the difference in phase angle $\Delta\Phi = \Phi_2 - \Phi_1$ we get

$$\Delta\Phi = \frac{2|\gamma|\mu m}{\hbar^2 2\pi} \delta \lambda \int H ds. \tag{5}$$

Here we assume small energy transfers, where the fields before and after the sample are chosen to be equal. If we further convert the change in wavelength $\delta\lambda$ to the energy transfer $\hbar\omega$ at the sample $\Delta\Phi$ reads

$$\Delta\Phi = \frac{2|\gamma|\mu m^2}{\hbar^3 \, 8\pi^3}\omega\lambda \int H ds. \tag{6}$$

The scattering experiment integrates over all energy transfers

$$P_f = P_i \int_{-\infty}^{+\infty} S(q, \omega) \cos(\omega t)\, d\omega \tag{7}$$

where the scattering function $S(q, \omega)$ is the probability that during scattering at a certain momentum transfer $\hbar q$ an energy change $\hbar\omega$ occurs. Furthermore, we have introduced the time variable

$$t = \frac{2|\gamma|\mu m^2}{8\pi^3 \hbar^3}\lambda^3 \int H ds = C_t \, H \, \lambda^3. \tag{8}$$

From Eq. (7) it is realized that the NSE is a Fourier method and essentially measures the real part of intermediate scattering function $S(q,t)$. The Fourier time thereby is proportional to λ^3 and the applied guide field H. A spin echo scan is performed by varying the guide field and thereby studying the intermediate scattering function $S(q, t)$ at different Fourier times. Finally, we remark that the use of a broad wavelength band introduces a further averaging process containing an integration over the incident wavelength distribution $I(\lambda)$

$$P_f = P_i \int_0^{\infty} I(\lambda)\, S(q, t(\lambda))\, d\lambda. \tag{9}$$

Table 1 Scaling variable x, lineshape parameter n, and characteristic frequency $\Omega(q)$ (see Eq. (14)) of the coherent dynamic structure factor of the Rouse and Zimm model (Γ: gamma function).

model	x	n	const	$\Omega(q)$
Rouse	$\dfrac{q^2 l^2}{3}\left(\dfrac{3k_B T}{f_0 l^2}t\right)^n$	1/2	$2/\sqrt{\pi}$	$\dfrac{1}{12}\dfrac{k_B T}{f_0} l^2 q^4$
Zimm	$\dfrac{q^2 l^2}{3}\left(\dfrac{k_B T}{6\sqrt{2\pi}\,\eta_0}t\right)^n$	2/3	$\dfrac{2^{2/3}}{\pi}\Gamma\left(\dfrac{1}{3}\right)$	$\dfrac{1}{6\pi}\dfrac{k_B T}{\eta_0} q^3$

This averaging process obscures somewhat the relationship between P_f and $S(q, t)$. For many relaxation processes, however, where the quasi-elastic width varies with power laws in q, the smearing of Eq. (9) is of no practical importance. E. g., for internal relaxation of polymers in dilute solution we have $S(q, t) = S(q^2 t^{2/3})$ (see Tab. 1). Since q varies with $1/\lambda$ and t with λ^3, the wavelength dependence drops out completely.

3 Microscopic Models for Segmental Diffusion and Related Coherent Scattering Laws

The most simple model for the internal chain relaxation on the scale of segmental diffusion is the Rouse model [4]. In this model a chain of freely jointed segments is considered. It is representative for Θ-conditions since the most favourable conformation obeys to Gaussian statistics. The individual segments are treated as friction points in a continuous surroundings. The internal relaxation modes on a length scale $1 < r < \langle R_g^2 \rangle^{1/2}$ (1 segment length) can be derived from a corresponding Langevin equation, where the inertial term is negligible and the friction forces are counterbalanced by entropic and stochastic forces. The spectrum of relaxation rates $1/\tau_j^R$ is given by

$$\frac{1}{\tau_j^R} = \frac{6\pi^2 k_B T}{\langle R_0^2 \rangle^2} \frac{l^2}{f} j^2, \quad j = 1, 2, \ldots N \tag{10}$$

(k_B Boltzmann constant, T temperature, $\langle R_0^2 \rangle$ mean square end to distance, f/l^2 friction coefficient per segment length squared, N total number of segments per chain).

A more complex model is the Zimm model [5]. There the Rouse dynamics is modified by the additional assumption that each moving segment creates a backflow in the solvent, which is felt by all the other segments of the chain. Introducing this hydrodynamic interaction into the Langevin equation via the preaveraged Oseen tensor, the related relaxation rates $1/\tau_j^Z$ are changed significantly compared to the predictions of the Rouse model

$$\frac{1}{\tau_j^Z} = 2\sqrt{3\pi} \frac{k_B T}{\langle R_0^2 \rangle^{3/2}} \frac{1}{\eta_0} j^{3/2}. \tag{11}$$

Other than in the Rouse model the relaxation rates, reduced with respect to coil dimensions, do not depend on polymer specific properties, but exhibit universal behaviour which is determined entirely by η_0, the viscosity of the pure solvent.

Another model, developed to describe segmental diffusion in concentrated solutions and melts, is the tube or reptation model [6 ... 9]. This model bases on the idea that the presence of neighboured molecules imposes topological constraints (entanglements) on the motion of each arbitrarily chosen chain. It starts from the assumption that these constraints restrict the diffusion process essentially to a tube along its own contour. Thus, on larger length scales the chain reptates like a snake, whereas on length scales, smaller than the tube diameter d_t, the segmental diffusion is Rouse like. As a consequence of the one-dimensional reptational motion along a tube, which itself has a coil conformation, the diffusion coefficient D and the viscosity η are predicted to scale with M^{-2} and M^3, respectively.

When the results of the different models are compared to viscosity, diffusion, or dynamic mechanical relaxation data one can see that the Zimm model provides an excellent description for the findings from dilute Θ-solutions [24]. In polymer melts and concentrated solutions the experimental data follow the predictions of the Rouse model only as long as the molar mass M and the concentration c do not exceed the so-called critical molecular mass M_c and critical concentration $c_e = \rho M_c/M$ (ρ polymer density). For $M > M_c$ and $c > c_e$, however, where $\eta \sim M^{3.4}$ and $D \sim M^{2.0}$ is found, the Rouse model fails completely and the reptation model, despite of the apparent discrepancies with respect to the power of the viscosity, is favoured [24, 25].

As outlined in Sect. 2 NSE gives direct access to the intermediate coherent scattering function, which is determined by the correlation function.

$$S(\vec{q}, t) = \frac{1}{N} \sum_j \sum_k \langle \exp\{-i\vec{q}\,\vec{r_j}(0)\} \exp\{i\vec{q}\,\vec{r_k}(t)\} \rangle. \tag{12}$$

$\vec{r_k}(t)$ and $\vec{r_j}(0)$ are the positions of the segments k and j at time t and time zero. The brackets $\langle \ldots \rangle$ denote the thermal average. In the so-called Gaussian approximation Eq. (12) can be written as

$$S(q, t) = \frac{1}{N} \sum_j \sum_k \exp\{-\frac{q^2}{2} \langle [r_k(t) - r_j(0)]^2 \rangle\}. \tag{13}$$

Thus, to compare the predictions of the microscopic models with the experimental scattering data, $\langle [r_k(t) - r_j(0)]^2 \rangle$ has to be calculated for the different models. In the limit of long times Eq. (13) can be approximated by

$$\frac{S(q, t)}{S(q, 0)} = \exp\{-\text{const}\,(\Omega(q)\,t)^n\}. \tag{14}$$

The parameters in (14) are the characteristic frequency in Fourier space $\Omega(q)$ and the lineshape parameter n.

The normalized intermediate coherent scattering functions of the Rouse [26] and the Zimm model [27] depend only on a single scaling variable x, which is given in Tab. 1 together with the three related quantities, defined by Eq. (14).

8

For the reptation model an explicit expression of the coherent dynamic structure factor is also available [28]. In the so-called regime of local reptation ($d_t < r < \langle R_g^2 \rangle^{1/2}$) it has the form

$$\frac{S(q, t)}{S(q, 0)} = (1 - \frac{q^2 d_t^2}{36}) + \frac{q^2 d_t^2}{36} \exp\left\{\frac{x^2}{36}\right\} \; [1 - \mathrm{erf}\left(\frac{x}{6}\right)], \tag{15}$$

where x is the scaling variable of the Rouse model and erf the error function.

4 NSE Experiments on Dilute Polymer Solutions

4.1 Segmental Diffusion in Θ-Solvents

In order to test the predictions of the Zimm model on a microscopic scale, NSE experiments were performed on PDMS (M $\approx$ 60.000 g/mol) in deuterated bromo-benzene [19, 20, 22], which is a Θ-solvent for this polymer at T = 357 K. The monomer concentration was 5%.

The experimental results are shown in Fig. 3, where $- \ln [S(q, t)/S(q, 0)]$ is plotted vs. the time t on a double logarithmic scale. The broken lines result from fitting the scattering law, calculated for the Zimm model, simultaneously to all data, using only XT/η_0 as adjustable parameter. The solid lines give the analogous fit result for the corresponding asymptotic form (s. Eq. (14) and Tab. 1).

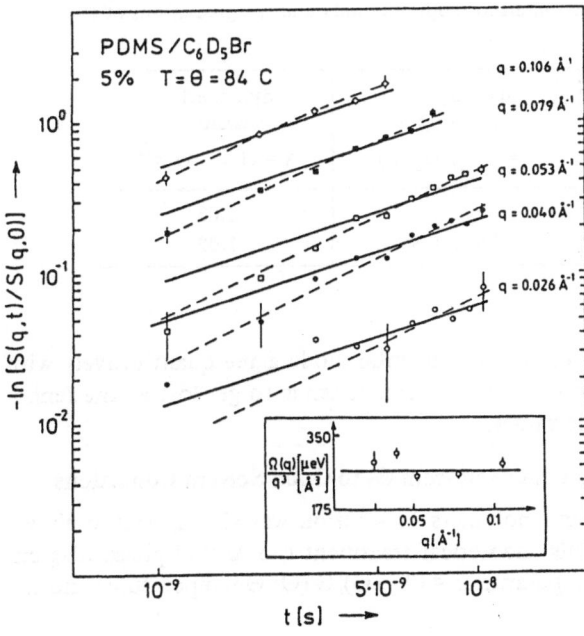

Fig. 3

Results of NSE experiments in the dilute Θ-system PDMS/d-bromobenzene at T = 357 K. Solid and broken lines result from fitting the asymptotic and non-asymptotic forms of the scattering law, calculated for the Zimm model, simultaneously to all correlation curves, using only one adjustable parameter. The insert presents the q-dependence of the characteristic frequencies Ω (q).

9

In the insert the q-dependence of the characteristic frequency $\Omega(q)$, as obtained from fitting the scattering law of the Zimm model to the experimental results at different q-values is shown.

It is obvious that the Zimm model is able to describe the experimental data very accurately both with respect to the lineshape and the q-dependence of the characteristic frequencies. The asymptotic regime is not yet reached in the covered q-t-regime. However, with increasing q a gradual tendency towards the asymptotic form is observable. Similar results were found for polystyrene (PS) in d-methyl-cyclohexane at T = 341 K [19, 20] and in d-cyclohexane at T = 311 K [16], respectively, which are also Θ-systems.

Concerning the absolute values of the characteristic frequencies $\Omega(q)$, the experimental PDMS data differ by a factor of X = 0.87 from the theoretical predictions (s. Tab. 2). However, the discrepancies become still increased by an additional factor of 0.85, if the non-preaveraged Oseen tensor is used to describe the hydrodynamic interactions.

In Tab. 2 the $\Omega(q)$ data, available for PDMS and PS from NSE and dynamic light scattering [29, 30] experiments, are summarized together with the related fundamental relaxation rates $1/\tau_1$, obtained from measurements of dynamic mechanical relaxation [31, 32]. All data are normalized to the predictions of the Zimm model with preaveraged Oseen tensor.

Table 2 Comparison between the characteristic frequencies and fundamental relaxation rates, as obtained by different microscopic and macroscopic methods, and the predictions of the Zimm model.

	NSE $X = \Omega(q)/\Omega_Z(q)$	dyn. light scattering $X = \Omega(q)/\Omega_Z(q)$	dyn. mech. relaxation $X = (1/\tau_1)/(1/\tau_1^Z)$
PDMS	0.87	–	1.03
PS	0.43 ... 0.50	0.85 ... 0.91	1.02

While the macroscopic results for the fundamental mode agree quantitatively with the prediction of the Zimm model, this agreement tends to get lost as the length scale is decreased to microscopic values.

4.2 Segmental Diffusion at the Transition from Θ- to Good Solvent Conditions

The analogy of ferromagnets and polymers in solution was already stated above. One of the consequences of this theoretical treatment is a related phase diagram [33, 34], where the reduced temperature $\tau = (T - \Theta)/\Theta$ (Θ: Θ-temperature) and the

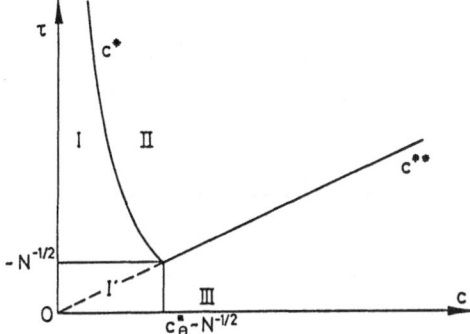

Fig. 4

Temperature-concentration diagram for polymer solutions [33, 34]. $\tau = (T - \Theta)/\Theta$ (Θ: Θ-temperature), c monomer concentration. (s. Tab. 3).

monomer concentration c enter as variables. The relevant part of this diagram ($\tau > 0$) is shown in Fig. 4.

The c-τ-plane is divided into 4 areas: The dilute regimes I' and I are separated from the semidilute regimes III and II, where the different polymer coils interpenetrate, by the so-called overlap concentration

$$c^* = \frac{M}{N_A \langle R_g^2 \rangle^{3/2}} \tag{16}$$

(N_A: Avogadro constant).

I' and III are the tricritical or Θ-regions. There the chain molecules exhibit an unperturbed random coil conformation. In contrast, I and II are the critical or good solvent regions, which are characterized by conformational fluctuations.

These fluctuations can be easily understood in the "blob" concept [35 ... 37]. E. g., in the semidilute region II on distances, smaller than a correlation length ξ_c, the monomers of a given chain mainly interact with like monomers of the same chain. Therefore it is expected that an individual chain behaves for $r < \xi_c$ as a chain in the dilute solution and that the conformation is determined by excluded volume interactions, characteristic for good solvent conditions. On distances $r > \xi_c$, however, the conformation of the chain is influenced by the interaction with other chains and the excluded volume interactions are screened.

Passing from I to II the conformation gradually changes from a self-avoiding random walk to an unperturbed coil conformation. The blobsize ξ_c thereby scales with $c^{-3/4}$.

An analogous behaviour is expected at the transition from I to I'. Within the chain, swollen by excluded volume interactions, blobs of Gaussian conformation are formed. Its blobsize ξ_τ scales with τ^{-1}. At the bordline between I and I', ξ_τ becomes equal to R_g and the whole chain performs an unperturbed random walk. In Tab. 3 the scaling predictions for $\langle R_g^2 \rangle$ and $\langle \xi^2 \rangle$ are summarized. These predictions later on were confirmed by small angle neutron scattering [34, 35, 38 ... 40].

Table 3

Scaling predictions for the mean square radius of gyration $\langle R_g^2 \rangle$ and the mean square correlation lengths $\langle \xi^2 \rangle$ in the different regimes (s. Fig. 4) of polymer solutions [33, 34].

regime	$\langle R_g^2 \rangle$	$\langle \xi^2 \rangle$
I'	N	—
I	$N^{6/5}\tau^{2/5}$	τ^{-2}
II	$N\,c^{-1/4}\tau^{1/4}$	$c^{-3/2}\tau$
III	N	c^{-2}

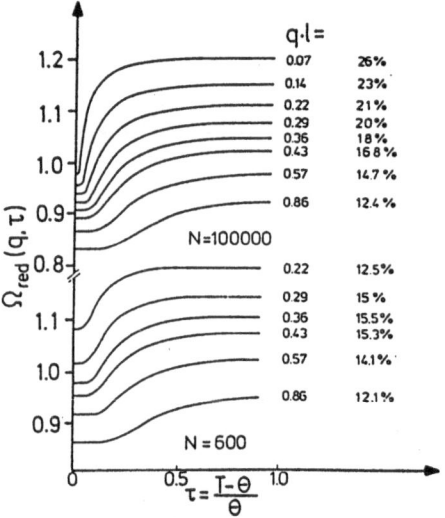

Fig. 5

Cross over from Θ- to good solvent conditions in dilute solutions. Calculated characteristic frequencies, normalized to Θ-conditions, as dependent on reduced temperature for two different chain length N at various values of $(q \cdot l)$. Right to the different curves the increase of $\Omega_{red}(q, \tau)$ between $\tau = 0$ and $\tau = 0.9$ is given.

The theoretical scattering function for polymer relaxation in the transition zone $(I \rightarrow I')$ and in good solvents are not available. However, another access to this problem [41], based on Kirkwood's generalized diffusion equation and the Mori projection operator technique, provides information on the initial slope

$$\Gamma = \lim_{t \rightarrow 0} \frac{\partial}{\partial t} - \ln \frac{S(q, t)}{S(q, 0)} \qquad (17)$$

of $S(q, t)$, which is proportional to the characteristic frequency $\Omega(q)$, introduced above. In this treatment the polymer properties enter alone by their static correlation functions.

Applying this method to infinite long chains in good solvents [42], $\Omega(q)$ it found to exhibit the same dependence on temperature, viscosity, and momentum transfer as in the Θ-state. This is in agreement with earlier scaling predictions [3]. In addition, the magnitude of $\Omega(q)$ increases by a factor of 1.34 compared to Θ-condi-

12

tions at the same temperature and viscosity. The transition behaviour of $\Omega(q, \tau)$ is shown in Fig. 5, where

$$\Omega_{\text{red}}(q, \tau) \equiv \frac{T}{\Theta} \cdot \frac{\eta_0(\Theta)}{\eta_0(T)} \cdot \frac{\Omega(q, \tau)}{\Omega_Z(q, 0)}$$

is plotted vs. τ for two different chain lengths. This plot shows:
(i) For large numbers of segments N and small values of $(q \cdot l)$ a sharp cross over from Θ- to good solvent conditions occurs; (ii) for a given chain length the sharpness of the cross over as well as the increase of $\Omega_{\text{red}}(q, \tau)$ is diminished with increasing $(q \cdot l)$ values; (iii) the reduction of N reduces considerably the increase of $\Omega_{\text{red}}(q, \tau)$ and smears out the sharpness of the transition.

Regarding exerimental results in good solvents (e. g. PDMS in d-benzene at 343 K or in d-chlorobenzene at 373 K) it was found that the lineshape of the scattering function agrees with that in Θ-solutions [15, 20, 22]. In addition, the characteristic frequencies $\Omega(q)$ follow the q^3-behaviour quite well (s. Fig. 8, 2 % and 5 %). The absolute values of $\Omega(q)$ are increased considerably compared to the predictions for Θ-conditions. However, the theoretically expected limits are not reached in any case. This was confirmed also by other authors [43, 44].

Fig. 6 shows experimental results, obtained from PDMS/d-bromobenzene in the transition regime from Θ- to good solvent conditions [19]. Although the observed cross over in $\Omega_{\text{red}}(q, \tau)$ exhibits quantitatively all the predicted features (s. Fig. 5, N = 600), quantitatively there are distinct discrepancies, both concerning the total amount of increase and the sharpness of the transition. Moreover, even larger quantitative disagreement was found for PS/d-cyclohexane in the range of reduced temperature $0 \leqslant \tau \leqslant 0.14$ [16].

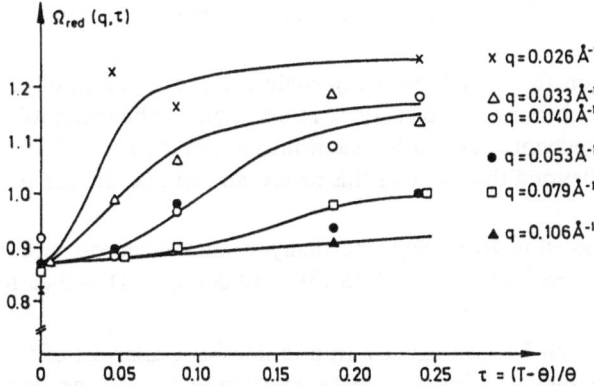

Fig. 6 Cross over from Θ- to good solvent conditions in PDMS/d-bromobenzene. Characteristic frequencies, normalized to Θ conditions, as a function of reduced temperature τ at different q-values. The various lines are guide lines for the eye only. For reasons of clarity the line related to $q = 0.053$ A^{-1} is omitted.

13

5 NSE Experiments on Semidilute Solutions Under Good Solvent Conditions

5.1 Transition from Single to Many Chain Behaviour

When the monomer concentration exceeds c* (s. Eq. (16) and Fig. 4), the different polymer molecules are no longer separated but interpenetrate each other forming a transient network of life time τ_g. At constant temperature this network structure is characterized by a concentration dependent correlation length $\xi(c) \equiv \langle \xi^2 \rangle^{1/2}$, which may be taken as the mean mesh size of the pseudo gel. In good solvents $\xi(c)$ is also the distance beyond which the hydrodynamic as well as the excluded volume interactions are screened [3]. Since the hydrodynamic interaction is introduced in the Zimm model as a typical intrachain effect, its screening can only be observed properly, if the internal dynamics of single labelled chains in the many chain system is considered. This will be done in the next section. Here, the dynamic behaviour of the many chain system itself will be the subject of discussion. From the analogy with permanent networks swollen by a solvent the scattering law for the many chain system ($r > \xi(c)$ or $q\xi(c) < 1$) on the time scale $t < \tau_g$ is given by [45]

$$\frac{S(q, t)}{S(q, 0)} = \exp(-D_c q^2 t). \tag{18}$$

D_c is the collective diffusion coefficient of the pseudo gel which governs the decay of density fluctuations. It can also be regarded as the blob diffusion coefficient.

For spatial dimensions $r < \xi(c)$ or $q\xi(c) > 1$, however, the network is not yet effective and single chain behaviour of Zimm type, as observed in dilute solutions on the whole intramolecular scale, is expected to prevail. The condition of a smooth cross over between regimes at $q\xi(c) \approx 1$ implies $D_c \sim \xi(c)^{-1} \sim c^{3/4}$ in good solvents.

The treatment of the transition from single to many chain behaviour in semidilute Θ-solutions (regime III) is much more complex since in addition to the intermolecular interactions, self-entanglements have to be taken into account [46, 47]. The discussion of this problem is beyond the scope of this review and will be the subject of a forthcoming paper [48].

NSE experiments on the transition from single to many chain relaxation under good solvent conditions were performed on PDMS ($M \approx 30.000$ g/mol) at 343 K [18].

Quantitatively, the cross over can be seen already from a lineshape analysis of the neutron scattering data (see Fig. 7). For the dilute solution ($c = 0.05$) the lineshape parameter n is equal to about 0.7 for all q values which is characteristic for Zimm behaviour. In the semidilute solution n-values of 0.7 are found only at larger q-values while n-values of about 1.0 are obtained at small q-values as predicted for the collective diffusion.

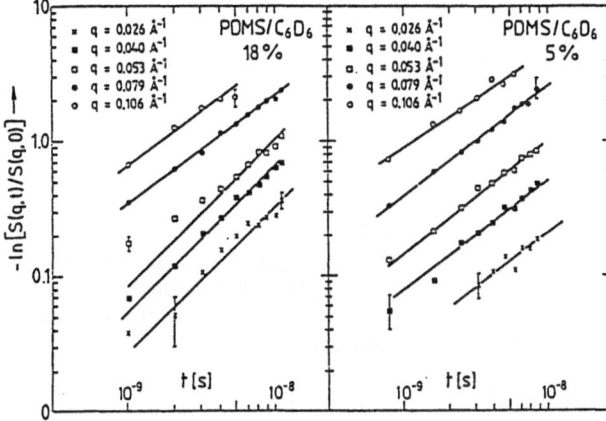

Fig. 7

Cross over from single chain to many chain relaxation at T = 343 K. Lineshape analysis for PDMS/d-benzene at c = 5 and 18 %; double logarithmic plot of $-\ln[S(q,t)/S(q,0)]$ vs. t/s.

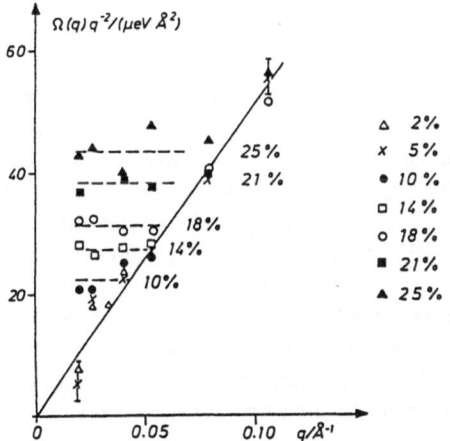

△	2%
×	5%
●	10%
□	14%
○	18%
■	21%
▲	25%

Fig. 8

Cross over from single chain to many chain behaviour in PDMS/d-benzene at T = 343 K. Characteristic frequencies $\Omega(q)$ divided by q^2 as a function of q.

Figure 8 presents the $\Omega(q)/q^2$ relaxation rates, obtained from a fit with the scattering law of the Zimm model, as a function of q. For both dilute solutions (c = 0.02 and c = 0.05) $\Omega(q) \sim q^3$ is found in the whole q-range of the experiment. With increasing concentrations a transition from q^3 to q^2 behaviour takes place at decreasing q. The position of the cross over point $q^*(c)$ is a direct measure of the dynamic correlation length $\xi(c)$ ($\xi(c) = 1/q^*(c)$). The plateau value at low q determines the collective diffusion coefficient D_c. A simultaneous fit of all low q

15

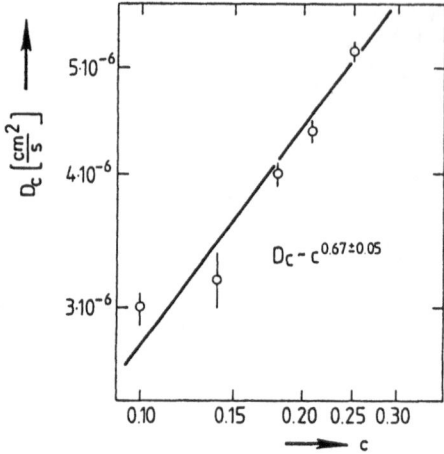

$D_c \sim c^{0.67 \pm 0.05}$

Fig. 9
Collective diffusion coefficient D_c in semidilute PDMS/d-benzene systems at T = 343 K as a function of the monomer concentration c.

spectra where a simple exponential decay was found let to the concentration dependence and the numerical values of D_c

$$D_c = (1.3 \pm 0.1)\ 10^{-5} \cdot c^{0.67 \pm 0.05}\ \frac{cm^2}{s}.$$ (19)

The result is shown in Fig. 9 where the different D_c values, obtained from separate fits at each concentration, are also presented. The dynamic correlation length $\xi(c)/\text{Å} = (3.4 \pm 0.7) \cdot c^{-0.67 \pm 0.05}$ is of the same order of magnitude as the value obtained from a static experiment on the PS/d-cyclohexane system [35]. The exponents for the concentration dependence of D_c and $\xi(c)$ are in agreement with the results of light scattering experiments [49] probing a much larger length scale. They are both smaller than theoretically predicted and found in static neutron scattering experiments. According to Weill and des Cloiseaux [50] this observation may not have to be attributed to a fundamental difference between static and dynamic scaling laws, but may result from the finite chain length which causes the dynamic properties to converge much more slowly to their asymptotic values.

5.2 Transitions in the Single Chain Behaviour due to the Screening of Hydrodynamic Interaction

The calculation of eigenmodes of single polymers in semidilute solutions from first principles is a demanding task requiring the solution of a complicated many-body problem [45, 51 ... 59].

However, with respect to the screening of hydrodynamic interactions one may obtain the spectrum of internal relaxations by a phenomenological approach [21, 22]. It bases on the Zimm model and on the additional assumption that the average hydrodynamic interaction in semidilute solutions is still of the same form as in the dilute case. Within this concept the effect of screening is taken into account re-

16

placing the solvent viscosity η_0 in the Oseen tensor by an effective phenomenological quantity $\eta_{eff}\,(r,\,c)$.

$$\eta_{eff}\,(r,\,c) = (\eta_0^{-1} - \eta_H\,(c)^{-1}) \cdot \exp\,\{-r/\xi_H\,(c)\} + \eta_H\,(c)^{-1}. \qquad (20)$$

Eq. (20) defines the hydrodynamic screening length $\xi_H\,(c)$ and introduces an additional parameter to account for residual hydrodynamic interactions on distances $r \gg \xi_H\,(c)$. Since the screening of excluded volume interactions and the screening of hydrodynamic interactions are attributed to the same origin, namely, the interchain interaction, $\xi_H\,(c)$ and $\xi\,(c)$ (s. Sect. 5.1) were predicted to exhibit the same scaling behaviour [45].

However, this conclusion is not generally accepted, leading to different predictions concerning the scaling laws and absolute values [52, 58, 59]. When Eq. (20) is incorporated into the Langevin equation, three different dynamic regimes (s. Fig. 10) can be distinguished:

(i) $q\,\xi_H\,(c) \gg 1$

$$\Omega\,(q) \equiv \Omega_z\,(q) = \frac{1}{6\pi}\,\frac{k_B T}{\eta_0}\,q^3, \qquad\qquad n = 2/3$$

unscreened Zimm relaxation (as in the dilute solution)

(ii) $q\,\xi_H\,(c) \ll 1$ and $q\,\xi_H\,(c) \gg \eta_0/\eta_H\,(c)$

$$\Omega\,(q) = \Omega_z\,(q)\,(1 - \frac{\eta_0}{\eta_H\,(c)}\,q\,\xi_H\,(c)), \qquad\qquad n = 1/2$$

enhanced Rouse relaxation

(iii) $q\,\xi_H\,(c) \ll 1$ and $q\,\xi_H\,(c) \ll \eta_0/\eta_H\,(c)$

$$\Omega\,(c) = \Omega_z\,(q)\,\frac{\eta_0}{\eta_H\,(c)}, \qquad\qquad n = 2/3$$

screened Zimm relaxation.

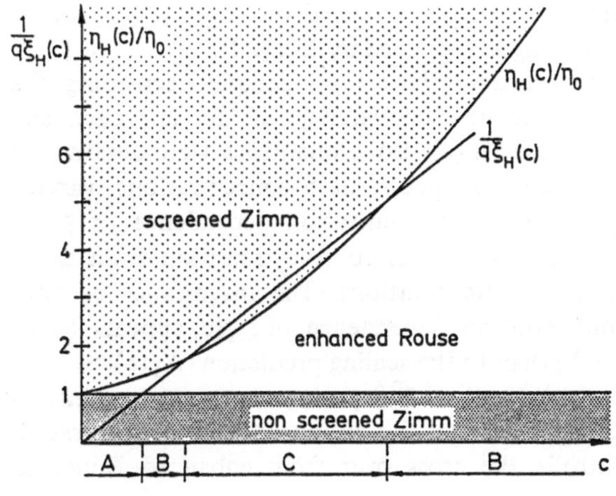

Fig. 10

Incomplete screening of hydrodynamic interactions in semidilute polymer solutions. Presentation of different regimes, which are passed with increasing concentration.

A, C unscreened and screened Zimm relaxation, respectively,

B enhanced Rouse relaxation.

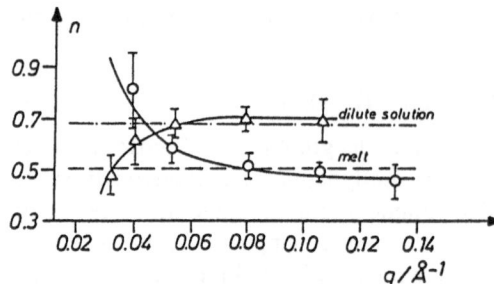

Fig. 11

Single chain behaviour in semidilute PDMS/d-chlorobenzene solutions. Lineshape parameter n as a function of q at the concentration c = 0.18 and c = 0.45, indicating the occurrence of two cross over effects, as predicted by the concept of incomplete screened hydrodynamic interactions. ($-\cdot-\cdot-$), ($---$) asymptotic Zimm and Rouse behaviour, respectively.

If a residual hydrodynamic interaction over large distances does not exist ($1/\eta_H (c) \equiv 0$), the regime of screened Zimm relaxation vanishes, and only the cross over from unscreened Zimm to enhanced Rouse relaxation remains.

The observation of single chain dynamics in semidilute solutions requires contrast matching and labelling. In the case of PDMS this can be achieved using a mixture of 95% deuterated and 5% protonated polymers, dissolved in d-chlorobenzene. Since the scattering length densities of d-PDMS and d-chlorobenzene are almost identical, for all monomer concentrations the dynamics of the single protonated chains can be observed. The NSE measurements were performed at T = 373 K and concentrations $0.04 \leqslant c \leqslant 0.45$ [21, 22].

Fig. 11 presents the results of the lineshape analysis for c = 0.18 and c = 0.45. In the first case the polymer relaxation is still determined by the Zimm modes at larger q-values while at smaller q the Rouse modes become dominant. Qualitatively, this behaviour is expected for the cross over from unscreened Zimm to enhanced Rouse relaxation. At c = 0.45 the q-dependence at n is inverted. Apparently, Zimm modes are active now over larger distances while the shorter spatial dimensions are governed by Rouse modes. This result, which is incompatible with the assumption that the hydrodynamic interactions are totally screened on larger length scales, supports the idea of incomplete screening as supposed by Eq. (20).

Employing the explicitly calculated scattering laws for incomplete screening, the quantitative data analysis demonstrated the consistency of the approach and revealed numerical results for $\xi_H (c)$ and $\eta_H (c)$. Both quantities were determined from a simultaneous fit of 25 experimental spectra at 5 concentrations, varying only two parameters. Fig. 12 presents the concentration dependence of $\xi_H (c)$ and compares it with $\xi (c)$, the correlation length for the transition from single chain to collective relaxation in semidilute solutions. The magnitudes of both lengths nearly coincide. The concentration dependence of $\xi_H (c)$ appears to be closer to the mean field ($\xi_H (c) \sim c^{-1}$) than to the scaling prediction ($\xi_H (c) \sim c^{-3/4}$) and the experimental findings for $\xi (c)$, ($\xi (c) \sim c^{-0.67}$). However, measurements on more concentrations are needed to determine the exponent. The concentration dependence of $\eta_H (c)$, which controlls the cross over from enhanced Rouse to

18

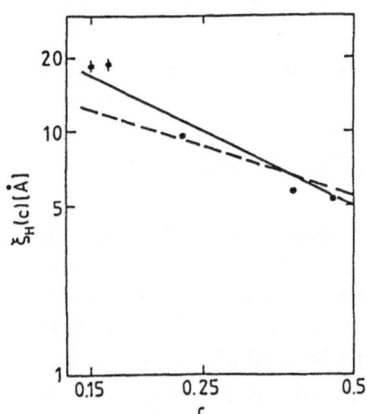

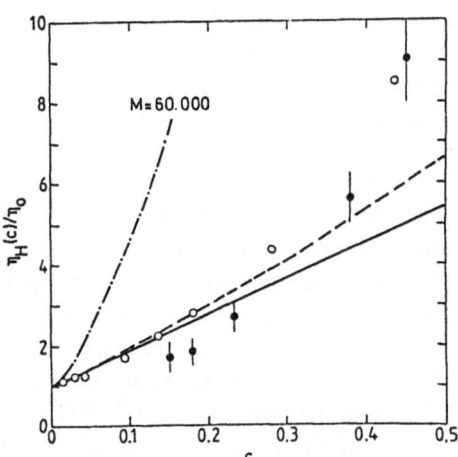

Fig. 12 Concentration dependence of the hydrodynamic screening length ξ_H (c). The solid line represents the result of the simultaneous fit, the dashed line the correlation length ξ (c) related to the transition from single to many chain behaviour.

Fig. 13 Concentration dependence of η_H (c)/η_0: • result of separate fits; o results of viscosity measurements on PDMS solutions (M_w = 7.400). The result of the simultaneous fit considering the linear term in η_H (c) = η_0 (1 + [η] c + K_H [η]^{2}c^2) is given gy the solid line; the inclusion of a quadratic term leads to the dashed line. The point-dashed line indicates the macroscopic viscosity for M = 60.000 g/mol.

screened Zimm relaxation, is shown in Fig. 13. Since no dependence of η_H (c) on the molar mass of the polymer matrix (M ≈ 60.000, 178.000 g/mol) was observed, η_H (c) can not be identified with the macroscopic viscosity of the polymer solution (s. Fig. 13).

On the other hand it was found that the microscopic parameter η_H (c) behaved very similar as the macroscopic viscosity η (c)/η_0 of a low molecular mass (M ≈ 7.400 g/mol) PDMS/d-chlorobenzene system at 373 K. For that low molar mass the terminal Zimm time τ_1^Z (s. Eq. (10)) is comparable to the time scale of the NSE experiment. Thus, the macroscopic viscosity can relax towards equilibrium within the available time window. This coincidence supports the idea that the cross over from enhanced Rouse to screened Zimm relaxation is determined by an effective time dependent viscosity.

6 NSE Experiments on Concentrated Polymer Solutions and Polymer Melts

When the monomer concentration becomes larger than $c_e = \rho M/M_c$, the influence of entanglements is supposed to govern the dynamic properties in these dense polymer systems [24, 25]. The experimental results on macroscopic diffusion and

19

zero shear viscosity ($D \sim M^{-2.0}$; $\eta \sim M^{3.4}$) support the reptation idea, although there are discrepancies concerning the molecular weight dependence of the viscosity.

With respect to the segmental diffusion the reptation or tube concept predicts a cross over from Rouse relaxation on distances, smaller than the tube diameter d_t, to a local reptation mechanism on scales larger than d_t. Since the tube diameter increases with decreasing monomer concentration, this cross over should become best observable in the melt.

The NSE measurements on a PDMS melt were performed at 373 K [17, 22]. Both the 5% labelled protonated and the 95% deuterated matrix molecules had a molar mass of 60.000 g/mol, well above the critical molar mass $M_c \approx 27.000$ g/mol of PDMS [24, 25]. The results are shown in Fig. 14. In this scaling graph, $\ln[S(q,t)/S(q,0)]$ is plotted vs. $(q \cdot l)^2 (Wt)^2/6$ ($W = 3k_B T/(fl^2)$), which is the universal variable of the scattering laws calculated for the Rouse model and the model of local reptation, respectively (cf. Tab. 1). All experimental data follow the master curve of the Rouse model and do not exhibit any indications for the q dependent splitting, to be expected for the reptational mechanism.

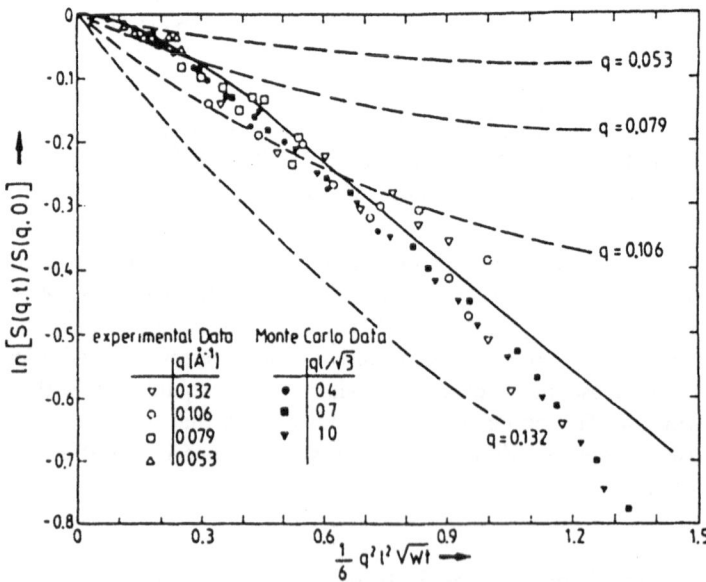

Fig. 14 Single chain relaxation in a PDMS/d-PDMS melt (M $\approx$ 60.000 g/mol) at T = 373 K. Scaling plot of data, obtained either from NSE measurements or Monte Carlo simulations.
────── Rouse model with $l^4W = 3.0 \cdot 10^{13}$ A^4 s^{-1},
------ tube model with $l^4W = 3.0 \cdot 10^{13}$ A^4 s^{-1} and tube diameter $d_t = 40$ A.

20

However, although the Rouse model, which was also confirmed by recent Monte Carlo calculations (s. Fig. 14) [17], is obviously superior to the tube model, it does not describe all details of the experimental data properly. This can be seen from Fig. 15, where the non-asymptotic and asymptotic form of scattering law, calculated for the Rouse model, were fitted simultaneously to all relaxation curves. While in the case of dilute solutions the asymptotic form of the Zimm scattering function was inferior to the non-asymptotic form, in the melt the situation is inversed. Here the asymptotic form (solid line) clearly provides the better agreement with the experimental data than the corresponding non-asymptotic form (broken line), which is actually appropriate to the time range of the experiment.

This effect, which was also observed in a polytetrahydrofuran melt [60, 61], is not restricted to labelled molecules of large molecular mass. It occurs as well for labelled PDMS chains of $M \approx 3.800$ g/mol in the matrix of $M \approx 60.000$ g/mol [22] but is not found, if the internal dynamics of large labelled chains $(M > M_c)$ in a macroscopically non-entangled matrix $(M < M_c)$ is studied [61]. Thus, the line-shape of the relaxation curves appears to be sensitive to the topological constraints build-up by the matrix molecules. This provides additional information on segmental diffusion which should stimulate further theoretical work.

Since the asymptotic Rouse and the local reptation model lead to nearly identical lineshape parameters, single relaxation curves are also well described on the basis of the reptation mechanism [61]. However, in contrast to Rouse relaxation local reptation predicts a stronger q-dependence of the relaxation rates $\Omega (q)$ and therefore does not fit the observed q-dependence of the spectra.

From the characteristic frequencies of the Rouse model the segmental friction coefficient f/l^2 was determined to $(5.3 \pm 0.6) \cdot 10^5$ g cm^{-2}s^{-1}. It is in satisfactory agreement with $7.1 \cdot 10^5$ cm^2 s g^{-1}, obtained from macroscopic viscoelastic measurements on PDMS at 373 K [24].

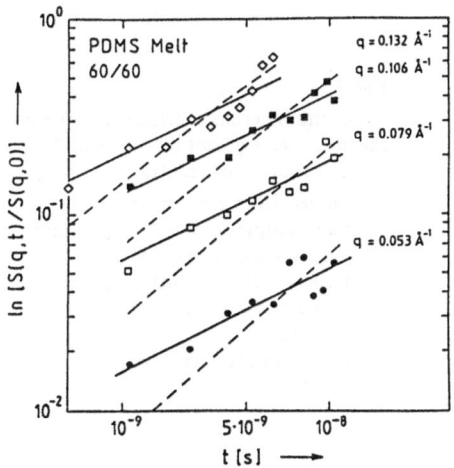

Fig. 15

NSE data from a PDMS/d-PDMS melt $(M \approx 60.000$ g/mol) at 373 K. The solid and broken lines result from fitting the asymptotic and non-asymptotic form of the scattering law, calculated for the Rouse model, simultaneously to all data, using only one adjustable parameter.

21

The scattering results on polymer melts, available up to now, do not exhibit the features of local reptation. This, however, does not yet demonstrate the invalidity of the model, since rather restrictive conditions have to be satisfied [46, 47]. These are $qd_t < 1$ and $t > (d_t/l)^4 W^{-1}$. If, for example, tube diameters of 40 to 50 Å, as estimated from the plateau moduli in viscoelastic experiments, are realistic [24, 25], it is impossible to reach these limts with the experimental conditions, accessible at present. Since no indications for local reptation were found on length scales reaching up to 40 Å, even in the melt, it is not surprising that such a mechanism is not observed in concentrated solutions ($0.5 \leqslant c < 1$). As in the melt all relaxation curves are well described by the asymptotic form of the Rouse scattering function. In Fig. 16 the derived segmental friction coefficients per segment length squared $f(c)/l^2$, as obtained by fitting all relaxations curves at fixed concentrations with the scattering law of the Rouse model, are plotted vs. the concentration on a double logarithmic scale.

In the case of non-entangled solutions the Rouse model relates $f(c)/l^2$ with the viscosity of the solution by

$$\eta(c) = \eta_0 + \frac{L}{36\,M} \langle R_0^2 \rangle^2 \, \rho \, \frac{f(c)}{l^2}. \tag{21}$$

In order to compare microscopic and macroscopic parameters, zero shear viscosity measurements on PDMS ($M \approx 7.400$ g/mol) in chlorobenzene were performed at T = 373 K. The results are also shown in Fig. 16.

Both $f(c)/l^2$ values, the microscopic ones, deduced from the scattering experiments which covered time scales of 10^{-8} to 10^{-9} s and length scales of 10 to 40 Å, and

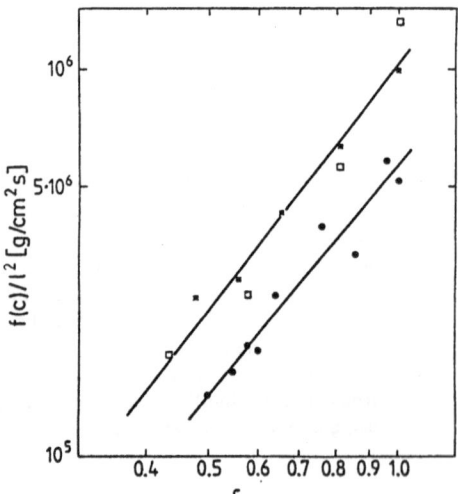

Fig. 16

Segmental friction coefficient $f(c)$ per mean square segment length l^2 of PDMS as a function of concentration:

● derived from NSE measurements on macroscopically entangled systems, using the Rouse model;

□ derived from viscosity measurements on macroscopically non entangled systems;

★ data from the literature [62].

22

those derived from macroscopic viscosity measurements, exhibit the same concentration dependence, which is close to c^2. The magnitudes differ by a factor less than 2. This agreement between macroscopic measurements on non-entangled solutions and melts, respectively, and microscopic measurements on macroscopically entangled systems may be taken as further evidence that entanglement constraints do not affect segmental diffusion on the time and length scales accessible to neutron spin echo.

7 Conclusions and Outlook

Applying high resolution neutron spin echo spectroscopy it became possible for the first time, to investigate internal relaxation processes of linear polymer molecules on the scale of segmental diffusion in the whole range of concentration. The aim of these investigations was to test the different microscopic models on microscopic length scales as well as to prove the predictions of universal behaviour.

For dilute solutions the neutron scattering experiments showed that the segmental diffusion within the polymer coils is governed by hydrodynamic interactions and described well by the Zimm model for Θ- as well as for good solvent conditions. The predicted universal behaviour for the characteristic frequencies $\Omega(q)$ is only confirmed for their q-dependence but not for their magnitude.

Despite of some quantitative discrepancies the internal dynamics at the transition from Θ- to good solvent conditions qualitatively exhibit all the features which are predicted by the scaling approach.

In semidilute solutions, where the different polymer coils interpenetrate each other and form a transient network of mean meshsize $\xi(c)$, the relaxation behaviour of single labelled chains as well as the collective relaxation of the many-chain system were studied. In the latter case a cross over from Zimm relaxation $(\Omega(q) \sim q^3)$ at short scales to collective diffusion $(\Omega(q) \sim q^2)$ at larger scales was observed. At the cross over also the lineshape of the relaxation spectra changes. The concentration dependence of correlation length $\xi(c)$ and collective diffusion coefficient D_c are close to the scaling predictions.

The segmental diffusion of single labelled chains in semidilute solutions exhibits a multiple transition behaviour, resulting from the incomplete screening of hydrodynamic interactions. With increasing concentration the regimes of unscreened Zimm, enhanced Rouse, and screened Zimm relaxation are passed. The experimental data allowed an evaluation of the hydrodynamic screening length $\xi_H(c)$. Its magnitude is close to that of $\xi(c)$, as required by the scaling ansatz of only one relevant correlation length. The viscosity parameter $\eta_H(c)$, which characterizes the residual long-range hydrodynamic interaction, cannot be identified with the macroscopic viscosity of the solution. It exhibits startling similarities with the viscosity of an equivalent solution of lower molecular weight, where the viscosity reaches its equilibrium value during the experimental observation time.

In concentrated polymer solutions and in polymer melts the Rouse model provides a general description of the segmental diffusion processes on length scales up to 40 Å and time scales in the 10^{-8} s region. In fact, deviations of the experimental lineshape from that of the Rouse model appear to permit conclusions on the presence of topological constraints build-up by the polymer matrix, but they cannot easily be attributed to local reptation.

The experimental data presented here demonstrate the uniqueness and the efficiency of the NSE method in the investigation of the internal dynamics of polymeric systems. The experiments were restricted to linear polymers. Further work on more complicated systems, as polymer stars [63] or polymer networks [64], is under progress now. However, as can be seen, e. g., from the measurements on polymer melts and concentrated solutions a promising treatment of fundamental problems, like the microscopic examination of the reptation concept, require the spatial range to be enlarged and the time scale to be extended to larger values. From the technical point of view these extensions are possible. Thus, the paper may also support the efforts to realize the second, improved generation of neutron spin echo in the near future.

Acknowledgement

The authors thank the Institut Laue-Langevin, Grenoble for making available the neutron spin echo spectrometer IN 11 and Dr. B. Lehnen (Universität Mainz) for providing the PDMS samples.

They are also indebted to Dr. A. Baumgärtner (KFA Jülich), Prof. Dr. K. Binder (Universität Mainz), Dr. B. Hayter (Oak Ridge), and Dr. B. Stühn (Universität Freiburg) for their contributions on this subject.

One of us (B. E.) gratefully acknowledges the financial support by the Bundesminister für Forschung und Technologie (BMFT), Bonn.

References

[1] *H. G. Elias*, Macromolecules 1 (Wiley, London, New York, Sydney, Toronto 1977)

[2] *P. J. Flory*, Principles of Polymer Chemistry (Cornell University Press, Ithaca, New York 1971)

[3] *P. G. de Gennes*, Scaling Concepts in Polymer Physics (Cornell University Press, Ithaca, London 1979)

[4] *P. E. Rouse*, J. Chem. Phys. **21**, 1272 (1953)

[5] *B. H. Zimm*, J. Chem, Phys. **24**, 269 (1956)

[6] *P. G. de Gennes*, J. Chem. Phys. **55**, 572 (1971)

[7] *S. F. Edwards* and *J. M. V. Grant*, J. Phys. A6, 1169 (1973)

[8] *M. Doi* and *S. F. Edwards*, J. Chem. Soc. Farad. Trans. 2 **274**, 1789, 1802, 1818 (1978)

[9] *M. Doi*, J. Polym. Sci., Polym. Phys. Ed. **18**, 1005 (1980); **21**, 667 (1983)

[10] *J. Des Cloiseaux*, J. Physique **36**, 281 (1975)

[11] *H. E. Stanley*, Introduction to Phase Transitions and Critical Phenomena (Oxford University Press, London 1972)

[12] *F. Mezei*, Z. Phys. **255**, 146 (1972)

[13] *J. B. Hayter and J. Penfold*, Z. Phys. **B35**, 199 (1979)

[14] *F. Mezei* (Ed.), Neutron Spin Echo (Springer, Heidelberg, Berlin, New York 1980)

[15] *D. Richter, F. Mezei, J. B. Hayter and B. Ewen*, Phys. Rev. Lett. **41**, 1484 (1978)

[16] *D. Richter, B. Ewen and J. B. Hayter*, Phys. Rev. Lett. **45**, 2121 (1980)

[17] *D. Richter, A. Baumgärtner, K. Binder, B. Ewen and J. B. Hayter*, Phys. Rev. Lett. **47**, (1981); **48**, 1694 (1982)

[18] *B. Ewen, D. Richter, J. B. Hayter and B. Lehnen*, J. Polym. Sci., Polym. Lett. Ed. **20**, 233 (1982)

[19] *B. Stühn*, Dissertation, Mainz (1983)

[20] *B. Ewen*, Pure & Appl. Chem. **56**, 1107 (1984)

[21] *B. Ewen, B. Stühn, K. Binder, D. Richter and J. B. Hayter*, Polymer Communication **25**, 135 (1984)

[22] *D. Richter, K. Binder, B. Ewen and B. Stühn*, J. Phys. Chem. **88**, 6618 (1984)

[23] *D. Richter and B. Ewen*, in: Scaling Phenomena in Disordered Systems, ed. by *R. Pynn* and *A. Skjeltorp*, NATO ASI Series B: Physics Vol. 133 (Plenum Press, New York 1985)

[24] *J. D. Ferry*, Viscoelastic Properties of Polymers (Wiley, New York 1980)

[25] *W. W. Graessley*, Adv. Polymer Sci. **12**, 1 (1973)

[26] *P. G. de Gennes*, Physics (USA), **3**, 37 (1967)

[27] *E. Dubois-Violette and P. G. de Gennes*, Physics (USA), **3**, 181 (1967)

[28] *P. G. de Gennes*, J. Chem. Phys. **72**, 4756 (1980); J. Physique **42**, 735 (1981)

[29] *C. C. Han and A. Z. Akcasu*, Macromolecules **14**, 1080 (1981)

[30] *Y. Tsunashima, N. Nemoto and M. Kurata*, Macromolecules **16**, 1184 (1983)

[31] *R. M. Johnson, J. L. Schrag and J. D. Ferry*, Polym. J. **1**, 742 (1970)

[32] *T. C. Warren, J. L. Schrag and J. D. Ferry*, Macromolecules **6**, 467 (1973)

[33] *M. Daoud and G. Jannink*, J. Physique **37**, 973 (1976)

[34] *J. P. Cotton, M. Nierlich, F. Boué, M. Daoud, B. Farnoux, G. Jannink, R. Duplessix and C. Picot*, J. Chem. Phys. **65**, 1101 (1976)

[35] *B. Farnoux, F. Boué, J. P. Cotton, M. Daoud, G. Jannink, M. Nierlich and P. G. de Gennes*, J. Physique **39**, 77 (1978)

[36] *M. Benmouna and A. Z. Akcasu*, Macromolecules **11**, 1187 (1978)

[37] *B. Nystroem and J. Roots*, Progr. Polymer Sci. **8**, 333 (1982)

[38] *M. Daoud, J. P. Cotton, B. Farnoux, G. Jannink, G. Sarma, H. Benoit, R. Duplessix, C. Picot and P. G. de Gennes*, Macromol. **8**, 804 (1975)

[39] *B. Farnoux, M. Daoud, D. Decker, G. Jannink and R. Ober*, J. Physique Lett. **35**, L122 (1975)

[40] *R. W. Richards, A. Machonnachie and G. Allen*, Polymer **19**, 266 (1978)

[41] *Z. Akcasu, H. Gurol*, J. Polym. Sci., Polym. Phys. Ed., **14**, 1 (1976).

[42] *M. Benmouna and A. Z. Akcasu*, Macromolecules **11**, 1187 (1978)

[43] *K. L. Nicholson, J. S. Higgins and J. B. Hayter*, Macromolecules **14**, 836 (1981)

[44] *J. S. Higgins, K. Ma, L. K. Nicholson, J. B. Hayter, K. Dodgson and J. A. Semelyen*, Polymer **24**, 793 (1983)

[45] *P. G. de Gennes*, Macromolecules **9**, 587, 594 (1976)

[46] *F. Brochard*, J. Physique **44**, 39 (1983)

[47] *F. Brochard* and *P. G. de Gennes*, Macromolecules **10**, 1157 (1977)

[48] *B. Ewen, D. Richter, B. Stühn* and *B. Farago*, to be published;

[49] *M. Adam* and *M. Delsanti*, Macromolecules **10**, 1229 (1977)

[50] *G. Weill, J. des Cloiseaux*, J. Physique **40**, 99 (1979)

[51] *S. F. Edwards* and *K. F. Freed*, J. Chem. Phys. **61**, 1189 (1974)

[52] *K. F. Freed* and *S. F. Edwards*, J. Chem. Phys. **61**, 3626 (1974)

[53] a) *K. F. Freed* and *S. F. Edwards*, J. Chem. Soc. Faraday Trans. 1 **71**, 2025 (1975)
      b) *K. F. Freed* and *S. F. Edwards*, J. Chem. Phys. **62**, 4032 (1975)
      c) *K. F. Freed*, in: Progress in Liquid Physics, ed. by *C. A. Croxton* (Wiley, London 1978), p. 343

[54] *K. F. Freed* and *H. Metiu*, J. Chem. Phys. **68**, 4604 (1978)

[55] *M. Muthukumar* and *K. F. Freed*, Macromolecules **10**, 899 (1977); **11**, 843 (1978)

[56] *K. F. Freed*, Ferroelectrics **30**, 277 (1980)

[57] *A. Perico* and *C. Cuniberti*, J. Polym. Sci., Polym. Phys. Ed. 1983 **17**, (1983)

[58] *K. F. Freed* and *A. Perico*, Macromolecules **14**, 1290 (1981)

[59] *M. Muthukumar* and *S. F. Edwards*, Polymer **23**, 345 (1982)

[60] *J. S. Higgins* and *L. K. Nicholson, J. B. Hayter*, Polymer **22**, 163 (1981)

[61] *J. S. Higgins* and *J. E. Roots*, J. Chem. Soc., Faraday Trans. 2, **81**, 757 (1985)

[62] *T. Kataoka* and *S. Ueda*, J. Polym. Sci. Polym. Lett. **4**, 317 (1960)

[63] *D. Richter, B. Stühn, B. Ewen* and *D. Nerger*, submitted for publication

[64] *R. Oeser, B. Ewen, D. Richter* and *B. Farago*, to be published

Temporal and Spatial Structures of Nonlinear Dynamical Systems

Hans-Georg Purwins, Günter Klempt, Jürgen Berkemeier

Institut für Angewandte Physik, Universität Münster,
D-4400 Münster, Federal Republic of Germany

Summary: The present article contains the description of simple experiments mounted for a demonstration of temporal and spatial structures of dissipative nonlinear dynamical systems. The systematic approach possible to systems with few degrees of freedom is described on an elementary level showing experiments on a rotating nonlinear oscillator. The temporal structure elements are those contained in the stationary, periodic, quasi-periodic, and chaotic motion. Structures of systems with many degrees of freedom are demonstrated by showing experiments on real spatially extended electronic circuits and gas discharge systems both described by reaction diffusion equations. Such systems have a complexity and richness of structures far beyond what can be described systematically by current techniques. However, for special cases a quantitative understanding is possible. Also filaments of rather well defined size and shape observed in our experiments can be considered as simple elements building up a variety of spatial patterns. We also show that noise is decisive in many cases for the formation and the nonreproducibility of stationary structures. Finally, we stress some features common to reaction diffusion systems and living beings.

1 Introduction

The idea of structure is of fundamental importance for all natural sciences. It implies that the complex system under consideration can be decomposed into simpler subsystems with mutual interactions:

$$(\text{element}) + (\text{interaction}) \to \text{structure}. \tag{1-1}$$

In physics the most important structures are those appearing in time and space; we therefore restrict ourselves to these two kinds. However, this does not mean that the elements are necessarily objects in time or ordinary space. Perhaps the most familiar elements in this context are the Fourier components which are defined in frequency and wave vector space. For the present talk also elements in phase space are important.

The term nonlinear dynamical system does not seem to be very well defined; however, among others the following properties are essential.

— The system is nonlinear with the consequence that the superposition principle does not apply.

— A given system can be conservative which has the advantage of dealing at least with one constant of motion or it can be dissipative with all the simplifications that arise from the contraction of phase space. In the present talk we consider exclusively dissipative systems.

27

— With increasing number of degrees of freedom the theoretical description becomes more and more difficult.

Examples of systems with structures being well understood are shown in Tab. 1a...c. We note that the concepts of linear superposition and weakly interacting quasi-particles play an important role. However, for the systems mentioned from Tab. 1d...g the description of temporal and spatial structures becomes somehow vague. Indeed today most of the structures found in daily life are even not classified.

In the presence of nonlinearities dynamical systems show a large variety of structures depending on the choice of parameters, boundary conditions, initital conditions, and perturbations. These features lead to considerable difficulties in experimental investigations and to formidable mathematical problems when treating the corresponding nonlinear equations. However, in recent time progress in experimental techniques, in the facilities of handling large amounts of data, in the geometrical and algebraic methods, and in the use of fast computers have made possible the discovery of many exciting and unexpected temporal and spatial structures. There are good reasons to believe that the whole field being still in its infancy will contribute in a substantial way to a deeper understanding of the overwhelming richness of structures seen almost everywhere in nature.

In this talk we report some important ideas used in the present understanding of temporal and spatial structures of nonlinear dynamical systems. This will be done

Table 1 Examples of systems with temporal and spatial structures

system	element	interaction	structure
a. linear information channel	pulses oscillations	linear super- position	information
b. solid body	atoms defects	electromagnetic exchange mechanical	crystallographic lattice defect structure
c. oscillating linear continuum	Fourier component	linear super- position	waves nodes frequency and wave vector Fourier spectrum
d. homogeneous chemical reactions	pulses oscillations	?	frequency Fourier spectrum
e. living beings	cells	chemical reactions (?)	arrangement of cells in the organ
f. fluids	vortex lines	?	vortex pattern
g. atmosphere	clouds	?	cloud pattern

by performing rather simple experiments from which we deduce the fundamental concepts. In chapter 2 we show a nonlinear rotating oscillator and demonstrate for systems with few degrees of freedom the concept of stability, attractors, and bifurcations of vector fields in phase space. This chapter is rather elementary and can be considered as an introduction to the field of nonlinear dynamical systems using an important model system as an example. Quasi-continuous and continuous systems are introduced in chapter 3, and various temporal and spatial structures are shown on electrical networks and a gas discharge system. We also draw the attention to semiconductor devices. These are examples of reaction diffusion systems which play an important role in chemistry, biology, and as is demonstrated also in physics. While chapter 2 presents well established concepts chapter 3 gives some very new ideas and experimental results. At the same time these systems are very suitable to demonstrate the richness of structure forming properties of systems with many degrees of freedom. In the course of the talk it will become clear that the understanding of systems with many degrees of freedom is rather poor. Finally in chapter 4 we make some conclusive remarks.

The literature in the field of nonlinear dynamical systems is rapidly growing, and it is beyond the scope of this article to present an exhaustive review. In what follows we therefore give references which reflect more or less the authors' approach to the subject. It is the hope that these references will help to fill in many details which cannot be dealt with in this presentation. A rather broad view of the properties of systems with few degrees of freedom is given in the introductory book of Thompson and Stewart [1]. The same point of view on an advanced level is treated in Guckenheimer and Holmes [2]. Of considerable value is a collection of important papers edited by Cvitanović [3]. Also the book of Schuster [4] may be mentioned. For systems with many degrees of freedom the largest amount of literature exists for topics related to hydrodynamics. A general survey of modern ideas is presented e.g. by Tritton [5] and edited by Swinney and Gollub [6]. Another broad field is the investigation of reaction diffusion systems where the books of Smoller [7], and Nicolis and Prigogine [8], and the work of Fife [9] and Nicolis [10] give an up to date account. From the biological point of view the review of Gierer [11] is of interest. Finally, we draw the readers' attention to the Springer series on synergetics which covers to some extent the whole field of nonlinear dynamical systems [12].

2 The Rotating Oscillator and Systems with Few Degrees of Freedom

2.1 The Rotating Oscillator

In this chapter we demonstrate experimentally how systems of few degrees of freedom can be described using the concept of stability, attractors, and bifurcations of vector fields in phase space. As experimental setup we use a mechanical nonlinear rotating oscillator. The details of the apparatus are shown in Fig. 1. A rotating copper wheel is subject to linear torques produced by a coiled spring, by eddy currents, and by a sinusoidally driven step motor. A nonlinear torque is produced by an additional mass.

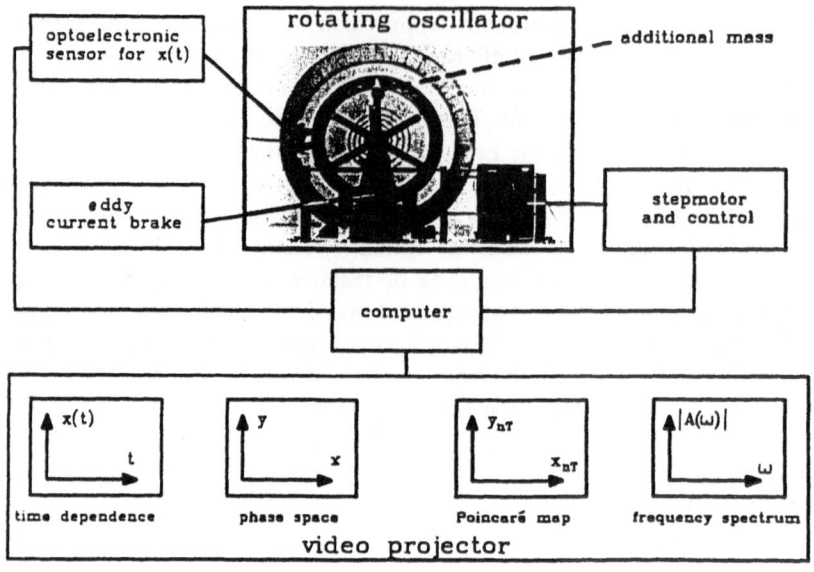

Fig. 1 Experimental setup of the rotating oscillator including the data handling and projection.

The equation of motion of the system has the form:

$$(I + mr^2) \frac{d^2\Phi}{dt^2} = -b \frac{d\Phi}{dt} - C\Phi + mgr \sin(\Phi) + A \sin(\Omega t), \qquad (2\text{-}1)$$

Φ = angular deflection, $\qquad$ I = moment of inertia for m = 0,
m = additional mass, $\qquad$ r = distance of m from axis,
b = damping constant, $\qquad$ C = spring constant,
g = gravitational constant, $\qquad$ A = amplitude of the driving torque,
Ω = driving frequency, $\qquad$ T = $2\pi/\Omega$ driving period,
I, m, r, b, C, A, $\Omega \geqslant 0$.

Transforming Eq. (2-1) we obtain with

$\Phi = x$, $\qquad\qquad$ $k = b/(I + mr^2)$, $\qquad$ $\omega_0^2 = C/(I + m r^2)$

$\mu = mgr/(I + m r^2)$, $\qquad$ $B = A/(I + mr^2)$,

$$\frac{d^2x}{dt^2} + k \frac{dx}{dt} + \omega_0^2 x - \mu \sin(x) = B \sin(\Omega t). \qquad (2\text{-}2)$$

30

This is the equation of a driven damped linear oscillator plus an additional nonlinear term governed by the nonlinearity μ. Eq. (2-2) can be transformed into an autonomous system of first order ordinary differential equations by writing $y = dx/dt$ and $t = z$:

$$\frac{dx}{dt} = y, \qquad\qquad\qquad\qquad\qquad\qquad\qquad\qquad\text{(a)}$$

$$\frac{dy}{dt} = -\omega_0^2 x + \mu \sin(x) - ky + B \sin(\Omega z), \qquad\qquad\text{(b)} \quad (2\text{-}3)$$

$$\frac{dz}{dt} = 1. \qquad\qquad\qquad\qquad\qquad\qquad\qquad\qquad\text{(c)}$$

For given initial conditions solutions of Eq. (2-3) are nonintersecting trajectories $(x(t), y(t), z(t))$ in 3-dimensional phase space. The ensemble of all possible trajectories in phase space defines a vector field which generates a flow in turn. In the sense of nonlinear dynamical theory the system is considered to be understood if the vector field in phase space is known for all parameter sets of interest.

By generalizing Eq. (2-3) we may state that a major goal in the field of nonlinear dynamical systems is to investigate physical, chemical, biological, or other objects which are described by an autonomous system of first order nonlinear ordinary differential equations of the kind:

$$\vec{x} = \vec{F}(\vec{x}), \quad \vec{x} = (x_1, \ldots, x_n), \quad \vec{F} = (f_1, \ldots, f_n). \qquad\qquad (2\text{-}4)$$

The system Eq. (2-4) is understood if the vector field in n-dimensional phase space in known. Because of the complexity of this field one often restricts the description to the global topological structure of the vector field and abandons the details of the trajectories. This will be discussed in length by looking at the rotating oscillator.

A detailed mathematical description of our apparatus has to take into account in Eqs. (2-1), (2-2), and (2-3) further terms, in particular static friction. We neglect these effects since our main interest is to show some most important nonlinear effects and to give an impression of the overall behaviour of nonlinear systems.

2.2 Point Attractors

First we investigate the freely rotating oscillator by switching off the driving torque ($B = 0$). The system is then described by a vector field in two-dimensional phase space resulting from Eq. (2-3a, b). Using in our experiment various initial conditions, we obtain a survey of the organization of the vector field. From Fig. 2a-d we note that the system can have stationary states which are solutions of Eq. (2-3a, b), with $(dx/dt, dy/dt) = (0,0)$. These solutions correspond to single points $(x^*, y^*) = \vec{x}^*$ in phase space which are called fixed points. Experimentally, we find for $\mu < 7.7\,\omega_0^2$:

$$\omega_0^2 > \mu \quad \vec{x}_0^* = (0,0) \qquad\qquad\qquad\qquad\qquad \text{stable,}$$
$$\omega_0^2 < \mu \quad \vec{x}_{1,2}^* = (\pm a, 0) \quad (a > 0: \omega_0^2\, a = \mu \sin a) \quad \text{stable,} \qquad (2\text{-}5)$$
$$\qquad\quad \vec{x}_3^* = (0, 0) \qquad\qquad\qquad\qquad\qquad \text{unstable.}$$

The stable fixed points have the property that after a small deviation the system is driven back to $\vec{x}*$ after long time. In the case of unstable fixed points the system is driven away.

The stable fixed points each have a basin of attraction that is a region in phase space from which all trajectories are pulled into the fixed point. Stable fixed points are therefore called point attractors. The basin of attraction for $\omega_0^2 > \mu$ is the whole phase space, and there is only one stationary state, namely, the stable state $\vec{x}_0^*$. We call the system monostable. For $\omega_0^2 < \mu < 7.7\,\omega_0^2$ the phase space is decomposed into two basins of attraction corresponding to the attractors $\vec{x}_1^*$ and $\vec{x}_2^*$ separated by the separatrix. Since in this range of parameters the system has two stable states, it is bistable which is a special case of multistability. From the experiments shown in Fig. 2a-d we see that there is a qualitative change of the vector field at $\omega_0^2 = \mu$. Apparently, this is related to a change of number and properties of fixed points; we call this a bifurcation, and a corresponding bifurcation diagram is shown in Fig. 2e. We conclude that for our case of a freely rotating oscillator fixed

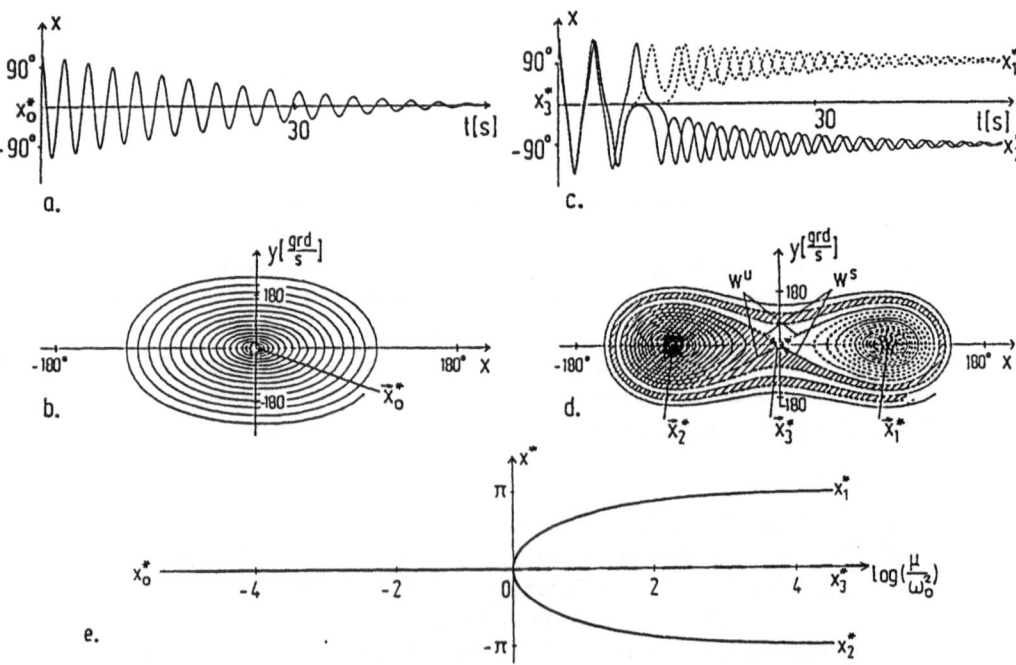

Fig. 2 Angular deflection x (a, c) and trajectories in phase space (b, d) of the rotating oscillator for B = 0, k $\approx$ 0.07 s^{-1}, $\omega_0^2 \approx 7.5$ s^{-2}, $\mu \approx 4.3$ s^{-2} (a, b), and $\mu \approx 11.5$ s^{-2} (c, d), respectively. Shaded area of (d) is the basin of attraction for $\vec{x}_2^*$, white area that for $\vec{x}_1^*$ and W^s and W^u of (d) are the stable and unstable manifold of $\vec{x}_3^*$. (e) represents the bifurcation diagram for the (stable) (———) and unstable (– – – –) fixed points using the nonlinearity as bifurcation parameter. Fixed points with $|x^*| > 180°$ are not shown.

32

points serve to organize the vector field in phase space. − For $\mu > 7.7\ \omega_0^2$ there are further bifurcations and fixed points with $|x^*| > \pi$ do appear in addition to those shown in Fig. 2e. This case is not considered in the present article. −

We generalize our observations in that we state that fixed points of Eq. (2-4) are essential for the organization of the vector field in phase space and that they belong to the most important objects investigated in the theory of nonlinear dynamical systems. Thereby one first determines some or all fixed points of Eq. (2-4). To characterize the fixed points one then linearizes Eq. (2-4) around the fixed points and determines the n eigenvalues and eigenvectors. These vectors span an n-dimensional eigenvector space. This space can be decomposed into n^s-, n^u- and n^c-dimensional stable, unstable, and center manifolds for each fixed point ($n^s + n^u + n^c = n$). Iterating the nonlinear equation out of fixed points by using as starting value the manifolds of the linear system one can obtain the stable, unstable, and center invariant manifolds W^s, W^u, and W^c of the whole nonlinear system with respect to a certain fixed point. W^s and W^u obtained in this way are unique; for W^c this is not necessarily the case. This is the essence of the important "center manifold theorem" of trajectories in phase space. The theorem allows to obtain important information of the whole flow in phase space by solving the linear system and calculating W^s, W^u, and W^c. This theorem and its extension to Poincaré maps (s. below) is therefore the privot of the analysis of almost all nonlinear dynamical systems with few degrees of freedom.

In the case of Fig. 2b, d the fixed points $\vec{x}_0^*$, $\vec{x}_1^*$, and $\vec{x}_2^*$ have a two-dimensional manifold W^s (W^u and W^c are empty) with the result that these points are attractors having each their basin of attraction. In contrast $\vec{x}_3^*$ has a one-dimensional stable manifold W^s and an unstable manifold W^u of the same dimensionality. Running out of $\vec{x}_3^*$ backward along W^s gives the separatrix, running foreward W^u gives the fixed points $\vec{x}_0^*$ and $\vec{x}_2^*$. Fixed points having no center manifold are called hyperbolic fixed points.

At $\omega_0^2 = \mu$ the fixed point $\vec{x}_0^* = 0$ has a one-dimensional center manifold, and consequently, this point is called nonhyperbolic. However, this is just the point where we observe a bifurcation. We therefore conclude that the center manifold plays an important role in the theoretical investigation of bifurcations. In fact it is precisely this manifold which determines exclusively the bifurcation behaviour. This in turn generates universal behaviour; all systems with the same center manifold exhibit the same bifurcation behaviour.

2.3 Periodic Attractors

We now investigate the driven oscillator as described by the full system of Eqs. (2-3a, b, c). For two sets of parameters we show in Fig. 3a, e x as a function of time and in Fig. 3b, f the corresponding trajectories projected into the x-y phase space plane after long time. From the latter we see that the system exhibits periodic motion with the frequency Ω of the driving motor. This is also clear from Figs. 3d, h where the Fourier spectrum is shown. The periodic motion represented in Fig. 3a...d

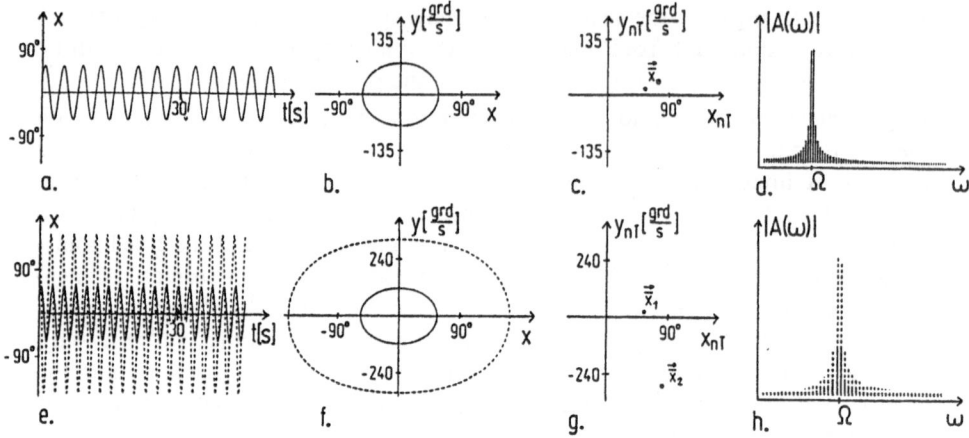

Fig. 3 Angular deflection x (a, e), projection of phase space into the (x-y)-plane (b, f), Poincaré map (c, g), and frequency spectrum (d, h) of the rotating oscillator for parameters $B \approx 2.3$ s^{-2}, $k \approx 0.34$ s^{-1}, $\omega_0^2 \approx 7.7$ s^{-2}, $\mu \approx 4.3$ s^{-2}, and $\Omega \approx 1.5$ s^{-1} for (a ... d), and $\Omega \approx 2.5$ s^{-1} for (e ... h), respectively.

seems to have the whole phase space as attractor. For the case of Fig. 3e...h we have two periodic attractors with basins of attraction which are difficult to represent.

We also plot $y_{nT} = y(nT)$ versus $x_{nT} = x(nT)$ in Fig. 3c, g. This stroboscopic representation of trajectories is an example of a large class of discrete maps called Poincaré maps. In this representation orbitals in phase space with period mT correspond to m points. In the case of Fig. 3a...d we have one such point $\vec{x}_0$ and in the case of Fig. 3e...h two, namely $\vec{x}_1$ and $\vec{x}_2$. $\vec{x}_0$ is a stable fixed point since the trajectories in the vicinity of the periodic orbital are pulled into the corresponding limit cycle of Fig. 3b. Consequently, the fixed point is an attractor having his proper basin of attraction also in the Poincaré map. Analytical arguments show that in our case there exists at least one unstable periodic orbital.

We now generalize our results: in addition to fixed points periodic and as can be shown quasi-periodic orbitals organize the vector field in phase space. To analyse the behaviour of systems with periodic orbitals the fixed points in the Poincaré map are investigated. In analogy to the fixed points in phase space there is a center manifold theorem for Poincaré maps. It tells us that we can find from the linearized system the stable, unstable, and center manifold of the nonlinear system. From these manifolds we can draw important conclusions concerning the whole phase space. Again the center manifold governs the bifurcations.

In Fig. 4 we show the resonance curve of our rotating oscillator for the nonlinear case. Apparently, the overhanging multivalued curve can be regarded as a sheared singlevalued curve of the linear case. Carrying out the experiment with Ω increasing

34

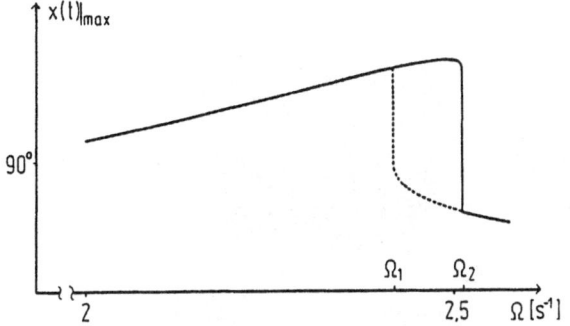

Fig. 4

Maximum amplitude $x(t)|_{max}$ of the angular deflection of the rotating oscillator as a function of the driving frequency changed in quasi-stationary manner for $B \approx 2.3 \text{ s}^{-1}$, $k \approx 0.34 \text{ s}^{-2}$, $\omega_0^2 \approx 7.7 \text{ s}^{-2}$, and $\mu \approx 4.3 \text{ s}^{-2}$ (increasing: ———, decreasing: — — — —).

quasi-stationarily, we follow the upper curve and for decreasing Ω the lower curve. At Ω_1 we have a bifurcation of one stable periodic orbital into two stable and one unstable periodic orbital, the latter being a result of theoretical considerations. At Ω_2 this bifurcation is reversed. The system is monostable for $0 \leqslant \Omega \leqslant \Omega_1$, $\Omega_2 \leqslant \Omega$ and bistable for $\Omega_1 < \Omega < \Omega_2$.

2.4 The Chaotic Attractor

We now show that in addition to point, periodic, and quasi-periodic attractors there is a fourth kind of attractors organizing the phase space in dissipative systems. To demonstrate this we measure the behaviour of the nonlinear driven rotating oscillator as shown in Fig. 5. Apparently, the time dependence of x in Fig. 5a has an irregular "chaotic" behaviour. This is also reflected by the trajectories of Fig. 5b in the phase space projection and the "strange" point set in the Poincaré map of Fig. 5c. The corresponding frequency spectrum of Fig. 5d is rather broad. We observe in Fig. 5a, b that trajectories being initially very near separate from each other after a short time interval demonstrating a sensitive dependence on initial conditions. From Fig. 5b, c we see that the trajectories are pulled into a certain region in phase space. We therefore conclude that there is a strange or chaotic attractor again having its proper basin of attraction. On the first glance the time dependence of x and the frequency spectrum may give the impression that we deal with a stochastic process of finite bandwidth. However, this is definitely not the case since the system under investigation is deterministic. For this reason we also speak of deterministic chaos. Effects related to strange attractors have been considered until recently as an annoying artefact. This opinion has changed by now and chaotic attractors are investigated by their own right because of the many interesting properties they have and because they are a common feature of nonlinear dynamical systems.

By generalizing the results from Fig. 5 we state some characteristics of strange attractors:

— There is an area in phase space and in the Poincaré map called the strange attractor which attracts the trajectories of some larger region.

35

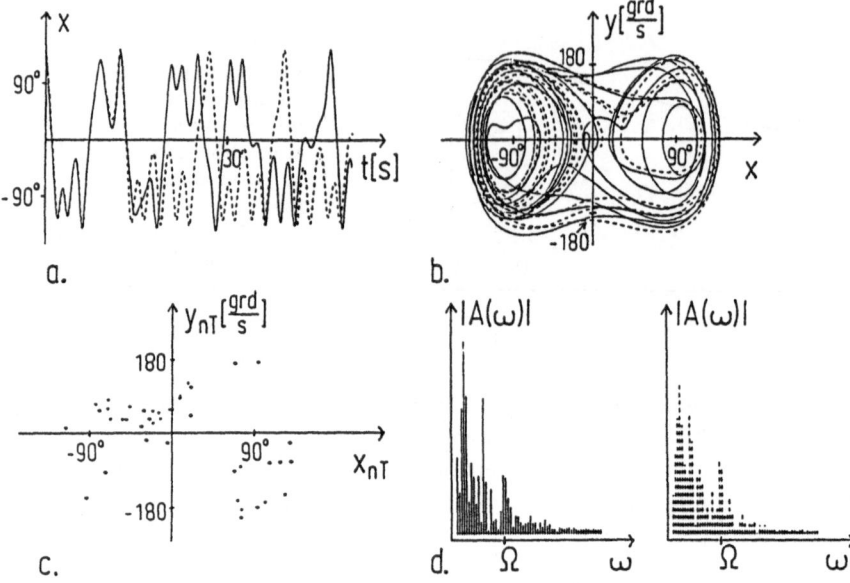

Fig. 5 Angular deflection x as a function of time (a), projection of the trajectories into to the (x-y)-plane (b), Poincaré map (c), and frequency spectrum (d) of the rotating oscillator for $B \approx 2.18 \text{ s}^{-2}$, $k \approx 0.33 \text{ s}^{-1}$, $\omega_0^2 \approx 7.3 \text{ s}^{-2}$, $\mu \approx 11.5 \text{ s}^{-2}$, and $\Omega \approx 2.5 \text{ s}^{-1}$. In (a) and (b) trajectories for slightly different initial conditions are denoted by ——— and – – – –.

- The motion of the system is deterministic so that there is no intersection of trajectories in full phase space.
- Initially neighbouring trajectories separate exponentially in the course of short time intervals.
- Because of the finite extension of the strange attractor exponentially diverging trajectories are folded back after finite time so that any trajectory on the attractor can approach any point of the attractor arbitrarily close.
- Small changes of parameters result in small changes of the attractor.

To demonstrate the property of back folding we measure in Fig. 6 the Poincaré map for various time shifts. It is clear from Eq. (2-3) that a time shift of the amount of the driving period does not change the Poincaré map. However, for smaller shifts there is a deformation of the attractor. By increasing the shift points of the attractor being initially very close can separate and those being far away can come close together. This complicated folding and mixing is clearly seen from Fig. 6 and is the main reason for the unusual topological properties of the chaotic attractor.

By now there are several quantities used to characterize chaotic attractors, a selection of them is given in Tab. 2. We draw particular attention to the dimensionality

36

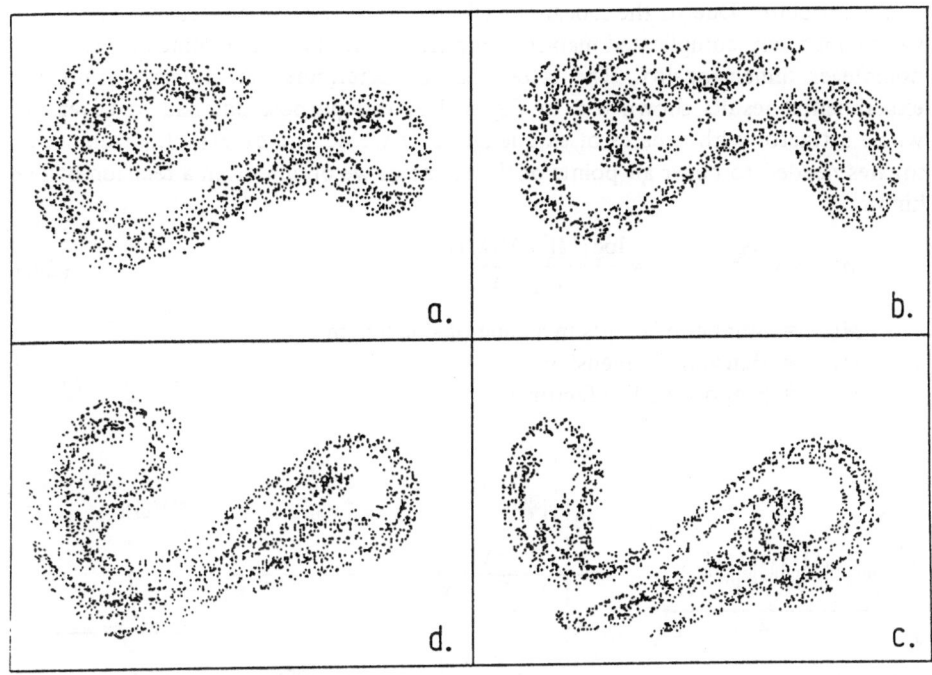

Fig. 6 Poincaré maps for time shift 0 (a), (1/8) T (b), (1/4) T (c), and (3/8) T (d) of the rotating oscillator with $B \approx 2.18 \text{ s}^{-2}$, $k \approx 0.15 \text{ s}^{-1}$, $\omega_0^2 \approx 7.3 \text{ s}^{-2}$, $\mu \approx 11.5 \text{ s}^{-2}$, and $\Omega \approx 2.5 \text{ s}^{-1}$.

Table 2 Some important quantities characterizing strange attractors

quantity	physical meaning	symbols, equations
invariant measure	density of points in the Poincaré map	$\rho(\vec{x})$, $\vec{x} \in \text{IR}^{n-1}$
Liapunov exponent	time average of the exponent of the divergence of initially nearby trajectories	$\lambda_1, \lambda_2, ..., \lambda_n$
Hausdorff dimension	exponent which gives the volume of a point set a finite measure	D_0
Kolmogorov entropy	average of information loss in time of an evolving system	$K = \int d^n x \rho(\vec{x}) \sum_i \lambda_i^+(\vec{x})$, λ_i^+ all λ_i with $\lambda_i > 0$

of the attractor. Due to the repeated back folding of the trajectories in phase space we obtain a very complicated manifold in phase space and in the Poincaré map with noninteger dimensionality. The latter may be determined in principle from our experimental results shown e.g. in Fig. 6. For this purpose one had to cover the whole attractor with squares of side length l. If we denote by M(l) the number of squares needed to cover all points of the attractor we obtain often a relation of the kind:

$$M(l) = \gamma \, l^{D_0} \qquad D_0 = \frac{\log\,(M(l)/M(l_0))}{\log(l/l_0)} \qquad (2\text{-}6)$$

$M(l)$ = number of points in a cube of side length l
D_0 = Hausdorff dimension
γ = proportionality factor

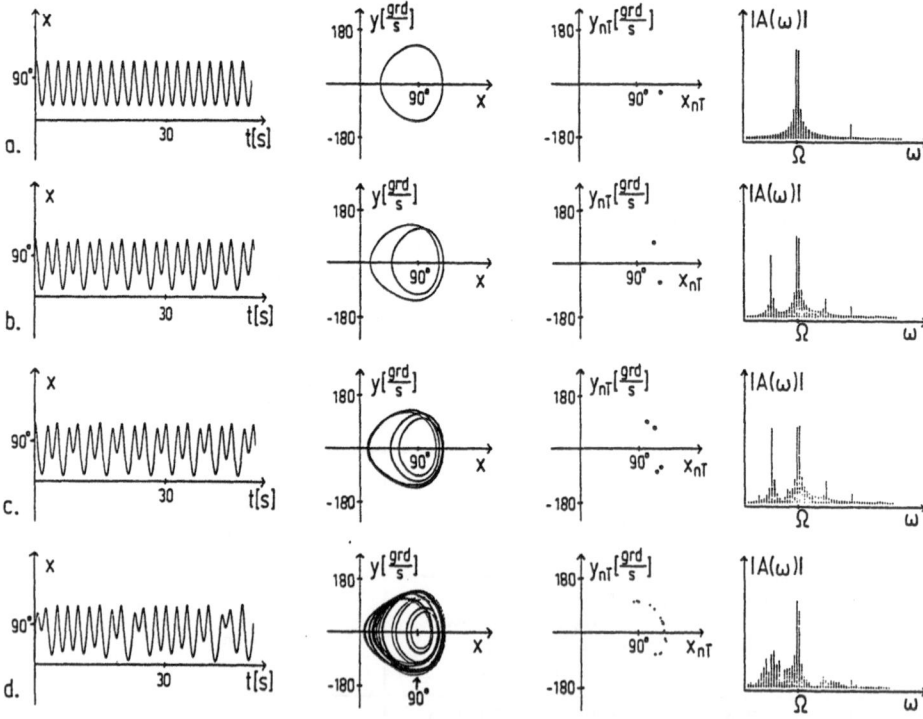

Fig. 7 Angular deflection x, projection of trajectories into the (x-y)-plane, Poincaré map, and frequency spectrum of the rotating oscillator for $B \approx 2.18\ s^{-2}$, $\omega_0^2 \approx 7.3\ s^{-2}$, $\mu \approx 11.5\ s^{-2}$, $\Omega \approx 2.5\ s^{-1}$ and $k \approx 0.88\ s^{-1}$ (a), $k \approx 0.75\ s^{-1}$ (b), $k \approx 0.70\ s^{-1}$ (c), and $k \approx 0.66\ s^{-1}$ (d).

where D_0 is the Hausdorff dimension having a constant value. Unfortunately, the stability of our apparatus concerning the time shift was not sufficient to get a reliable result for D_0. The point sets of Fig. 6 are similar to those called Cantor sets in mathematics. Recently, Grassberger and Procaccia [13] have shown that it is useful to define several in general not much differing dimensions. Also it has been shown by Halsey et al. [14] and Jensen et al. [15] that attractors can have a spectrum of dimensions.

We now turn to the question of how a system can bifurcate into chaotic behaviour. For this purpose we look at the experimental results shown in Fig. 7. With decreasing damping k we observe a cascade of period doublings ending up in a chaotic attractor. Because of experimental resolution we can measure only a small number of period doublings. However, we are rather sure that we are confronted with an infinite sequence breaking up into chaos at a finite value of $k = k_\infty$. Also as more precise measurements would show the value k_n, where the n-th period doubling occurs, scales in the following way

$$\frac{(k_\infty - k_n)}{(k_\infty - k_{n+1})} = \delta = 4.66 \ldots, \qquad n \gg 1. \tag{2-7}$$

This result for δ is observed in many physical systems showing the route of period doubling bifurcations into chaos. Remarkably enough, the control parameter can have very different physical meaning. Apparently, we deal with some kind of universality common to a whole class of nonlinear dynamical systems. If we remember what has been said in Chap. 2.2, we conjecture that the topological reason is again the existence of the same center manifold governing the bifurcation route for the whole class.

The period doubling route to chaos is not the only one. Also the Manneville-Pomeau and the Ruelle-Takens-Newhouse routes are observed. In addition very many other routes are possible. A survey of the properties of the three routes investigated most at present time is given in Tab. 3.

2.5 Discrete maps

Of course in principle the theoretical investigation of nonlinear dynamical systems with few degrees of freedom could be limited to the investigation of the corresponding differential equations of type Eq. (2-4). However, this can be very hard mathematically, and it is tentative to look for concepts which are simpler from the mathematical point of view. One idea is to describe the evolution of the system in discrete time steps. We have already met such a description when defining the Poincaré map. A much simpler example is the quadratic map where the integer i can be interpreted as a discrete time variable

$$X_{i+1} = f_\Lambda (X_i) = 1 - \Lambda X_i^2 . \tag{2-8}$$

Table 3 Three frequently observed routes to chaos

type	system	bifurcation scheme
Feigenbaum	Rayleigh-Bénard convection Taylor-Couette convection nonlinear driven oscillator Belousov-Zhabotinsky reaction optical bistability	infinite sequence of period doublings: $$T = \frac{2\pi}{\Omega} \to 2T \to 4T \to 8T \to \ldots$$
Manneville-Pomeau	Rayleigh-Bénard convection Josephson junction Belousov-Zhabotinsky reaction laser	intermittency: quasi-periodic motion with nearly periodic time intervals interrupted by strongly nonperiodic motion
Ruelle-Takens-Newhouse	Rayleigh-Bénard convection Taylor-Couette convection nonlinear conductor	3-step bifurcation route: • (point attractor) → ◯ (periodic attractor) → ◯ (attractor consisting of two incomensurable periodic oscillations) → chaotic attractor

This map has a rather rich structure (see e.g. Schuster [4] and Cvitanović [3]), and some features are illustrated in Fig. 8. We now search for fixed points X* defined by

$$X^* = f_\Lambda^k (X^*) = f_\Lambda^{k-1} [f_\Lambda (X^*)], \quad k = 2^n, \quad n = 0, 1, 2, \ldots. \tag{2-9}$$

and observe that there are ranges Λ_k where Eq. (2-8) has $k = 2^n$ solutions. In terms of Eq. (2-8) this means that after 2^n time steps the system returns to the same value of X. This is shown schematically in Fig. 8. With increasing Λ we, therefore, observe a cascade of period doublings which ends up in a chaotic behaviour at a finite value of Λ_∞. But this is just what we assume to observe in the experiment described in Chap. 2.4. Also we note that e.g. $\lim (\Lambda_k/\Lambda_{2k}) = 4.66 \ldots$ is a universal constant. Besides this and various other universal properties there are many additional interesting features of Eq. (2-8) which are indicated in Fig. 8 and which are observed in many real systems.

40

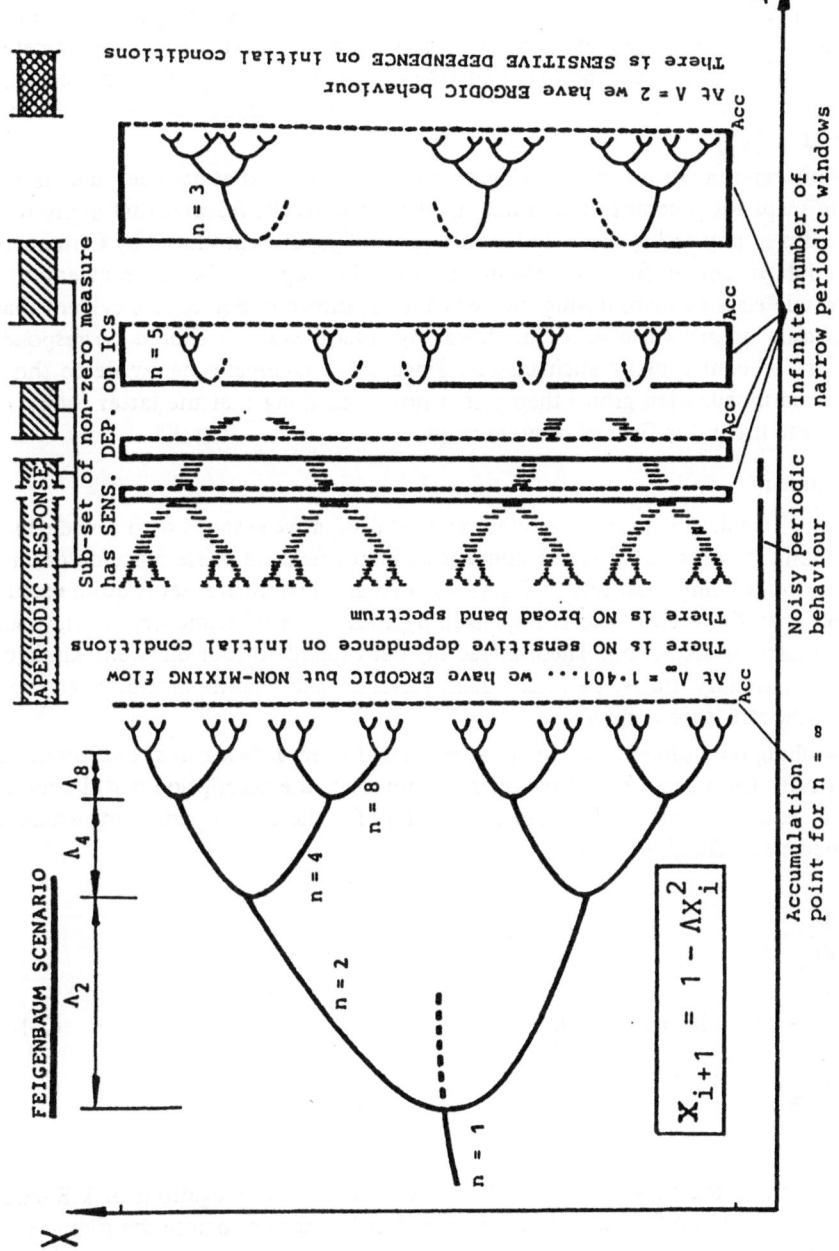

Fig. 8 Behaviour of the quadratic map $X_{i+1} = 1 - \Lambda X_i^2$ (after Thompson and Stewart [1]).

41

From these considerations we conclude that discrete maps like Eq. (2-8) describe important features of real physical systems. Indeed there are striking examples how systems exhibit properties of discrete maps near bifurcation points in parameter space. Nice cases are the experiment of Gwinn and Westervelt [16] on doped Germanium at low temperature and the discussion of a Rayleigh-Bénard experiment by Jensen et al. [15].

From our experiment on the rotating oscillator and from what has been said about discrete maps we generalize that these maps have universal features that apply to a whole class of physical systems. Indeed in a series of papers reproduced by Cvitanović [3] it could be shown first by Feigenbaum that the map Eq. (2-8) is representative for a whole class of maps having the behaviour as shown in Fig. 8. It is evident that to this class of maps a class of nonlinear dynamical systems should correspond. There are of course many such classes. Since these arguments resemble to those used in renormalization group theory it is not astonishing that the latter is now in wide-spread use in the field of discrete maps.

2.6 Summary

In Chapt. 2 we discuss in an elementary manner dissipative systems with few degrees of freedom. It turns out that the concept of vector fields in phase space describing all trajectories being solutions of Eq. (2-4) for a given parameter set is quite useful. We find that the vector field is organized by four types of attractors having their proper basins of attraction. These attractors correspond to four different kinds of motion: stationary states, periodic, quasi-periodic, and chaotic motion with their proper time structure elements.

By describing the number and kind of attractors, the vector field can be characterized qualitatively for a given set of parameters. However, the description is still complicated in parameter space. This can be seen e.g. for the driven pendulum which is described by Duffing's equation

$$\frac{dx}{dt} = y,$$

$$\frac{dy}{dt} = -x^3 - ky + B \cos(z),$$ (2-10)

$$\frac{dz}{dt} = 1.$$

This is a simplified version of Eq. (2-3). The results of investigations in k,B-parameter space by Ueda [17] are shown in Fig. 9. It is amazing to note the richness of possible vector fields compared to the case where Eq. (2-10) is changed to the ordinary driven harmonic oscillator replacing the cubic term by a linear term. In the latter case the whole k,B-plane would contain a single period one orbital.

42

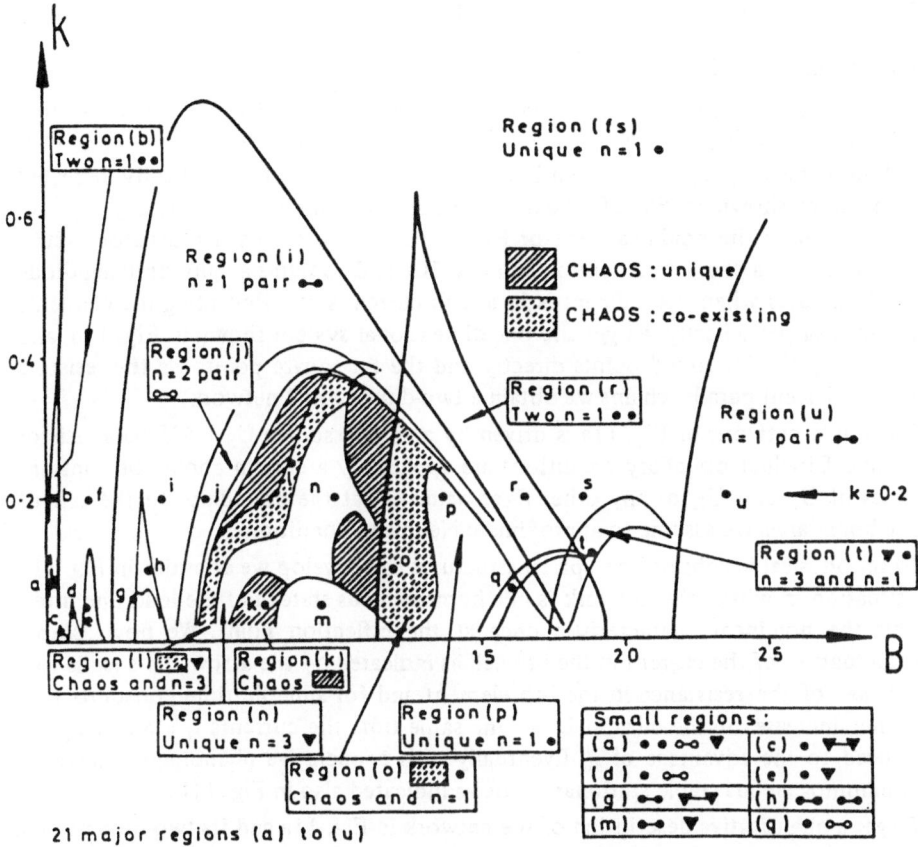

Fig. 9 Parameter plane spanned by the damping constant k and the driving amplitude B of the Duffing oscillator describing the qualitative behaviour of the vector-field (•, ○, ▼ period 1, 2, and 3 oscillations) (after Ueda [17]).

Though a systematic approach to systems with few degrees of freedom is possible there are many important unsolved problems. We want to mention only four.

— The visualization of vector fields in four or higher dimensional space is increasingly difficult.

— A satisfying characterization of chaotic attractors does not exist.

— The bifurcation of vector fields governed by center manifolds with higher dimension than one is not well understood.

— It is difficult to find the correct discrete map for a given physical system.

3 Reaction Diffusion Systems and Many Degrees of Freedom

3.1 The Electrical Network

We now describe an electrical network being the discretized version of a reaction diffusion system and which has been investigated in detail by Berkemeier et al. [18] and Berkemeier [19]. The network consists of an array of periodically repeated elements as shown in Fig. 10. Apart from the linear circuit elements R_V, R_D, L, and C we have the nonlinear resistor R(I) which is realized by a transistor circuit. The latter has a characteristic shown in Fig. 10b and contains a light emitting diode which radiates when the element is in a high current state. Repeating the elements of Fig. 10a periodically we get the one-dimensional system shown in Fig. 11a, and connecting the fat round points directly and the fat square points via the resistors R_D to adjacent parallel chains we obtain a two-dimensional network.

The whole network of Fig. 11a is driven by a voltage source $U_S + \Delta U$ via a resistor R_0 and Dirichlet boundary conditions are applied by a certain choice of constant values of U_L and U_R or any other fixed value U_b at the boundary. In the case of free boundaries we assume to approximate Neumann conditions.

To get physical insight of how spatial structures can develop we consider in Fig. 11b a situation in which the network is in a homogeneous state, and the load line intersects the nonlinear characteristic once at the inflection point. We now assume a fluctuation of the current in the i-th cell as indicated by the arrow. This leads to a decrease of the resistance in the i-th element and for suitable time constants to a further increase of the current I_i. At the same time the currents I_{i-1} and I_{i+1} are reduced to the advantage of I_i. Eventually, this cooperative phenomenon leads to an inhomogeneous stable stationary state as indicated also in Fig. 11b.

To get a quantitative description of the network in Fig. 11a and its two-dimensional extension we apply Kirchhoff's rules and write for the two-dimensional network:

$$\frac{dU_{i,j}}{dt'} = \frac{1}{C}\left[\frac{1}{R_D}(U_{i-1,j} + U_{i+1,j} + U_{i,j-1} + U_{i,j+1} - 4U_{i,j}) - \right.$$

$$\left. - \frac{1}{R_V}U_{i,j} - I_{i,j} + \frac{1}{R_V + NR_0}\cdot\left(U_S + \Delta U + \frac{R_0}{R_V}\sum_{i,j}U_{i,j}\right)\right],$$

$$\frac{dI_{i,j}}{dt'} = \frac{1}{L}[U_{i,j} - S(I_{i,j})], \tag{3-1}$$

$$S(I_{i,j}) = R(I_{i,j})\,I_{i,j} \approx U_0 - \chi'(I_{i,j} - I_0) + \Phi'(I_{i,j} - I_0)^3,$$

$$i = 1 \ldots N_i, j = 1, \ldots N_j, N = N_i N_j.$$

The one-dimensional network is described by dropping the index j. We note that qualitatively the structural behaviour of the network is not effected by the details of the s-shaped characteristic of Fig. 10b if only the general behaviour is retained. We therefore use the simple polynomial form of Eq. (3-1) for $S(I_{i,j})$ when writing down equations.

44

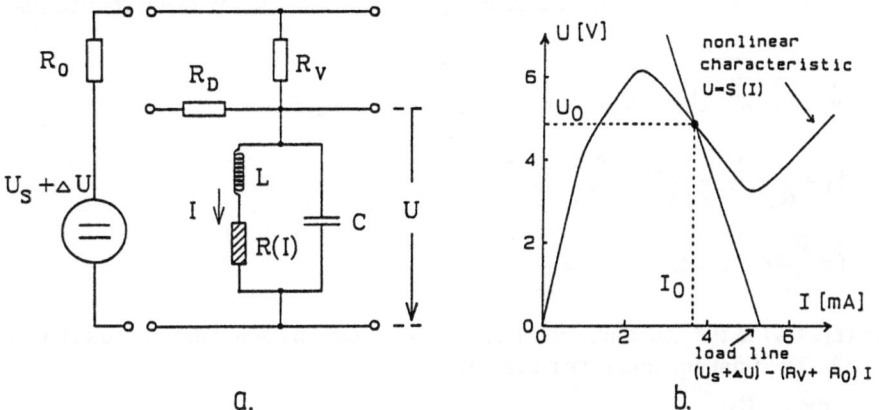

Fig. 10 Isolated element of the electrical network (a) and characteristic of the nonlinear resistor R (I) including the load line (b). (R_V = 3 kΩ, L = 1 mH, C = 820 pF, R_0 = 0 in general and R_D = 70 Ω ... 430 Ω or 4 kΩ throughout this paper.)

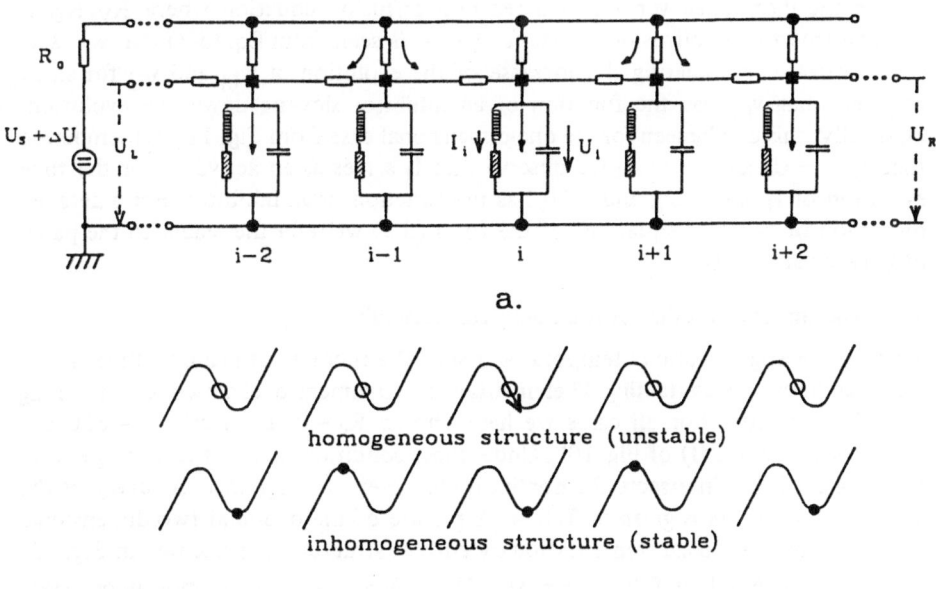

Fig. 11 Periodic electrical network (a) and the principle of formation of inhomogeneous stable stationary structures (b). ∘, • denote an unstable and a stable situation.

To see the mathematical structure of Eq. (3-1) we make the following transformations:

$$w_{i,j} = \sqrt{\frac{\Phi'}{R_V}} (I_{i,j} - I_0), \qquad v_{i,j} = \sqrt{\frac{\Phi'}{R_V^3}} (U_0 - U_{i,j}),$$

$$I_0 = \frac{1}{R_V + NR_0} \left(U_S + \frac{NR_0}{R_V} U_0 \right) - \frac{U_0}{R_V}, \tag{3-2}$$

$$t = \frac{R_V}{L} t', \quad \epsilon = R_V^2 \frac{C}{L}, \quad \lambda = \frac{\chi'}{R_V}.$$

Here (I_0, U_0) is the inflection point of the nonlinear characteristic as indicated in Fig. 10b. The final equations have the form

$$\epsilon \frac{dv_{i,j}}{dt} = \frac{R_V}{R_D} (v_{i-1,j} + v_{i+1,j} + v_{i,j-1} + v_{i,j+1} - 4v_{i,j}) +$$

$$+ w_{i,j} - v_{i,j} + \frac{R_0}{R_V + NR_0} \cdot \sum_{i,j} v_{i,j} - \sqrt{\frac{\Phi'}{R_V}} \cdot \frac{\Delta U}{R_V + NR_0}, \tag{3-3}$$

$$\frac{dw_{i,j}}{dt} = \lambda w_{i,j} - w_{i,j}^3 - v_{i,j}.$$

This is the discretized version of a reaction diffusion equation where R_V/R_D is the analogue to the diffusion constant. Also it is seen from Eq. (3-3) that $w_{i,j}$ acts as an activator stimulating the increase of the evolution of $w_{i,j}$ and $v_{i,j}$ for small $w_{i,j}$ and that $v_{i,j}$ has the function of an inhibitor slowing down the evolution. Physically, this can be seen for the one-dimensional case from Fig. 11. Remembering that $v_i \sim - U_i$ and $w_i \sim I_i$ we observe that I_i serves as an activator for the time evolution of I_i and $- U_i$, and $- U_i$ has the function of an inhibitor. For a detailed discussion of types of equations of the form (3-3) we refer the reader to the paper of Purwins et al. [20].

3.2 Experimental Results for the Electrical Network

We now investigate various temporal and spatial structures of real one-dimensional electrical networks containing 33 elements and two-dimensional networks containing 31 × 31 elements. For all cases we have chosen $R_0 = 0$, $L = 1$ mH, $C = 820$ pF, $R_V = 3000 \ \Omega$, and S(I) of Fig. 10b. Under these conditions and in the homogeneous state the load line intersects the nonlinear characteristic only once. A survey of the experimental results is given in Tab. 4. A picture of the one- and two-dimensional network showing typical stable stationary current structures is presented in Fig. 12.

In experiment No. 1 of Tab. 4 $U_S + \Delta U = U_V$ is such that in the homogeneous state the load line of Fig. 10b intersects the nonlinear characteristic in the inflection point. For the one-dimensional network we have plotted the voltage U_i of each net-

Table 4 Temporal and spatial structures of the one- (IR^1) and twodimensional (IR^2) networks under various conditions. U_b = const means at least one element of the boundary is kept fixed and const' means that another constant value is switched on. Arrows indicate the switching sequence. Fast switching mode denotes that the switching time is much smaller than the characteristic time constant of the network which is of the order $1/\sqrt{LC} \approx 10^{-5}$ s. From experiment No. 1 to 5 R_D is 70 ... 430 Ω, for No. 6 4 kΩ.

No. of experiment	network	switching mode		structure	remark
		U_b	$U_S + \Delta U$		
1	IR^1, IR^2	const	→ fast on	inhomogeneous stationary	see Figs. 12, 13, 14
2	IR^1, IR^2	free	→ fast on	homogeneous oscillatory	
3	IR^1, IR^2	free → const	→ fast on	inhomogeneous stationary	boundary serves as nucleation center
4	IR^1, IR^2	const → const'	→ fast on → fast off → fast on	inhomogeneous stationary	multistability
5	IR^1, IR^2	free	→ slow on	inhomogeneous in general stationary	structure determined by inhomogeneities
6	IR^1	special		inhomogeneous oscillating areas	see Fig. 16

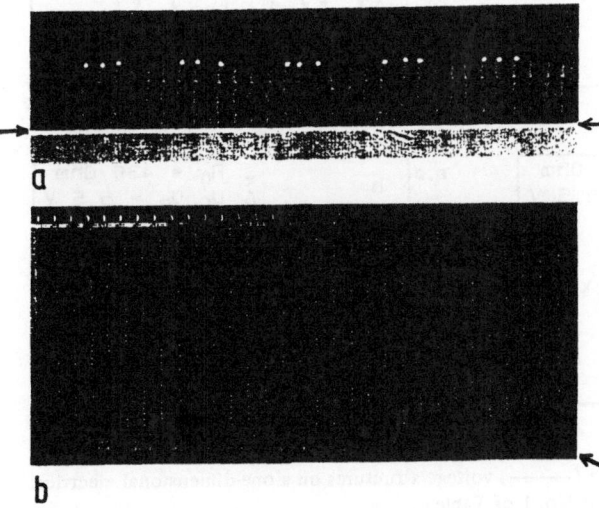

Fig. 12
The one- (a) and two-dimensional (b) electrical network with fixed boundaries at the points marked by arrows. Stable stationary inhomogeneous structures are indicated by light emitting diodes.

47

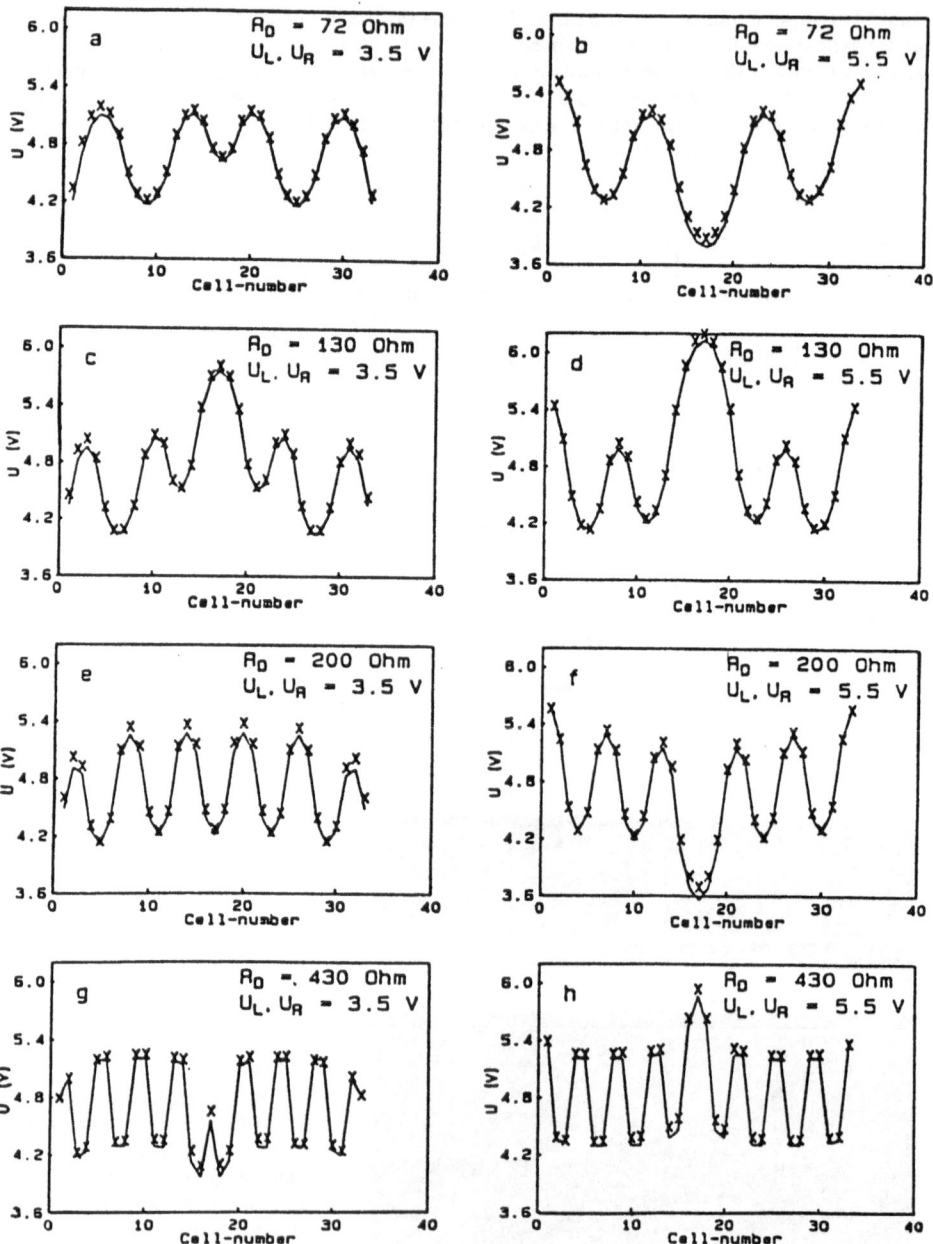

Fig. 13 Measured (x) and calculated (———) voltage structures on a one-dimensional electrical network corresponding to experiment No. 1 of Table 4.

work element in Fig. 13. This has been done for various coupling resistors R_D. We note that the smaller R_D the greater is the spatial extension of the voltage filaments observed on the network. This is to be expected from the correspondence of R_V/R_D to the diffusion constant. Small diffusion constant means small diffusive spread. The values of the continuous curves taken at a given cell number represent the calculated voltage. We find good agreement between calculation and experiment. Typical measured spatial structures of the two-dimensional network are shown in Fig. 14a. Also in the two-dimensional case good agreement with the calculated structures of Fig. 14b is obtained. Of course to any voltage structure there is a current structure with regions of high current where the voltage is low. Due to the lack of current diffusion (see below) the slopes of the current filaments are infinitely steep. − Experiment No. 2 demonstrates that homogeneous oscillatory temporal structures can be stable. − No. 3 shows that the latter can be changed to an in-homogeneous stable stationary state by applying a Dirichlet boundary condition afterwards at any boundary point. The latter seems to serve as a nucleation center from which the inhomogeneous structure can grow. − No. 4 demonstrates the high multistability of the system. Choosing U_b = const and switching on the network we obtain a certain stable stationary structure. Changing to another value U_b = const$'$ does not effect much the structure in general. Now, switching $U_S + \Delta U$ off and on again we obtain in most cases a very different structure. − A particular difficult

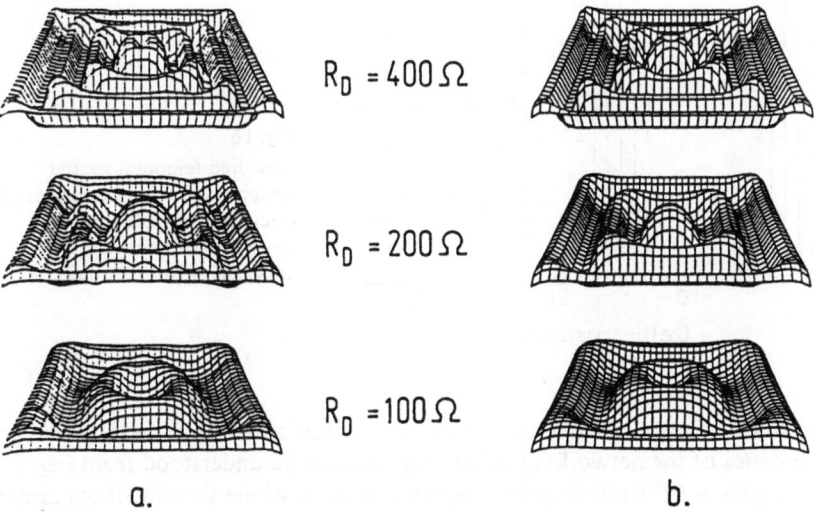

$R_D = 400\,\Omega$

$R_D = 200\,\Omega$

$R_D = 100\,\Omega$

a.   b.

Fig. 14 Measured (a) and calculated (b) voltage structures on a two-dimensional electrical network corresponding to experiment No. 1 of Table 4 with homogeneous boundary voltage of 4.5 V.

49

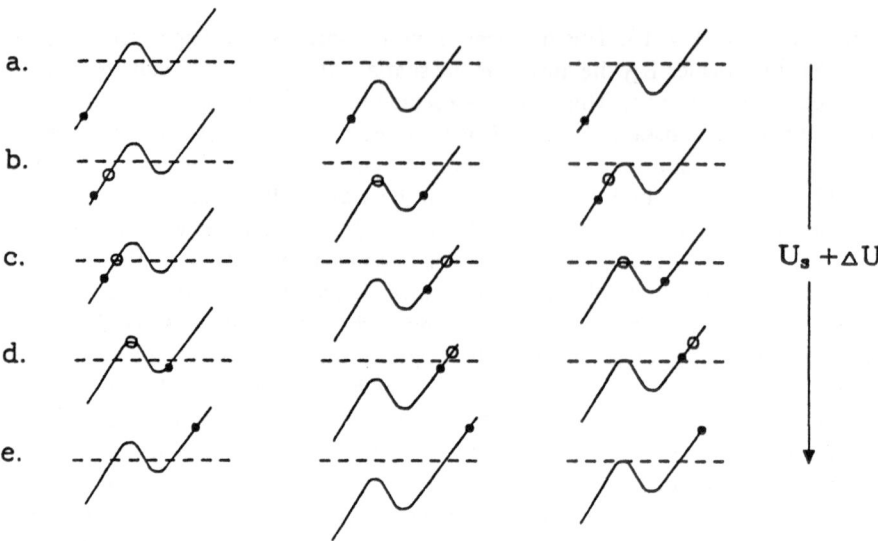

$U_s + \Delta U$

Fig. 15 Effect of inhomogeneities on the structure formation of electrical networks by increasing slowly the driving voltage $U_S + \Delta U$ from a to e. ○, ● denote an unstable and a stable situation.

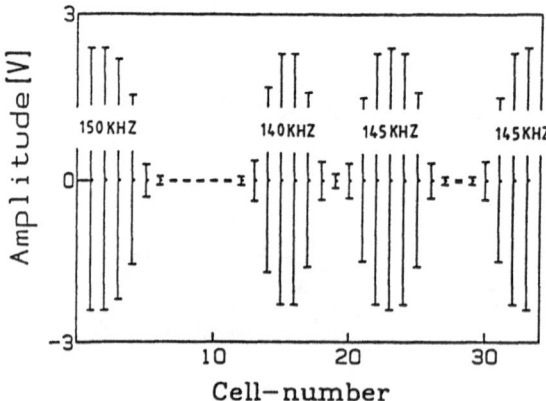

Fig. 16

Measured temporal spatial structure on a one-dimensional electrical network with $R_D = 4 \, k\Omega$ corresponding to experiment No. 5 of Table 4.

situation arises in experiment No. 5. Apparently, the structure is determined by inhomogeneities of the network. Qualitatively, this can be understood from Fig. 15. By increasing $U_S + \Delta U$ from zero we come to a situation where the lowest maximum of a nonlinear characteristic is reached (open points in Fig. 15b). This leads to an instability, and the system undergoes a transition to a stationary inhomogeneous current and voltage distribution. This kind of transition is repeated from Fig. 15c ... e until all cells are in the high current state.

50

In experiment No. 6 a particular interesting temporal spatial structure is observed which is shown in Fig. 16. There are regions of cells oscillating with exactly the same frequency within one region separated by stationary regions. However, the different regions have different frequencies. For more details see Dirksmeyer et al. [21].

3.3 The Electrical Network as an Equivalent Circuit for a Continuous Material

It is instructive to interpret the two-dimensional extension of the electrical network of Figs. 10, 11 as an equivalent circuit for a nonlinear composite material as shown in Fig. 17. In the linear upper part of the network R_V and R_D allow for a vertical and horizontal current flow. This part of the network can be considered as an equivalent circuit for a linear material L as shown in the upper part of Fig. 17. The lower part of the network corresponds to a nonlinear material with s-shaped characteristic. Since the current flow in the nonlinear part of the network is vertical, we have to choose a in Fig. 17 sufficiently small so that the horizontal current can be neglected with respect to the vertical current. The details are discussed by Radehaus et al. [22], Radehaus [23], and Purwins et al. [20]. Performing the transition from Eq. (3-1) to the continuum and adding a current diffusion term because of the diffusion of charge carriers in the x,y-plane we obtain for the current density $j(x, y, t)$ and the potential $U(x, y, t)$ in the L-N-interface

$$\frac{\partial U}{\partial t'} = \frac{1}{C_s} \left[\frac{b}{4\rho_L} \Delta' U - \frac{1}{\rho_L b} U - j + \frac{1}{\rho_L b} U_V \right] ,$$

$$\frac{\partial j}{\partial t'} = \frac{1}{L_s} \left[D\Delta' j + U - s(j) \right] ,$$

$$\frac{U_V}{\rho_L b} = \frac{U_S + \Delta U - R_0 I}{\rho_L b} \tag{3-4}$$

$$= \frac{1}{\rho_L b + R_0 dl} \left[U_S + \Delta U + \frac{R_0}{\rho_L b} \int_0^d \int_0^l U \, dx' \, dy' \right] ,$$

$$s(j) = U_0 - \chi(j - j_0) + \Phi(j - j_0)^3 ,$$

ρ_L = specific resistivity of N, C_s = effective specific capacity,
$\quad$D = current diffusion constant, L_s = effective specific inductance.

Under favourable circumstances C_s is related directly to the capacity of the composite material, whereas L_s has to do with the microscopic relaxation processes.

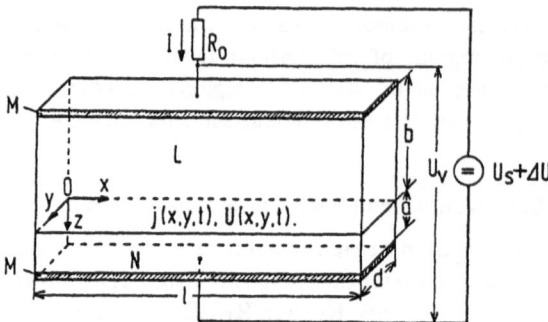

Fig. 17

Composite material having the two-dimensional extension of the network of Figs. 10 and 11 as equivalent circuit. (M = metallic contacts, L = linear, N = nonlinear material.)

Again to see the mathematical structure of Eq. (3-4) we make the following transformation:

$$v = \sqrt{\frac{\Phi}{\rho_L^3 b^3}} \, (U_0 - U),$$

$$w = \sqrt{\frac{\Phi}{\rho_L b}} \, (j - j_0),$$

$$j_0 = \frac{1}{\rho_L b + R_0 dl} \left[U_S + \frac{R_0 dl}{\rho_L b} U_0 \right] - \frac{U_0}{\rho_L b},$$

$$\lambda = \frac{\chi}{\rho_L b}, \quad \epsilon = \frac{\rho_L^2 b^2 C_s}{L_s}, \quad \sigma = \frac{4D}{\rho_L b^3},$$

$$\kappa_1 = \sqrt{\frac{\Phi}{\rho_L b}} \, \frac{\Delta U}{\rho_L b + R_0 dl}, \quad \kappa_2 = \frac{R_0 b^2}{4(\rho_L b + R_0 dl)}$$

$$t = \frac{\rho_L b}{L_s} \, t', \quad x = \frac{2}{b} x', \quad y = \frac{2}{b} y', \quad \zeta = 2\frac{d}{b}, \quad \xi = 2\frac{l}{b},$$

(3-5)

and obtain

$$\epsilon \frac{\partial v}{\partial t} = \Delta v + w - v - \kappa_1 + \kappa_2 \int_0^\zeta \int_0^\xi v \, dx \, dy,$$

$$\frac{\partial w}{\partial t} = \sigma \Delta w + \lambda w - w^3 - v,$$

$$\epsilon, \lambda, \sigma, \kappa_1, \kappa_2 \geqslant 0,$$

(3-6)

v = inhibitor ($\sim -U$), ϵ = relative relaxation time of v to w,

w = activator ($\sim I$), σ = relative diffusion constant of v to w.

A comprehensive discussion of Eq. (3-6) is given by Purwins et al. [20] and Radehaus [23]. We merely note that Eq. (3-6) is a diffusion equation because of the
simple partial derivative with respect to time and because of the Laplace operator.
It is also a reaction equation because of the terms v and w appearing at the right
hand side. For our purpose we may say that the complete integro-differential
equation behaves qualitatively as the equation with vanishing integral part ($\kappa_2 = 0$).
The mechanism of pattern formation is dominated by the inhibitor v damping the
production of v and w and the activator w stimulating the increase of v and w.
Thereby the nonlinear term in f(w) prevents the system from exploding. Also we
remark that the inverse of the coupling resistance $1/R_D$ is formally related to a
voltage diffusion. Consequently, large values of $1/R_D$ support the smearing out of
voltage structures.

For the physical realization of the composite material of Fig. 17 we need a linear
high ohmic material and a nonlinear s-shaped material. Radehaus et al. [22] have
shown that such devices are approximated by pin-diodes as they are investigated
by Jäger et al. [24] and Baumann et al. [25]. Also the present ideas should apply to
thyristor-like structures and other devices. In the following section we describe a
particular simple device consisting of a gas discharge system.

3.4 The Gas Discharge System

In Fig. 18 we show the experimental realization of the gas discharge system as
investigated by Radehaus et al. [26]. A high voltage generator is connected via a
resistor R_0 and a broad metallic contact to a thin linear semiconductor electrode.
The latter is surrounded by a He-atmosphere with 10 % O_2 of variable pressure.
The counterelectrode is a thin metallic plate. The nonlinear material is the He-O_2-
gas having an s-shaped characteristic. We remark that in contrast to all experiments
discussed for electrical networks we have in the present setup the situation that for

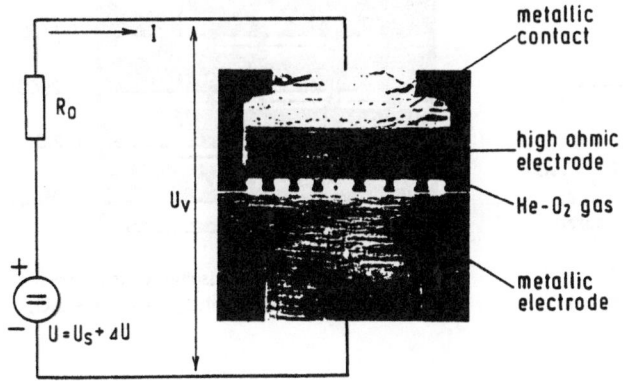

metallic
contact

high ohmic
electrode

He-O_2 gas

metallic
electrode

Fig. 18
Experimental setup for the
gas discharge system.

53

appropriate values of $U_S + \Delta U$ and homogeneous current and voltage distributions the load line can intersect the nonlinear characteristic three times. – The high ohmic electrode has been cut from Au-doped Si with resistivity of 1,6 kΩcm. The dimensions of the system corresponding to Figs. 17, 18 are a = 1.5 mm, b = 15 mm, d = 0.3 mm, and l = 30 mm. The other parameters in particular those of the nonlinear characteristic are not known yet. Therefore, the experimental results are discussed only qualitatively. – It is obvious from the experimental setup that we work with free boundary conditions.

In Fig. 19 we present the current versus voltage characteristic of the gas discharge system corresponding to Fig. 18. We note that the characteristic exhibits well pronounced jumps. Also there is hysteresis showing that the system is multistable. Finally, it is observed that the whole curve can be reproduced in its general shape but not in its details. It can be shown that this is not an experimental insufficiency but a characteristic feature indicating a high degree of multistability.

In Fig. 20 we demonstrate the formation of stable stationary inhomogeneous current filament structures when increasing the external supply voltage $U = U_S + \Delta U$.

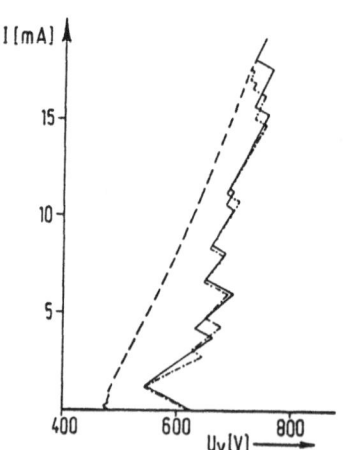

Fig. 19 I-U-characteristic (U = U_S + ΔU) for the device of Fig. 17 realized as a gas discharge system Fig. 18 measured in a quasi-stationary manner (——— increasing U, – – – decreasing U, –.–.– repetition with increasing U).

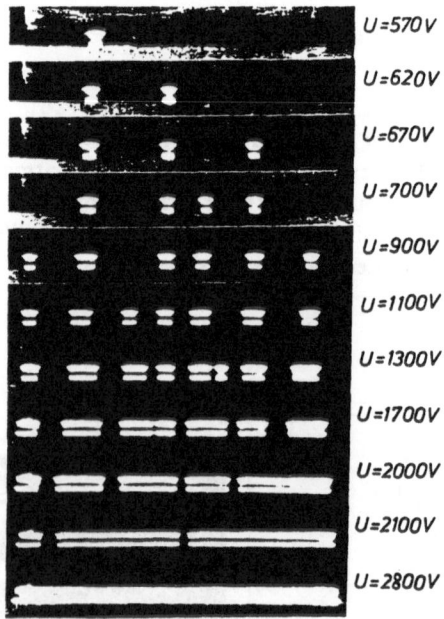

Fig. 20 Photos of gas discharge structures for increasing driving voltage U = U_S + ΔU.

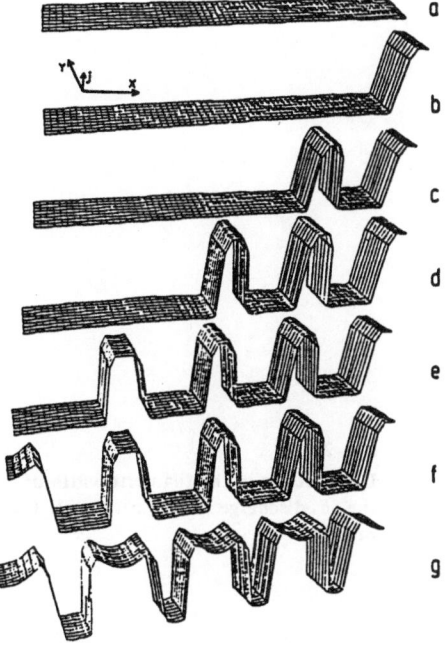

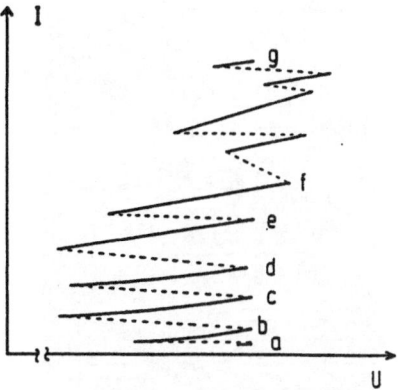

Fig. 21
Calculated spatial current structures and I-U-characteristic calculated from Eq. (3-4) for parameters as discussed in Purwins et al. [20]. The letters indicate the correspondence of spatial structures to branches in the I-U-characteristic.

Obviously, in order to allow a larger total current I the system prefers to form new filaments in a discontinuous manner instead of changing much their size or geometrical form. This is observed between zero voltage and approximately 900 V. In this range one may speak of a solitary type of behaviour of the current filaments. From U = 1100 V on the filaments seem to interact and become rapidly broader until the structure merges into a single filament extending over the total width of the electrodes. The experiments show that any formation of a new filament is accompanied by a jump in the current-voltage characteristic. Also for the spatial current-filament patterns we observe hysteresis and nonreproducibility of details as there are e.g. the exact position of the filaments or the exact number at a certain value of the driving voltage.

In Fig. 21 we present typical solutions of Eq. (3-4) for the current-density distribution and the I-U-characteristic by Purwins et al. [20] for a quasi-stationary increase of the driving voltage and small thickness d of the device Fig. 17. We note that all qualitative features of the current-filament patterns and the I-U-characteristic observed experimentally are reproduced by the calculations. This is true in particular for the solitary-like behaviour of the filaments, the discontinuous jumps, and the hysteresis. Also calculations have shown that due to the high degree of multistability noise can be decisive for the sequence of stable stationary states realized in the course of the quasi-stationary increase of the driving voltage. Finally, as one may expect, also spatial inhomogeneities determine the details of the structures. For suf-

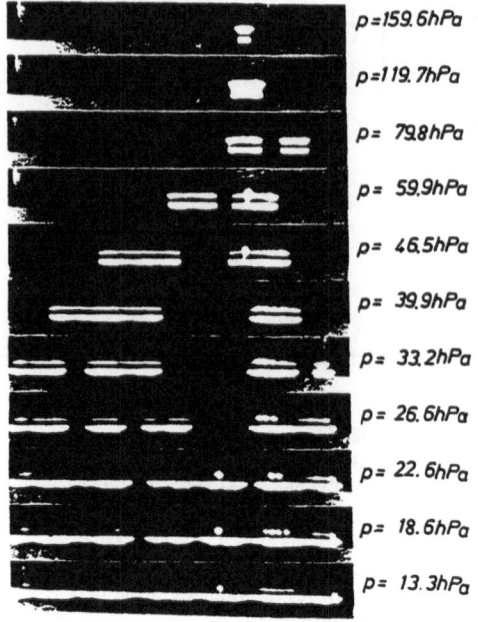

p = 159.6 hPa

p = 119.7 hPa

p = 79.8 hPa

p = 59.9 hPa

p = 46.5 hPa

p = 39.9 hPa

p = 33.2 hPa

p = 26.6 hPa

p = 22.6 hPa

p = 18.6 hPa

p = 13.3 hPa

Fig. 22

Dependence of the filament width of the gas-discharge system on the He-O_2-pressure.

ficiently large inhomogeneities the effect of the noise may be ruled out, and the reproducibility is improved.

From the physical meaning of the diffusion constant D in Eq. (3-4) one expects that an increase of D leads to an increase of the filament width. Increasing D is easily realized by lowering the pressure of the He-O_2 atmosphere. The results obtained by Radehaus et al. [26] are shown in Fig. 22. Again our model allows a prediction of the qualitative behaviour. However, since a change of D effects the stability and choice of possible stationary states there may be a change of the number of filaments. — We stress that the agreement between calculation and experiment is qualitative so far. Quantitative investigations are in progress.

An interesting feature of the gas discharge system is also the fact that by changing the polarity of the system we still observe filaments; however, they appear and disappear in an irregular manner in the course of time. We are convinced that this observation has to be attributed to what has to be called "temporal spatial chaos", a term which is still not properly defined.

3.5 A Microscopic Model

In chapters 3.1 ... 3.4 we have discussed a phenomenological description of models for pattern formation in electrical conducting devices. Kardell et al. [27] and Kardell [28] have shown that under appropriate conditions stable stationary multifilament space charge and conduction electron density structures can be found in semiconductors on the basis of a microscopic model. Starting point is a semiconductor

56

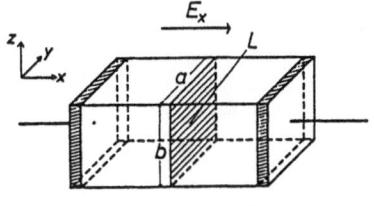 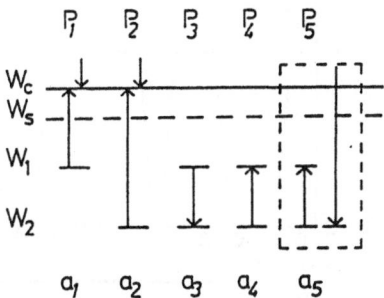

Fig. 23 Semiconductor device (a) allowing for the kinetic processes P_1 to P_5 (b) related to the shallow donor levels W_S and the deep double donor levels W_1 and W_2. The kinetics of P_1 to P_5 are described by the reaction coefficients a_1 to a_5.

with metallic contacts as shown in Fig. 23a which allows the kinetic processes as shown in Fig. 23b. The semiconductor has shallow donor levels corresponding to W_s in Fig. 23b and deep double donor levels indicated by W_1 and W_2. The transitions considered in the model are labeled by $P_1, ..., P_5$. There are also occupied acceptor levels not shown in the figure.

We consider a situation where a $\ll$ b and much smaller than the fundamental wave length of the spatial structure. The space-charge and conduction-electron densities of the device shown in Fig. 23a can then be calculated for the plane L being far from the metallic contacts by solving the following system of equations:

$$0 = 1 - \nu - \nu_1 - \frac{a_3 \nu_1}{a_2 \nu + a_4} + \frac{\partial^2}{\partial \xi^2} \ln \nu,$$

$$\frac{\partial \nu_1}{\partial \tau} = \nu_1 \left\{ -a_1 \nu - a_3 + \frac{a_3 \left[a_4 + a_5 (2N_D + N_D' - N_A - \nu_1) \right]}{a_2 \nu + a_4} \right\} + d \frac{\partial^2}{\partial \xi^2} \nu_1,$$

$$\rho' = 1 - \nu - \nu_1 - \frac{a_3 \nu_1}{a_2 \nu + a_4}$$

$$(\nu, \nu_1, \rho') = \frac{(n, n_1, \rho/e)}{2N_D + N_D' - N_A}, \quad \tau = \Gamma_0 t, \quad \xi = \frac{y}{L_D}, \quad d = \frac{D'}{D}. \tag{3-7}$$

n	= conduction-electron density,	N_D, N_D', N_A = concentration of shallow donors, deep donors, and acceptors,
$n_{1,2}$	= electron densities corresponding to $W_{1,2}$,	Γ_0 = dielectric relaxation frequency,
ρ	= space charge density,	L_D = Debye length,
e	= electron charge,	a_i = reaction coefficients depending on electrical field.
D', D	= diffusion constant of electrons within W_1-levels and for free electrons,	

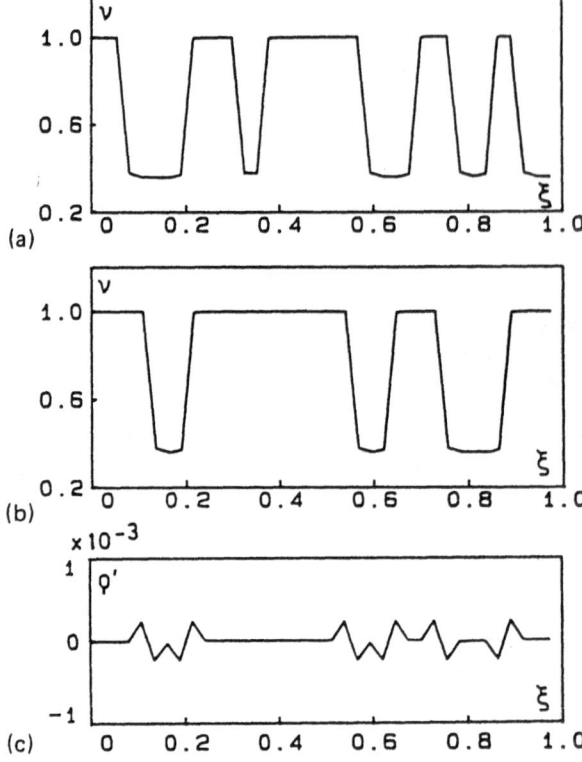

Fig. 24

Two possible stable stationary states for the normalized conduction electron concentrations distribution (a, b). The space charge density corresponding to (b) is shown in (c). Calculations are done for
$a_1 = 0.03$,
$a_2 = 0.06$,
$a_3 = 0.005$,
$a_4 = 0.009$,
$a_5 = 0.63$, and
$2N_D + N' - N_A = 0.69$.

Again Eq. (3-7) is a reaction diffusion equation, however, this time with component dependent diffusion constant.

Examples of stable stationary multifilament structures of the normalized conduction-electron concentration and space charge ν and ρ' being solutions of Eq. (3-7) are given in Fig. 24a, b, c. We note that there are two stable states for the electron density indicating the multistability of the system. — It would be of considerable interest to find a semiconductor material having the transition scheme Fig. 23b.

3.6 Summary

The physical objects investigated in Chap. 3 are all examples of reaction-diffusion systems. With the exception of the system treated in chapter 3.5 diffusion enters via a linear coupling of neighbouring elements which is the simplest form of spatial coupling. The reaction terms enter via an activator and an inhibitor. It is the interplay of these terms which in connection with the nonlinearities gives rise to interesting temporal, spatial, and temporal-spatial structures. As has been shown in discussing the electrical network, the pattern formation can easily be visualized. We believe that the systems discussed in Chap. 3 belong to the simplest types one could possibly imagine having a wealth of interesting temporal, spatial, and temporal-spatial structures. At the same time these systems are of great importance in chemistry, biology, and as we have demonstrated also in physics. This makes reaction-diffusion systems to a candidate being ideally suited for fundamental investigations of pattern formation in nature.

In the present article we have shown how real physical systems of reaction-diffusion type can be built and investigated. Thereby we get fundamental insight into the possible structures, their stability, and their dependence on parameters, boundary conditions, initial conditions, noise, and inhomogeneities. In particular we find that due to the high degree of multistability noise has the effect that the observed structures are nonreproducible though the overall structure is. This is a feature common to most of the chemical and biological structures faced in daily life. Spatial inhomogeneities in contrast have the effect to improve the reproducibility to the expense of the richness of possible structures and complexity of the system.

Also it is easy to see that reaction-diffusion systems are of considerable practical interest. We can imagine that patterns discussed in this article can be used as the basis for memories, switches, modulators, pulse generators, and analogue-digital converters. Also one may ask e.g. to which extent the boundary of a two-dimensional continuous field can be used to write predefined patterns. Finally, we are convinced that the subject is strongly related to the field of pattern recognition. It is e.g. an interesting question to which extent a system can complete a pattern which is given only partially due to lack of knowledge.

Concerning electrical networks described in the present article one may ask whether a system following Kirchhoff's rules is an interesting physical object. As we have seen, even the very simple network discussed in the present paper has a degree of complexity which demands for new concepts. We are convinced that such concepts can best be found by investigating the system theoretically and experimentally at the same time. To do this with an object where we can be rather sure that we have the correct equations is a considerable advantage. It is a general experience in physics that even if the description of a system is known in principle, many interesting effects and discoveries have been made experimentally. Also theoretical problems frequently occur which cannot be solved numerically in reasonable time or a numerical solution is impossible due to computer noise. Finally, nonlinear networks will be of increasing importance when progressing in miniaturization. Already today there are indications that nonlinear effects appearing in this connection due to the spatial coupling are used by their own right in circuit design.

4 Conclusive Remarks

The aim of the present paper is to demonstrate experimentally the richness of temporal and spatial structures of dissipative nonlinear dynamical systems and to describe approaches for their systematic understanding. As an example for systems with few degrees of freedom, we show experiments on the driven anharmonic rotating oscillator and as an example for many degrees of freedom reaction-diffusion systems in the form of electrical networks and gas discharge systems. These systems were so simple that they could easily be mounted as a demonstration apparatus suitable for an impressive visualization of the richness of possible structures.

It is fair to say that a systematic approach to systems with few degrees of freedom seems to be possible. The fundamental temporal structure elements are those contained in the stationary, periodic, quasi-periodic, and the chaotic motion being well described in low dimensional phase space. We observe that to any of these structures there corresponds a wealth of structures in spatially extended systems with a high degree of multistability in general, and only in favourable circumstances a reduction to a small number of degrees of freedom is possible. These features together with the problems related to the high dimensionality of the involved phase space are the reasons why the understanding of systems with many degrees of freedom is so poor at present.

On the other hand there are interesting starting points for an improved understanding also for systems with many degrees of freedom. We mention that in special cases structures can be calculated quantitatively. We also note that the filaments discussed in Chap. 3 are just the elements that we are looking for when returning to our original definition of structure. Also our understanding of the influence of noise and inhomogeneities on the structure formation has been improved.

Finally, we remark that already systems with few degrees of freedom can have an incredible complexity being well demonstrated e.g. by the Poincaré maps of the chaotic motion or the complex interlocking of basins of attraction. In addition, we were faced with the problem of reproducibility when discussing the sensitive dependence of trajectories of chaotic motion on initial conditions. The complexity is still far richer for spatially extended systems and even in nonchaotic behaviour we are confronted with the problem that the details of the structures cannot be reproduced in many cases though the overall structures can be in general. These features are also striking characteristics of living structures. We conjecture that the field of temporal and spatial structures of nonlinear dynamical systems iş going to provide a deeper understanding of one of the most fundamental problems in natural science, namely, the understanding of living structures.

Acknowledgement

The authors are grateful to J. Kondakis and the many collaborators of the Institut für Angewandte Physik for their enthusiastic support of this work.

References

[1] *J.M.T. Thompson* and *H.B. Stewart*, Nonlinear Dynamics and Chaos (John Wiley and Sons, New York 1986)

[2] *J. Guckenheimer* and *P. Holmes*, Nonlinear Oscillations, Dynamical Systems and Bifurcations of Vector Fields (Springer, Berlin 1983)

[3]   *P. Cvitanović*, Ed., Universality in Chaos (Adam Hilger, 1984)

[4]   *H. G. Schuster*, Deterministic Chaos (Physik-Verlag, Weinheim 1984)

[5]   *D. Tritton*, Physical Fluid Dynamics (Van Nostrand Reinhold, New York 1977)

[6]   *H. Swinney* and *J. Gollub*, Hydrodynamic Instabilities and the Transition to Turbulence, Topics in Appl. Phys. **45** (Springer, Berlin 1981)

[7]   *J. Smoller*, Shock Waves and Reaction-Diffusion Equation (Springer, Berlin 1983)

[8]   *G. Nicolis* and *J. Prigogine*, Self-Organisation in Nonequilibrium Systems (John Wiley and Sons, New York 1977)

[9]   *P. C. Fife*, Current Topics in Reaction-Diffusion Systems, in Nonequil. Coop. Phenom. in Phys. and Related Fields, Nato Advanced Study Inst. Series B Phys., Vol. 116, 1983.

[10] *G. Nicolis*, Rep. Prog. Phys. **49**, 873 (1986)

[11] *A. Gierer*, Prog. Biophys. molec. Biol. **37**, 1 (1981)

[12] *H. Haken*, Synergetics, 2nd Edition, Springer Series in Synergetics (Springer, Berlin 1978) and other volumes of this series

[13] *P. Grassberger* and *I. Procaccia*, Physica **9D**, 189 (1983)

[14] *T. C. Halsey, M. H. Jensen, L. P. Kadanoff, I. Procaccia*, and *B. I. Shraiman*, Phys. Rev. **A33**, 1141 (1986)

[15] *M. H. Jensen, L. P. Kadanoff*, and *A. Libchaber*, Phys. Rev. Lett. **55**, 2798 (1985)

[16] *E. G. Gwinn* and *R. M. Westerveld*, Phys. Rev. Lett. **57**, 1060 (1986)

[17] *Y. Ueda*, in: New Approches to Nonlinear Problems in Dynamics, ed. by P.J. Holmes (SIAM, Philadelphia 1980)

[18] *J. Berkemeier, T. Dirksmeyer, G. Klempt*, and *H.-G. Purwins*, Z. Phys. B Condensed Matter **65**, 255 (1986)

[19] *J. Berkemeier*, Thesis, University of Münster 1987

[20] *H.-G. Purwins, C. Radehaus*, and *J. Berkemeier*, submitted for publication in Phys. Rev. A

[21] *T. Dirksmeyer, M. Bode, J. Berkemeier*, and *H.-G. Purwins* to be published

[22] *C. Radehaus, K. Kardell, H. Baumann, D. Jäger*, and *H.-G. Purwins*, Z. Phys. B Condensed Matter **65**, 515 (1987)

[23] *C. Radehaus*, Thesis, University of Münster 1987

[24] *D. Jäger, H. Baumann*, and *R. Symanczyk*, Phys. Lett. **A117**, 141 (1986)

[25] *H. Baumann, T. Pioch, H. Dahmen*, and *D. Jäger*, Scanning Electron Microscopy 1986 II, 441 (1986)

[26] *C. Radehaus, T. Dirksmeyer, H. Willebrand*, and *H.-G. Purwins*, submitted for publication in Phys. Lett. A

[27] *K. Kardell, C. Radehaus, R. Dohmen*, and *H.-G. Purwins*, submitted for publication in **Phys. Rev. Lett.**

[28] *K. Kardell*, Thesis, University of Münster 1987

Festkörperprobleme 27 (1987)

Localization Effects in Disordered Systems

Dieter Vollhardt

Max-Planck-Institut für Physik und Astrophysik — Werner-Heisenberg-Institut für Physik —
D-8000 München 40, Federal Republic of Germany
From October 1, 1987: Institut für Theoretische Physik, Rheinisch-Westfälische Technische
Hochschule Aachen, D-5100 Aachen, Federal Republic of Germany

Summary: The basic concepts involved in the physics of localization in disordered systems are discussed on an elementary level. In the case of weak disorder localization effects may be understood as a coherent wave phenomenon. Initially developed to describe electronic transport in disordered metals, localization theory has now found wide application in other areas related to disordered systems. The article is intended to explain how and why localization has recently experienced such an explosive growth. We discuss localization effects in the propagation of various wave-like quantities, quantum oscillations in different geometries (hollow cylinders, networks, rings), the developments concerning mesoscopic systems, as well as the effects of universal fluctuations in such systems. An extensive list of references is given.

1 Introduction

The concept of "localization" due to disorder originates from the work of Anderson in 1958 [1]. He investigated the motion of a quantum mechanical particle on a three-dimensional lattice with randomly varying site-energies. The "disorder" in the problem was thus given by the magnitude of the energy fluctuations from site to site on the lattice. Assuming a particle on a site j at time $t = 0$ he calculated the return probability P for the particle in the limit $t \to \infty$. Below a critical value of the disorder he found $P = 0$, i.e. the particle had diffused away and had disappeared in the system. The particle is thus described by an *extended* state. For larger disorder one finds $P > 0$, indicating that the particle did not disappear but remained within a certain region around the site j. This corresponds to a *localized* state with a certain spatial extent (localization length). There is then a critical strength of the disorder where a sharp transition ("Anderson-transition") distinguishes an extended and a localized regime. In other words: if the energy E of the particle lies below a certain critical energy E_c it is localized, while for $E > E_c$ the energy fluctuations of the system will not be able to dominate the particle such that it is described by an extended state. In the first case one deals with an insulator, in the second one with a metal.

We note that the localization is caused by fluctuations imposed on the wave function and does *not* mean some kind of trapping or local binding to a particular site. Hence it is the coherence of the wave function which is important in this problem.

The disorder discussed above is due to the randomness of the on-site energies of the lattice. Alternatively and equivalently the disorder may enter via a random *spatial* distribution of scattering centers, off which a quantum mechanical particle (with constant energy E) is scattered elastically [2]. The latter approach ("Edwards model") is particularly suited for a perturbation-theoretical approach studying weak disorder.

Then, in the end of the seventies, the investigation of localization in two-dimensional systems, i.e. of the question whether very thin films may have a metallic conductivity at zero temperature or not, led to an explosive development in this field of condensed matter physics [3]. New approaches to the problem of Anderson localization and the metal-insulator transition were devised [4], which — together with perturbational methods — clarified the situation and demonstrated the absence of true metallic conductivity (or of a minimal metallic conductivity [5], for that matter) in two dimensions. In particular, the perturbational treatment of the effects of weak disorder by diagrammatic means [6 ... 8] ("maximally crossed diagrams" [9]) and the subsequent interpretation of the underlying physics [10 ... 12] in the beginning of the eighties led to a significantly new understanding of transport in disordered systems. It was soon realized that the localization effects known from weakly disordered electronic systems ("weak localization") were not peculiar to quantum mechanical particles as such but rather were due to the *wave-nature* of quantum mechanical particles and were thus a general phenomenon common to any wave-propagation. The basic physics of weak localization, namely the coherent backscattering of electrons, is therefore shared by all wave-like transport and is not connected to quantum mechanics or particle statistics. This notion will be discussed in Sect. 2 in the context of the diffusion of electrons in a weakly disordered metal. Consequently, localization effects are in principle also observable in the propagation of light (i.e. electromagnetic radiation) and any kind of sound (phonons). This is indeed the case and will be discussed in Sects. 3 and 4, respectively. Yet different systems and geometries discussed in the context of localization will be mentioned in Sect. 5.

The magnetic field dependence of weak localization led to the prediction of *macroscopic quantum oscillations* in multiply connected, normal-conducting geometries with a periodicity of half the flux quantum known from the Aharonov-Bohm effect. This will be discussed in Sect. 6.

The subsequent experimental investigation of these effects involved the study of *metallic networks*, composed of a macroscopic number of small loops, where quantum oscillations could be well observed (Sect. 7).

On the other hand, the investigation of single rings, i.e. *mesoscopic systems*, led to the discovery of quantum oscillations of the flux both with period $\frac{hc}{e}$ and $\frac{hc}{2e}$, which involve very different physics (Sect. 8).

In the course of the investigation of such mesoscopic systems unexpected *universal fluctuations* were discovered, which will be addressed in Sect. 9.

64

2 Localization in Disordered Electronic Systems

We will first give a rather detailed discussion of the physics of "weak localization"*
in disordered electronic systems. (An introduction to this topic in the general
context of the metal-insulator transition in disordered systems can be found in
[13].) The basic ideas [14] may then easily be applied to comprehend localization-
effects involving other wave-like quantities.

Weak Disorder and Weak Localization

Concentrating on the case of weak disorder, we consider (i) non-interacting, quantum mechanical particles, which (ii) are scattered by point-like randomly distributed
scattering centers of equal strength. The scattering in turn leads to a diffusive
motion of the particles. We are then interested in the conductivity σ or the diffusion coefficient D of such a disordered system. The disorder is measured by a
dimensionless parameter γ with $\gamma \sim n_i V_0^2$, i.e. γ is essentially given by the impurity
concentration n_i and the scattering strength V_0^2 of the scatterers. The case of very
weak disorder corresponds to $\gamma \ll 1$. The starting point is the *metallic* regime, which
is characterized by a finite dc-conductivity

$$\sigma_0 = \frac{e^2 n}{m} \tau, \tag{1}$$

where e and m are the charge and the mass of the particles (e.g. non-interacting
electrons), respectively, n is the density and τ is an average collision time between
successive scatterings; τ is related to the mean free path ℓ by $\ell = v_F \tau$ ($v_F = \hbar k_F/m$
is the Fermi velocity). The quantity σ_0 is often called "Boltzmann-conductivity",
because Eq. (1) is a simple result of the Boltzmann transport theory.

In the following we want to understand how a small concentration of impurities
affects the metallic behavior.

Weak disorder means that the mean free path ℓ is much greater than the average
particle distance $a \approx k_F^{-1}$, i.e. $k_F \ell \gg 1$. We will therefore choose

$$\gamma = \frac{1}{\pi k_F \ell} \tag{2}$$

as our (small) perturbation parameter. Starting from the metallic regime we intend
to consider the precursor effects of localization, i.e. the correction $\delta\sigma$ to the metallic conductivity

$$\sigma = \sigma_0 + \delta\sigma, \quad |\delta\sigma| \ll \sigma_0. \tag{3}$$

These perturbational effects are commonly called "weak localization". The correction $\delta\sigma$ depends on external parameters like the system's size L, the frequency
ω, the temperature T, or the magnetic field H.

---

* The discussion follows the presentation of Altshuler, Aronov, Khmelnitskii, and Larkin
[10]; see also [11].

As already mentioned, the dc-conductivity $\sigma_0 \sim 1/\gamma$ is a direct consequence of Boltzmann transport theory. In this theory consecutive collisions of particles are assumed to be independent of each other, i.e. collisions are uncorrelated. This implies that multiple scattering of a particle at a particular scattering center is not taken into account. Consequently, if there is a finite probability of the repeated occurrence of such multiple scatterings, the basic assumption of the independence of scattering events breaks down and, the validity of the result for σ in Eq. (1) becomes, at least, questionable.

To investigate this fundamental point we consider the diffusive behavior of a particle in a d-dimensional disordered system.

If at $t = 0$ a particle is located at some point $\vec{r}_0$ then, after some time $t \gg \tau$, the solution of the diffusion equation implies that the particle will have diffused into a smooth volume $V_{diff} \approx (D_0 t)^{d/2}$ around $\vec{r}_0$, where $D_0 = v_F^2 \tau/d$ is the diffusion constant. This is an entirely classical result. To understand the differences in the diffusive behavior of classical and quantum ("wave") mechanical particles, we take a look at the path of a particle diffusing from point A to point B (Fig. 1). This transport can take place via different trajectories (in Fig. 1 four examples are shown). The trajectories, or "tubes", have a typical width given by the Fermi wavelength

$$\lambda_F = \frac{h}{v_F m}. \tag{4}$$

In the classical case ($h = 0$) these paths are arbitrarily sharp ($\lambda_F = 0$) — in the quantum mechanical case, however, one has $\lambda_F = 2\pi/k_F \approx a$, i.e. the tubes have a finite diameter due to the wave nature of the electrons. We assume that the temperature is low enough such that inelastic processes, characterized by an inelastic scattering time τ_{in}, occur only very rarely ($\tau_{in} \gg \tau$).

Since the transport from A to B may take place along different trajectories, there is a probability amplitude A_i connected to every path i. The total probability W to reach point B from A is then given by the square of the magnitude of the sum of all amplitudes:

$$W = |\sum_i A_i|^2 \tag{5a}$$

$$= \sum_i |A_i|^2 + \sum_{i \neq j} A_i A_j^*. \tag{5b}$$

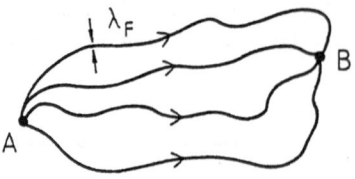

Fig. 1
Typical paths for an electron (or wave) diffusing from A to B.

66

The first term in Eq. (5b) describes separate, i.e. non-interfering paths — this is the classical result, in which the tubes are infinitely sharp. On the other hand, the second term represents the contribution due to *interference* of the path-amplitudes. It originates from the wave nature of the electron and is therefore an exclusively quantum mechanical effect. In Boltzmann theory these interference terms are neglected. In most cases this is in fact justified: since the trajectories have different lengths, the amplitudes A_i carry different phases. On the average this leads to destructive interference. Hence the quantum mechanical interferences in Fig. 1 are generally unimportant.

There is, however, *one* particular exception to this conclusion, namely, if point A and B coincide (Fig. 2). In this case starting-point and end-point are identical, such that the path in between can be traversed in two opposite directions: forward and backward. The probability to go from A to B is then nothing but the *return*-probability to the starting-point. If paths 1 and 2 in Fig. 2 are indeed equal (this is only the case if time reversal invariance of single particle states, i.e. the equivalence of states with $\vec{k}$ and $-\vec{k}$, is valid) the amplitudes A_1 and A_2 have a coherent phase relation. This leads to *constructive* interference, such that the contribution to W from interference becomes very important. Eq. (5b) then tells us that for $A_1 = A_2 \equiv A$ the classical return probability (due to the neglect of the interference terms) is given by $W_{class} = 2|A|^2$, while the quantum mechanical case yields $W_{qm} = 2|A|^2 + 2A_1A_2^* = 4|A|^2$. Hence one obtains

$$W_{qm} = 2 \cdot W_{class}. \tag{6}$$

The probability for a quantum mechanical particle to return to its starting point is hence seen to be *twice* that of a classical particle. One might say "quantum-diffusion" is slower than classical diffusion because in the first case, where the electron is described by a wave, there exists a constructive interference for backscattering. In other words: quantum mechanical particles in a disordered medium are (at low temperatures) less mobile than classical particles. This in turn leads to a correspondingly lower conductivity σ.

It should be stressed that the factor of 2 in (6) is simply a consequence of constructive wave interference of the two time reversed paths in Fig. 1. In the case of electrons its origin is quantum mechanical only because the wave nature of electrons is an inherently quantum mechanical effect. In general, any wave propagation in a disordered medium will lead to a qualitatively identical result. Any wave

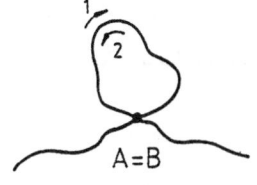

Fig. 2
Time reversed paths on a closed loop.

will do. For example, shouting into a forest* (we assume a naturally grown forest, where trees are irregularly spaced ...) will yield the same kind of enhancement ("echo") into the backward direction as will result from shining light into white paint [15]. Localization involving classical wave propagation has been discussed by Anderson [15], who also gave a number of examples for related electromagnetic and acoustic phenomena.

Correction to the conductivity

To estimate this effect on σ we consider the change $\delta\sigma$ relative to the metallic conductivity σ_0, i.e. $\delta\sigma/\sigma_0$. Because of the expected lowering of σ, the sign of $\delta\sigma/\sigma_0$ will be negative. Furthermore, the change will be proportional to the probability to find a particle in a closed tube, i.e. for the trajectory to intersect itself during the diffusion. Let us therefore have a look at a d-dimensional tube (Fig. 3) with diameter λ_F, i.e. cross-section λ_F^{d-1}. During the time interval dt the particle moves a distance $d\ell = v_F dt$, such that the corresponding volume element of the tube is given by $dV = v_F dt \, \lambda_F^{d-1}$. On the other hand, the maximally attainable volume for the diffusing particle is given by $V_{diff} \approx (D_0 t)^{d/2}$. The above mentioned probability for a particle to be in a closed tube is therefore given by the ratio of these two volumes. We find

$$W = \int_{\tau}^{\tau_{in}} \frac{dV}{V_{diff}} = v_F \, \lambda_F^{d-1} \int_{\tau}^{\tau_{in}} \frac{dt}{(D_0 t)^{d/2}} \tag{7}$$

where we have integrated over all times $\tau \lesssim t \lesssim \tau_{in}$; τ is the microscopic time for a single elastic collision, while τ_{in} is the shortest inelastic relaxation time in the system; it determines the maximal time during which coherent interference of the path-amplitudes is possible. Because of $D_0 \sim 1/\gamma$ and $\lambda_F \sim \hbar$ we obtain

$$\frac{\delta\sigma}{\sigma_0} \sim -\gamma \cdot \begin{cases} (\tau_{in}/\tau)^{1/2}, & d = 1 \\ \hbar \ln(\tau_{in}/\tau), & d = 2 \\ \hbar^2 (\tau_{in}/\tau)^{-1/2}, & d = 3. \end{cases} \tag{8}$$

If we assume that for $T \to 0$ the inelastic relaxation rate vanishes with some power of T, i.e. $1/\tau_{in} \sim T^p$, where p is a constant,

cross section $\sim \lambda_F^{d-1}$

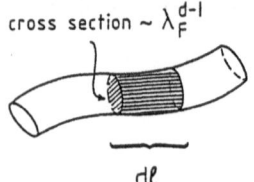

dℓ

Fig. 3
Part of a trajectory of a quantum mechanical particle.

---

* This example was mentioned to me by G. Bergmann.

Eq. (8) is given by

$$\frac{\delta\sigma}{\sigma_0} \sim -\gamma \cdot \begin{cases} T^{-p/2}, & d = 1 \\ \hbar\frac{p}{2} \ln\left(\frac{\hbar/\tau}{k_B T}\right), & d = 2 \\ \hbar^2 T^{p/2}, & d = 3. \end{cases} \tag{9}$$

We observe the following: (i) the conductivity decreases for decreasing temperature, (ii) the relative correction $\delta\sigma/\sigma_0$ is linear in the disorder parameter $\gamma \ll 1$ (lowest order in γ), (iii) except for $d = 1$, these corrections are of quantum mechanical origin, i.e. they disappear for $\hbar \to 0$. (In the case $d = 1$ the "tube" in Fig. 3 has no finite diameter – just as in the classical situation; furthermore, since in $d = 1$ there is only forward and backward scattering all paths are trivially "closed".)

In $d = 2$ one therefore obtains a *logarithmic* temperature dependence of the conductivity correction $\delta\sigma$, which has been confirmed in many experiments (Fig. 4). We note that the elastic scattering due to the disorder in principle leads to a *divergent* temperature behavior of $\delta\sigma$ in $d \leqslant 2$. For the initial assumption $|\delta\sigma| \ll \sigma_0$ to remain valid, the results in Eq. (9) for $d \leqslant 2$ may therefore not be used at too low temperatures. In particular, Eq. (9) does not allow to draw conclusions about $\delta\sigma$ at exactly $T = 0$.

We should like to stress once more that Eqs. (8) and (9) are based on the explicit consideration of backscattering effects, i.e. multiple scattering and the correlation of consecutive collisions. Thus they cannot be obtained within the framework of the Boltzmann transport theory. Also CPA ("coherent potential approximation" [17]), is not able to obtain these results because it makes similar assumptions as the Boltzmann theory ("single site approximation", etc.).

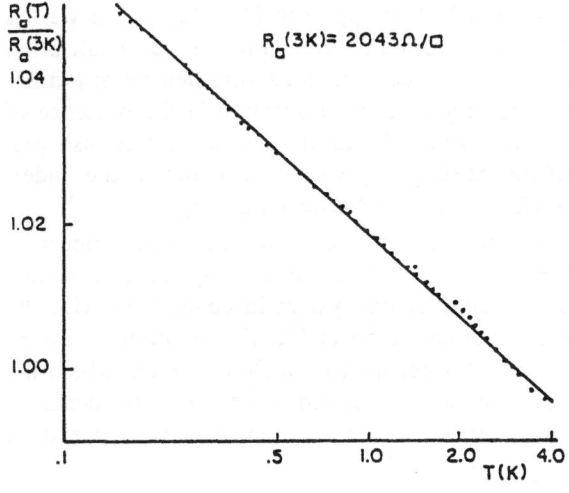

Fig. 4

Logarithmic temperature dependence of the resistivity of a thin palladium film [16].

The preceding discussion was limited to the so-called "normal" scattering, i.e. scattering by non-magnetic impurities. Therefore the spin of the particles was unimportant. However, in the case that the impurities carry a magnetic moment, spin scattering will occur, causing the spin of the particles to flip. Therefore the particles experience something similar to a fluctuating magnetic field. Time reversal invariance is then destroyed, and the weak-localization picture is no longer valid. Field theoretical investigations [18, 19] have shown that even in this new situation the conductivity acquires a logarithmic correction in d = 2. However, now the prefactor goes like γ^2 instead of γ, i.e. the correction is even smaller than in the case of normal scattering. It has not yet been possible to understand this result by means of the simple probability arguments used before in the case of normal scattering.

Impurities with a heavy nucleus lead to yet another type of scattering, namely to spin-orbit scattering of the particles. Theoretical investigations [19, 20] have again predicted a logarithmic correction for σ — but this time with a *positive* sign. The conductivity therefore *increases* with decreasing temperature. A simple quantum mechanical explanation of this effect in terms of multiple scattering (time reversal invariance holds in this case) and experimental results fully supporting these findings have been given by Bergmann [12].

Beyond weak localization

The intuitive picture of constructive interference of waves, propagating on time reversed paths, only allows for an estimation of the lowest order correction to the conductivity or the diffusion coefficient of the metallic regime. Higher order corrections or the Anderson transition itself cannot be studied in this way. For this purpose more powerful theoretical methods have to be employed.

It was Wegner [18, 21] who first realized that Anderson localization shared many properties with the problem of critical phenomena and who accomplished a mapping to a suitable field theoretical model. This approach [18, 22, 23] as well as other field theoretical methods [19, 24], allowed for conclusions about localization and the Anderson transition in various physical situations unrivaled by any other approach. This is particularly true for the case of localization in the presence of spin-flip scattering or magnetic fields, where localization in d = 2 was also predicted to occur as in the case of normal impurity scattering, although the underlying physics is necessarily quite different (see the discussion above).

A different approach to Anderson localization is due to Abrahams, Anderson, Licciardello, and Ramakrishnan [6]. They constructed a one-parameter scaling theory for the conductance g of a d-dimensional system in connection with the first diagrammatic, perturbative calculation of $\delta\sigma$ in Eq. (3). Assuming g to be the only relevant parameter these authors constructed a flow diagram which led to the conclusion that for d ≤ 2 all states of a disordered system are localized, irrespective of the strength of disorder, while for d > 2 an Anderson transition occurs at a finite critical disorder γ_c. For $\gamma < \gamma_c$ the system is metallic, for $\gamma > \gamma_c$ it is insulating.

Yet another approach to the Anderson transition is based on a self-consistent calculation of the diffusion coefficient $D(\omega)$ or conductivity $\sigma(\omega)$. Within this concept, which was first introduced by Götze [25, 26], one attempts to express $\sigma(\omega)$ or $D(\omega)$ by means of a non-trivial, generally approximate relation which itself involves this quantity. So one wants to find an equation of the form

$$D(\omega) = \mathscr{F}[D(\omega)].\qquad(10)$$

whose "self-consistent" solution then yields $D(\omega)$ for all ω and all disorder-parameters γ. For this to be successful it is necessary to start from known limiting cases (e.g. the perturbation theory for $\gamma \ll 1$) such that the theory can be anchored to an exact result [27, 28]. The self-consistency is then used to go beyond perturbation theory, i.e. to the transition itself (and even further). It is therefore used as a substitute for an (untractable) perturbation theory to infinite order.

Since $D(\omega)$ vanishes at the transition, its inverse $D_0/D(\omega)$ correspondingly diverges at that point. Within a diagrammatic perturbation theory Vollhardt and Wölfle [27, 28] showed that a self-consistent calculation of the latter quantity can be performed by summing up the largest (i.e. most divergent) contributions of perturbation theory [29]. In this way a self-consistent equation is derived. It has the simple structure

$$\frac{D_0}{D(\omega)} = 1 + \frac{1}{\pi N_F} \int \frac{d\vec{k}}{(2\pi)^d} \frac{1}{-i\omega + D(\omega)k^2},\qquad(11)$$

where $D_0/D(\omega)$ is given by the integral over a diffusion pole involving the diffusion coefficient $D(\omega)$ rather than the diffusion constant D_0. (Eq. (11) actually involves the particle-particle diffusion pole obtained from the "maximally crossed diagrams" [6 ... 8], which, in the case of time reversal invariance, can be related to the diffusion coefficient $D(\omega)$ of particle-hole diffusion.) This relation can also be derived by other methods [30 ... 32]. Its solution can easily be obtained: One finds that for $d \leqslant 2$ the dc-conductivity $\sigma(0)$ is always zero, irrespective of how small the disorder is (insulating behavior). However, in dimension $d = 2$ the localization length ξ is exponentially large for $\gamma \ll 1$ [27, 28]: $\xi \sim \exp(1/2\gamma)$. For $d > 2$ there exists a critical value of the disorder below which $\sigma(0)$ is finite (metallic regime), while for larger values it vanishes (insulating regime). Since the limit $\omega \to 0$ can be explicitly performed within this theory, one obtains results which go beyond the range of applicability of the scaling theory described above. Besides that one obtains complete agreement [33] with the results of scaling theory.

A field theoretical analysis of the problem by Hikami [24], which involves the solution of the Callan-Simanzyk equation, yields exactly the same relation for $D(\omega)$ as in Eq. (11). Hence, if perturbation theory is valid at all, Eq. (11) is an exact relation at least close to two dimensions.

Most recently a full renormalization-group treatment of the density-density correlation function by Abrahams and Lee [34] yielded the scaling behavior of the diffusion coefficient $D(\omega, \vec{q})$. For $v_F |\vec{q}| < \omega$ the result is again identical to the self-consistent equation (11).

3 Weak Localization of Light (Photons)

By 1983 the interpretation of weak localization in disordered electronic systems as an interference phenomenon of waves had been widely recognized. On the other hand, a very similar effect had already been discussed much earlier by Watson [35] and de Wolf [36] with respect to scattering of electromagnetic waves from fluctuations in a plasma and general turbulent media, respectively. These investigations were related to questions concerning radar scattering from ionized or neutral gases, which, for example, arise in the remote probing of the atmosphere. Since they originated in a subject very different from condensed matter physics, these findings went unnoticed by the disorder community. In fact, in 1984 Kuga and Ishimaru [37] and Tsang and Ishimaru [38] presented experimental and theoretical results, respectively, for the scattering of electromagnetic waves from a random distribution of discrete scatters, which clearly showed an enhancement in backscattering. The authors [38] used second-order multiple scattering theory to explain their results. Therefore their interpretation was based on the same physical idea, i.e. constructive interference in the backward direction due to *multiple* scattering, which had already been identified as the cause for weak localization in disordered systems. A connection with electron localization was not made. The same interference effects involving light where simultaneously discussed by Golubentsev [39].

It was Anderson [15] who discussed the phenomenon of localization from a general point of view and who explicitly addressed the question of classical wave localization.

Inspired by the weak localization effects known from disordered electronic system van Albada and Lagendijk [40] and Wolf and Maret [41] convincingly showed that coherent backscattering equally applies to the propagation of light in a disordered medium. Shining light into a highly concentrated aqueous suspension of sub-micron size polystyrene spheres, (also used in [37]) these two groups measured the scattered intensity and found a striking enhancement in the backscattering direction within a narrow cone. This enhancement comes from the constructive interference of light-waves travelling on closed, time-reversed paths just as explained in the case of weak localization. Note that the explicit condition of *static* disorder necessary for weak localization is fulfilled even in these experiments, since the thermal motion in the liquid is much slower than the propagation of the light wave along any relevant closed path in the medium. Ideally, i.e. assuming isotropic scattering and scalar waves, the backscattered intensity should be enhanced by the factor of 2 in Eq. (6) relative to the incoherent background. This would require that the starting and endpoint of the loops really coincide (A = B in Fig. 2). Otherwise interference cannot be complete, resulting in a reduced enhancement. This effect

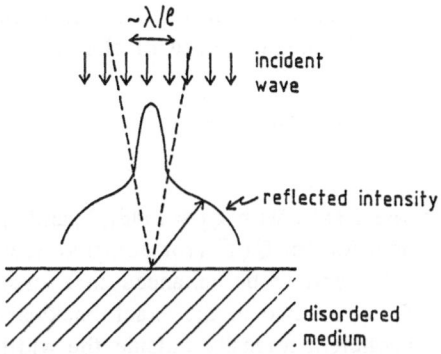

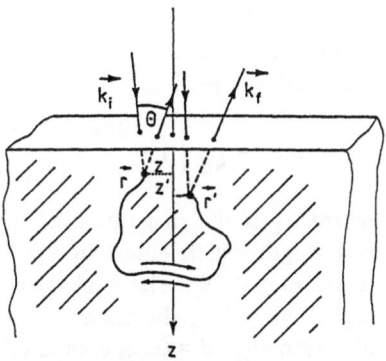

Fig. 5 Schematic view of the backscattering intensity of a (light) wave reflected from a disordered medium.

Fig. 6 Interference of light waves propagating on time reversed paths in a disordered medium (after [43]).

leads to a limitation of the angular width within which the enhancement is visible (see below), as indicated in Fig. 5. A detailed theoretical analysis of the effect [42] and of the experiments, in particular of the observed peak line shape, was given by Akkermans, Wolf, and Maynard [43]. Fig. 6 illustrates the wave propagation in the disordered system. Here $\vec{k}_i$ and $\vec{k}_f$ are the wave vectors of the incident and the emerging light wave (with wave length λ) and θ is their relative angle. The figure makes clear that coherent backscattering is only possible within an angular width $\delta\theta \approx \lambda/|\vec{r}_i - \vec{r}_f|$, where $\vec{r}_i$, $\vec{r}_f$ are the positions of first and last scattering in the medium, respectively. Since $|\vec{r}_i - \vec{r}_f| \gtrsim \ell$, the enhancement is only visible in a narrow cone of width $\delta\theta = \lambda/\ell \ll 1$.

The intensity of the reflected relative to the incident light, the so-called albedo $\alpha(\vec{k}_i, \vec{k}_f)$, is given by [43]

$$\alpha(\hat{k}_i, \hat{k}_f) = \frac{c}{4\pi\ell^2} \int dz\, dz'\, d^2\rho\, \exp\left(-\frac{z}{p_i\ell}\right) \cdot$$
$$\cdot \{1 + \cos[\vec{q} \cdot (\vec{r}_i - \vec{r}_f)]\}\, Q(\vec{r}, \vec{r}')\, \exp\left(-\frac{z'}{p_f\ell}\right) \tag{12}$$

where c is the wave velocity, z and z' are the z-components of $\vec{r}$ and $\vec{r}'$ (see Fig. 6), $\vec{q} = \vec{k}_i + \vec{k}_f$, $\vec{\rho}$ is the projection onto the surface, and $p_{i,f} = \hat{k}_{i,f} \cdot \hat{z}$. The two exponentials in Eq. (12) are attenuation factors associated with the scattering events which take place at distances z and z' from the interface inside the disordered medium. The factor $1 + \cos[\vec{q} \cdot (\vec{r}_i - \vec{r}_f)]$ corresponds to the usual $(1 - \cos\vartheta)$ term in transport theory, e.g. appearing in the expression for the relaxation time, which favors *large*-angle scattering (note, that $\vartheta = \pi - \vec{q} \cdot (\vec{r} - \vec{r}')$). The Green function $Q(\vec{r}, \vec{r}')$ describes the wave propagation from $\vec{r}$ to $\vec{r}'$. Owing to the disorder this transport is diffusive in nature, such that $Q(\vec{r} - \vec{r}')$ is determined by the three-

73

dimensional diffusion equation with proper boundary conditions. For isotropic scattering Eq. (12) may be evaluated to yield the angular shape of the relative scattered flux [43]

$$\alpha(\theta) = \frac{3}{8\pi} \left\{ 1 + \frac{2z_0}{\ell} + \frac{1}{(1 + q_\perp \ell)^2} \left[1 + \frac{1 - \exp(-2q_\perp z_0)}{q_\perp \ell} \right] \right\} \qquad (13)$$

where $q_\perp$ is the magnitude of $\vec{q}$ normal to the z-axis, with $q_\perp \approx 2\pi|\theta|/\lambda$, and z_0 is a length which enters in the boundary condition for $Q(\vec{r})$ (for pointlike scatterers $z_0 \approx 0.7\ell$). The backscattering is indeed found to be enhanced in a small regime $\delta\theta \sim \lambda/\ell$, the enhancement being a factor of two in the exact backward direction ($q_\perp$, $\theta = 0$) as compared with the incoherent intensity outside the width $\delta\theta$. $\alpha(\theta)$ has a peculiar shape: for small angles it varies linearly with θ, i.e. has a triangular shape with a sharp tip at $\theta = 0$. Using this result Akkermans, Wolf, and Maynard [43] were able to describe the experimental curves for $\alpha(\theta)$ of [41] without any adjustable parameter. Convoluting their theoretical result for $\alpha(\theta)$ with the given instrumental profil they found a truely remarkable quantitative agreement (Fig. 7).

These authors also pointed out that for small angles θ all diffusional trajectories of size $L \lesssim \lambda^2/(\ell\theta^2)$ contribute to the coherent backscattering; hence, the smaller θ is, the larger the size of the maximal loops are. The quantity $\lambda^2/(D \cdot \theta^2)$ therefore corresponds to the inelastic scattering time τ_{in} acting as a cut-off in Eqs. (8), (9). The sharp tip of $\alpha(\theta)$ will always be rounded unless loops of arbitrary size are included.

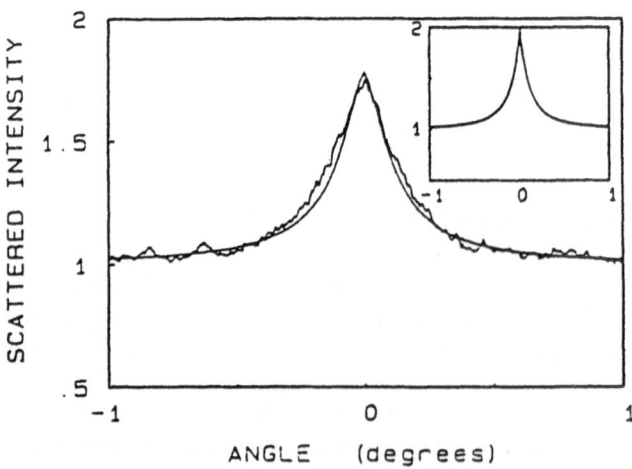

Fig. 7 Line shape of the intensity of coherently backscattering light [43]. The measured curve by Wolf and Maret [41] is compared with the theoretical results of Akkermans, Wolf, and Maynard [43] after folding with the experimental profil. Insert: bare theoretical curve.

74

In the case of light the backscattered intensity is polarization dependent [37, 40, 41], i.e. it depends on the relative orientation between incident and reflected polarization (parallel or perpendicular). Depending on the instrumental profil of the apparatus an enhancement in the former case of up to 1.96 ± 0.02 [44] has been observed. For perpendicular orientation only smaller enhancements (≈ 1.3) have been found. In [43] a maximal enhancement of 2 and ≈ 1.5, respectively, has been calculated for the two different cases. A detailed theoretical investigation of the polarization dependence of the backscattering was given by Stephen [45] and Stephen and Cwilich [46].

Localization effects in light scattering from a disordered *solid* have also been observed [47, 48]. In contrast to a liquid, a rigid disordered medium leads to large-amplitude fluctuations in the scattered intensity which has to be subtracted (by ensemble averaging) to reveal the backscattering peak. (In the liquid this is automatically done by the thermal motion of the atoms.) Localization effects caused by the reflection of light from random gratings [49] or random layered systems [50] have also been considered. The critical behavior of electromagnetic absorption, i.e. the "photon mobility edge" was studied within a renormalization group theory by John [51].

4 Localization of Acoustic Waves (Phonons) and other Sound

The wave nature of acoustic phenomena may in principle lead to similar localization effects in a disordered medium as discussed in the case of electronic transport or light propagation. On the other hand, as pointed out by Anderson [15], there will be fewer systems available than in the case of light where localization effects can actually be expected to be seen. (This is due to the problem of having an inhomogeneous mixture of two propagating media, where different waves will be excited within the system; see, however, Sect. 5.) He suggested to use expanded silica gel filled with a denser liquid to study acoustic localization. Most recently the observability of acoustical and optical localization was analyzed by Condat and Kirkpatrick [52]. Using the self-consistent equation (11) of Vollhardt and Wölfle [27, 28] to investigate the transition to the localized state, they conclude that in $d = 3$ acoustic localization will not be observed unless the scatterers are more efficient than hard spheres. On the other hand the optical localization transition is found to be not far from the conditions used in the weak localization experiments [37, 40, 41, 47].

A different kind of "acoustic" localization experiment in a one-dimensional system (where weak localization can never really be observed because there is no true extended wave behavior) has been reported by He and Maynard [53]. Disorder was introduced in a wire by either varying the size of periodically positioned small masses along a wire (alloy-type disorder) or their position itself (liquid-type disorder). The frequency response of the system, i.e. of a transverse wave generated in the wire, was then measured, where an additional electron-phonon interaction was simulated by means of a longitudinal strain in the wire. Studying the eigen-frequencies of the wire extended and localized states were clearly observed.

Concerning theoretical work the study of phonon localization was the first to follow the respective investigations of electronic systems [54, 55]. Using a field-theoretical formulation John, Sompolinsky, and Stephen [55] showed that a "phonon mobility edge" should occur above $d = 2$. A diagrammatic theory, based on a weak-localization, i.e. perturbational, approach to the localization of phonons, was presented by Akkermans and Maynard [56]. Although in spirit and technique very similar to the corresponding diagrammatic theory of the localization of electrons, there are important differences due to the absence of Fermi-Dirac statistics and because of the strong frequency dependence of characteristic quantities.

Localization of acoustic (sound) waves in a random array of hard scatterers in $d = 1$ [57] and in $d = 2, 3$ [58], has also been investigated within the self-consistent theory, Eq. (11). Agreement with the field-theoretical results [55] (which were derived for $d = 2 + \epsilon$) are obtained even in $d = 3$ where $\epsilon = 1$.

Localization of sound modes different from the conventional acoustical sound ("first" sound), namely, of "third" sound (i.e. surface modes in a superfluid such as ^{4}He) was also studied [59...61] using the self-consistent theory. It appears that in such a system the localization length can be continuously varied from practically infinity to a few millimeters. This would allow for investigations of localization in $d = 2$, unfeasible in electronic systems.

5 Other Localizing Media and Waves

Studying localization one usually considers the random scatterer to be uncor-related. The more complicated situation where wave propagation and localization takes place in a random potential having a long range correlation has been in-vestigated by John and Stephen [62]. Using an appropriate field-theoretical model they again find that for $d \leqslant 2$ all states are localized and that the mobility edge in $d = 2 + \epsilon$ is characterized by the same critical exponents as for spatially uncor-related disorder.

Localization of waves in a fluctuating plasma was studied by Escande and Souil-lard [63]. They find that, in the absence of dissipation, density fluctuations in a plasma may lead to exponential localization of electron plasma waves. The cor-responding localization transition is expected to be easily observable since the strength of the disorder can be readily varied.

Localization caused by surface roughness has been discussed in the context of electrons in thin films [64]. A conceptually similar but nonetheless different effect, namely, localization of surface plasmon polaritons (SPP) and its role in surface enhanced optical phenomena, was addressed by Arya, Su, and Birman [65]. After having been excited by a photon the SPP can propagate parallel to the metal surface but will be scattered elastically by spatial fluctuations in the dielectric function near the otherwise smooth metal-vacuum interface. Using the

76

self-consistent theory of Vollhardt and Wölfle [27, 28] an equation for the re-normalized diffusion coefficient is derived, showing that localization effects occur over a certain frequency range, in particular, if radiation losses are small.

The same approach has been employed to study scalar wave localization in a two-component composite [66] (see the remarks made in the context of acoustic effects). Localization is predicted if the impedance contrast of the medium (i.e. the ratio of the respective indices of refraction) exceeds a certain minimum value.

6 Localization and Magnetic Fields

In the case of normal impurity scattering weak localization is due to the constructive interference of waves on time-reversed paths. Therefore this effect is very sensitive to any kind of disturbance of time reversal invariance of the momentum states $\vec{k}$ and $-\vec{k}$. Such a perturbation is, for example, caused by a magnetic field. In its presence a state is no longer characterized by a momentum $\vec{k}$, but rather by the electromagnetic momentum $\vec{k} - 2e\vec{A}$. Here, $\vec{A}$ is the vector potential and the factor 2e (instead of simply e) is due to the correlation of *two* particles just as in superconductivity. If we now let $\vec{k}$ go into $-\vec{k}$, the momentum states, i.e. the paths 1 and 2 in Fig. 2, are no longer equivalent. Mathematically speaking this is a consequence of the fact that now the amplitudes A_1 and A_2 carry field dependent phase factors [10, 11], determined by the magnetic flux $\Phi = \oint d\vec{\ell} \cdot \vec{A}$, such that $A_1 \to Ae^{i\varphi}$, $A_2 \to Ae^{-i\varphi}$, where

$$\varphi = 2\pi \frac{\Phi}{hc/e} . \tag{14}$$

The magnetic flux is given by $\Phi = H \cdot S$, where H is the magnetic field and S is the area of the closed path in Fig. 2 (c = velocity of light). Since the motion of the particles is diffusive, S is given by $S = D_0 t$. The return probability W_H of a particle to its starting point in the presence of a magnetic field is again given by Eq. (5). One therefore obtains

$$W_H = 2|A|^2 \left[1 + \cos \frac{2eHD_0 t}{\hbar c} \right]. \tag{15}$$

The conductivity correction in the presence of a magnetic field, $\delta\sigma(H)$, is determined by the return probability W_H. The total change of the conductivity due to a magnetic field, $\Delta\sigma(H) = \delta\sigma(H) - \delta\sigma(0)$, therefore depends on the probability difference $W = W_H - W_{H=0}$, such that

$$\Delta\sigma(H) = -\int_\tau^{\tau_{in}} dt \frac{v_F \lambda_F^{d-1}}{(D_0 t)^{d/2}} \left[\cos \frac{2eHD_0 t}{\hbar c} - 1 \right]. \tag{16}$$

In $d = 2$ Eq. (16) can be written as $\Delta\sigma(H) = e^2 F(x)$, where $x = 2eHD_0 \tau_{in}/\hbar c$. The function $F(x)$ has the limits

$$F(x) = \begin{cases} x^2, & x \ll 1 \\ \ln x, & x \gg 1. \end{cases} \tag{17}$$

For weak magnetic fields, $x \ll 1$, one therefore finds

$$\Delta\sigma(H) \sim H^2 \tau_{in}^2 \tag{18}$$

while stronger fields, $x \gg 1$, give rise to a logarithmic field dependence

$$\Delta\sigma(H) \sim \ln(H\tau_{in}). \tag{19}$$

In any case, $\Delta\sigma$ is always positive ($\Delta R < 0$), so the resistance *decreases* with increasing magnetic field ("anomalous magnetoresistance") [19]. The reason lies in the disturbance of the phase coherence by the magnetic field, leading to a weakening of the localization effects. The "critical" field H_c, determined by $x = 1$, at which the change from the H^2 to the $\ln H$ behavior occurs, depends on τ_{in} and thereby on temperature. At temperatures commonly used in experiments, H_c is of the order $H_c \approx 100 \ldots 500$ Gauss ($\hat{=} 10 \ldots 50$ mT). This should be contrasted with the classical result $\Delta\sigma(H)/\sigma_0 \approx -(\omega_L\tau)^2$, which is not only many orders of magnitude smaller but also has a different sign (ω_L is the Larmor frequency)! So we see that even very small magnetic fields have a drastic influence on localization.

Oscillation effects

As first observed by Altshuler, Aronov, and Spivak [67] the phase dependence of the electron wave function leads to a novel kind of quantum oscillation in the magnetoresistance of a multiply connected geometry, e.g. a torus made by wrapping up a thin, disordered metallic film (Fig. 8). The total change in phase, $\Delta\varphi_{loc}$, of the two oppositely traversed paths in the cylinder (Fig. 9(b)), is given by (14)

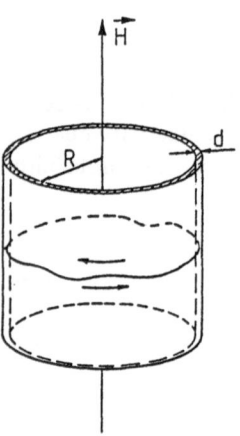

Fig. 8

Geometry used for detecting quantum oscillations in the magnetoresistance with period hc/2e.

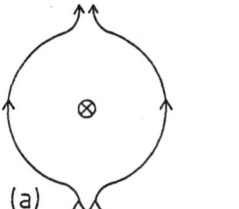 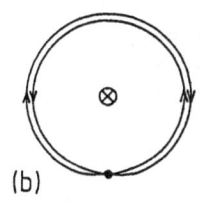

Fig. 9

Paths of interfering waves around screened magnetic field (⊗); (a) Usual Aharonov-Bohm effect, (b) geometry of Altshuler, Aharonov, and Spivak [67].

$$\Delta\varphi_{loc} = 2\varphi = 2\pi \frac{\Phi}{\Phi_0}, \tag{20}$$

where $\Phi_0 = hc/2e$ is the flux quantum known from superconductivity (although we are here in a normal-conducting situation!). The weak localization correction to the conductivity is thus given by a straightforward extension of Eq. (16), i.e. [10, 11],

$$\Delta\sigma(H) = \int_\tau^{\tau_{in}} dt \left[W_0 + 2 \sum_n W_n(t) \cos\left(2\pi n \frac{\Phi}{\Phi_0}\right) \right], \tag{21}$$

where the W_n are the return-probabilities for an electron after having traversed the loop n times. Clearly, $\Delta\sigma$ is an oscillatory function of the flux with periodicity Φ_0. This finding must be contrasted with the well-known Aharonov-Bohm effect (Fig. 9(a)) where electrons passing a coil enclosing a magnetic field, will only acquire a phase change of *half* the change given by Eq. (14). This is so because in the Aharonov-Bohm effect each electron only samples half the flux Φ, because it only passes through half the loop (Fig. 9(a)). This yields a total phase change of

$$\Delta\varphi_{AB} = 2 \cdot \frac{\varphi}{2} = 2\pi \frac{\Phi}{hc/e} \tag{22}$$

leading to oscillations of the flux with period $hc/e = 2\Phi_0$. While the Aharonov-Bohm effect in a cylinder geometry is similar to Dingle-oscillations [68], the effect predicted in [67] with period Φ_0 is reminiscent of Parks-Little oscillations [69] in a superconducting geometry.

To observe the effect it is important that the inelastic diffusion length $L_{in} = \sqrt{D_0 \tau_{in}}$ be larger than the circumference $2\pi R$ of the cylinder (Fig. 8), because otherwise the coherence is destroyed. (Note that both L_{in} and R can be much larger than the mean free path!).

The predicted oscillations were first measured by Sharvin and Sharvin [70] (Fig. 10), who found full agreement with theory. These experiments were done on thin Mg-films; other experiments used Li (where spin-orbit scattering is negligible) and also yielded very good agreement with theory [71].

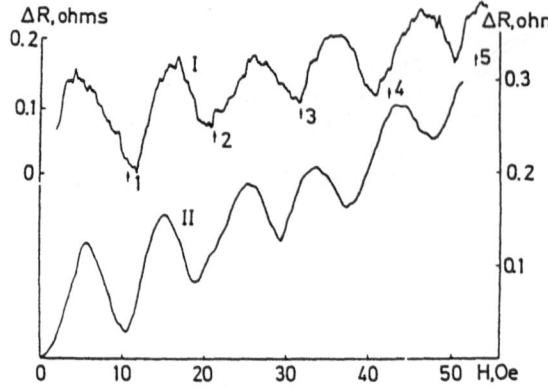

Fig. 10
Oscillations of the magnetoresistance of two different cylindrical Mg-films as measured by Sharvin and Sharvin [70].

7 Networks

In the attempt to reproduce the experiments of [70] other multiply-connected geometries than a single hollow cylinder where studied. This led to the investigation of networks of loops, e.g. of samples containing about $2.7 \cdot 10^6$ identical hexagonal loops forming a regular, two-dimensional honeycomb-network [72], or of ladders of 1000 little squares in series with 50 ladders in parallel [73]. The magnetoresistance of these new geometries were measured, and oscillations with period $\Phi_0 = hc/2e$ were found. A detailed theory of interference effects and quantum oscillations in the magnetoresistance of normal metal networks (loops, ladders, lassos, fractal networks, etc.) was worked out by Doucot and Rammal [74, 75] who also found remarkable agreement with the experiment (Fig. 11). This work is

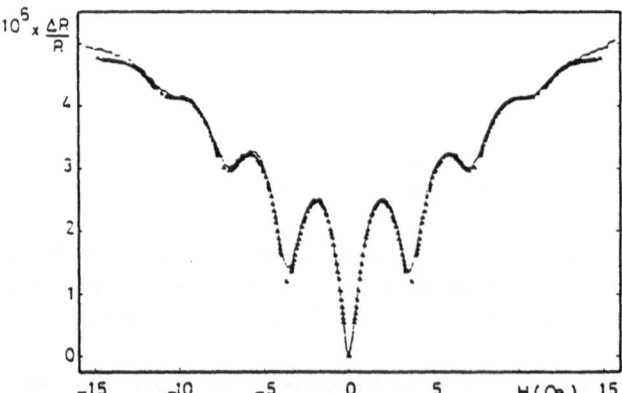

Fig. 11 Comparison between the theoretical results [74, 75] and experimental data [74] of Doucot and Rammal for magnetoresistance oscillations measured with a copper network with honey comb structure.

80

closely related to similar investigations of *superconducting* networks [76], where fascinating physics is known to occur (frustration, fractional number of flux quanta per unit cell of the network, fractal fine structure of the upper critical field line due to interference effects between adjacent loops, etc.). In contrast, static properties of normal-conducting networks do not show such a fine structure because of an inherent regularization of the otherwise complicated spectrum [74, 75].

8 hc/e Versus hc/2e Oscillations

Both from an experimental and theoretical point of view a single ring should be about the simplest geometry to observe the above-mentioned oscillations in the magnetoresistance. Experimentally, the opposite was true since, at first, the Φ_0-oscillations could not be found. Later, both $2\Phi_0 = hc/e$ and $\Phi_0 = hc/2e$-oscillations were detected in individual, micron size, normal metal rings [77]. The different temperature and field dependence clearly distinguishes between the two effects and their physical origin. At low fields the localization induced hc/2c-effect is seen, while at higher fields, when localization is suppressed, the hc/e-effect is visible. Most recently both types of oscillations were also measured in samples made up of N such rings in series [78]. It was found that, on averaging, the amplitude of the hc/e oscillations showed a $1/\sqrt{N}$ decrease, while the hc/2e-effect was independent of N. This has also been verified theoretically [79]. It clarifies the role of *ensemble averaging* in calculating corrections to the conductivity. This kind of averaging is canonically employed in the framework of weak localization (yielding hc/2e-oscillations) but *not* in the calculation of the transmission coefficient [80] in metal rings, where only the hc/e-effect is found. So, to obtain the hc/2e-effect, ensemble averaging is necessary.

9 Mesoscopic Systems and Universal Fluctuations

As an unexpected byproduct, the investigations of quantum oscillations in (sub-) micron structures ("mesoscopic" systems) led to the discovery of anomalously large, *universal* fluctuations [81 ... 83]. These fluctuations, known from experiment [84] and numerical simulations [85], are not due to time dependent noise or finite-size effects. They are a consequence of quantum interference, are reproducible and occur if the temperature is low enough such that the inelastic diffusion length L_{in} exceeds the sample dimension. In this case the conductance of a small sample shows fluctuations as a function of magnetic field, chemical potential, or impurity configuration whose r.m.s. value is approximately e^2/h, *independent* of sample size, i.e. is universal. The effect depends only weakly on dimensionality and the strength of (weak) disorder and is, of course, much larger than expected classically. It has clearly been observed in the experiment [86]. As discussed by Lee and Stone [82] this behavior is compatible with one-parameter scaling of Anderson localization [6] where in the scaling regime the conductance is essentially length independent and of order e^2/h such that its fluctuations must also be large and scale independent. The quantum interference of randomly diffusing electrons, which leads to the large fluctuations, implies an extraordinarily large sensitivity of the conductance on the impurity configuration. Indeed, the displacement of a *single* impurity by only λ_F (de Broglie wave length) affects essentially all quantum mechanical paths and hence changes the conductance by a universal, i.e. sample size independent amount [87, 88]. There are many other unusual phenomena (e.g. asymmetries [89] in the magnetoconductance [90], ability to rectify alternating currents [89]). A comprehensive discussion of the effects of finite temperatures, interactions, and magnetic fields on the universal conductance fluctuations, as well as of the physical assumptions underlying the ergodic hypothesis has been presented by Lee, Stone, and Fukuyama [91], who also relate theory to experiment.

References

[1] *P. W. Anderson*, Phys. Rev. 109, 1492 (1958)

[2] *S. F. Edwards*, Phil. Mag. 8, 1020 (1958)

[3] For a review, see *P. A. Lee* and *T. V. Ramakrishnan*, Rev. Mod. Phys. 57, 287 (1985)

[4] Anderson Localization, ed. by *Y. Nagaoka* and *H. Fukuyama*, Springer Series in Solid State Sciences, Vol. 39 (Springer, Berlin 1982)

[5] *N. F. Mott*, Phil. Mag. 26, 1015 (1972)

[6] *E. Abrahams, P. W. Anderson, D. C. Licciardello*, and *T. V. Ramakrishnan*, Phys. Rev. Lett. 42, 673 (1979)

[7] *E. Abrahams* and *T. V. Ramakrishnan*, J. Non-Cryst. Solids 35, 15 (1980)

[8] *L. P. Gor'kov, A. I. Larkin*, and *D. E. Khmelnitskii*, Zh. Eksp. Teor. Fiz. Pis'ma Red. 30, 248 (1979) [JETP Lett. 30, 248 (1979)]

[9] *J. S. Langer* and *T. Neal*, Phys. Rev. Lett. 16, 984 (1966)

[10] *B. L. Altshuler, A. G. Aronov, D. E. Khmelnitskii*, and *A. I. Larkin*, in: Quantum Theory of Solids, ed. by *J. M. Lifshitz* (MIR Publishers, Moscow, 1983)

[11] *D. E. Khmelnitskii*, Physica 126 B + C, 235 (1984)

[12] *G. Bergmann*, Phys. Rep. 107, 1 (1984)

[13] *D. Vollhardt*, in: Solid State Physics, Lecture Notes of the NORDITA Spring School, Tvärminne, Finland (April 1984), Vol. I; p. 73 (Nordita, Copenhagen)

[14] For a discussion of weak localization based on the quasi classical theory of electrons, see *S. Chakravarty* and *A. Schmid*, Phys. Rep. 140, 193 (1986)

[15] *P. W. Anderson*, Phil. Mag. B52, 505 (1985)

[16] *W. C. McGinnis, M. J. Burns, R. W. Simons, G. Deutscher*, and *P. M. Chaikin*, Physica 107B, 5 (1981)

[17] See, for example, *R. J. Elliot, J. A. Krumhansl*, and *P. L. Lieb*, Rev. Mod. Phys. 46, 465 (1974)

[18] *F. Wegner*, Z. Phys. B35, 207 (1979)

[19] *S. Hikami, A. I. Larkin*, and *Y. Nagaoka*, Prog. Theor. Phys. 63, 707 (1980)

[20] *R. Oppermann* and *K. Jüngling*, Phys. Lett. 76A, 449 (1980)

[21] *F. Wegner*, in [4]

[22] *L. Schäfer* and *F. Wegner*, Z. Phys. B38, 113 (1980)

[23] *R. Oppermann* and *F. Wegner*, Z. Phys. B34, 327 (1979)

[24] *S. Hikami* in [4]

[25] W. Götze, Solid State Comm. 27, 1393 (1978)

[26] W. Götze, Phil. Mag. 43, 219 (1981)

[27] D. Vollhardt and P. Wölfle, Phys. Rev. Lett. 45, 482 (1980)

[28] D. Vollhardt and P. Wölfle, Phys. Rev. B22, 4666 (1980)

[29] P. Wölfle and D. Vollhardt in [4]

[30] P. Prelovsek, in: Recent Developments in Condensed Matter Physics, ed. by J. T. Devreese (Plenum Press, New York 1981), Vol. II, p. 191

[31] D. Belitz, A. Gold and W. Götze, Z. Phys. B44, 273 (1981)

[32] C. S. Ting, Phys. Rev. B26, 678 (1982)

[33] D. Vollhardt and P. Wölfle, Phys. Rev. Lett. 48, 699 (1982)

[34] E. Abrahams and P. A. Lee, Phys. Rev. B33, 683 (1986)

[35] K. M. Watson, J. Math. Phys. 10, 688 (1969)

[36] D. A. de Wolf, IEEE Trans. Antennas Propag. 19, 254 (1971)

[37] Y. Kuga and A. Ishimaru, J. Opt. Soc. Am. A8, 831 (1984)

[38] L. Tsang and A. Ishimaru, J. Opt. Soc. Am. A8, 836 (1984)

[39] A. A. Golubentsev, Zh. Eksp. Teor. Fiz. 86, 47 (1984) [Sov. Phys. JETP 59, 26 (1984)]

[40] M. P. van Albada and A. Lagendijk, Phys. Rev. Lett. 55, 2692 (1985)

[41] P. E. Wolf and G. Maret, Phys. Rev. Lett. 55, 2696 (1985)

[42] E. Akkermans and R. Maynard, J. Physique Lett. 46L, 1045 (1985)

[43] E. Akkermans, P. W. Wolf, and R. Maynard, Phys. Rev. Lett. 56, 1471 (1986)

[44] U. Adler, Diploma-thesis, Ludwig-Maximilians Universität München, 1986 (unpublished)

[45] M. J. Stephen, Phys. Rev. Lett. 56, 1809 (1986)

[46] M. J. Stephen and G. Cwilich, Phys. Rev. B34, 7564 (1986)

[47] S. Etemad, R. Thompson, and M. J. Andrejco, Phys. Rev. Lett. 47, 575 (1986)

[48] M. Kaveh, M. Rosenbluh, I. Edrei, and I. Freund, Phys. Rev. Lett. 57, 2049 (1986)

[49] A. R. McGurn, A. A. Maradudin, and V. Celli, Phys. Rev. B31, 4866 (1985)

[50] E. Bouchaud and M. Daoud, J. Physique 47, 1467 (1986)

[51] S. John, Phys. Rev. Lett. 53, 2169 (1984)

[52] C. A. Condat and T. R. Kirkpatrick, Phys. Rev. Lett. 58, 226 (1987)

[53] S. He and J. D. Maynard, Phys. Rev. Lett. 57, 3171 (1986)

[54] J. Jäckle, Solid State Comm. 39, 1261 (1981)

[55] S. John, H. Sompolinsky, and M. J. Stephen, Phys. Rev. B27, 5592 (1983)

[56] E. Akkermans and R. Maynard, Phys. Rev. B32, 7850 (1985)

[57] C. A. Condat and T. R. Kirkpatrick, Phys. Rev. B32, 495 (1985)

[58] T. R. Kirkpatrick, Phys. Rev. B31, 5746 (1985)

[59] S. M. Cohen and J. Machta, Phys. Rev. Lett. 54, 2242 (1985)

[60] C. A. Condat and T. R. Kirkpatrick, Phys. Rev. B32, 4392 (1985)

[61] C. A. Condat and T. R. Kirkpatrick, Phys. Rev. B33, 3102 (1986)

[62] S. John and M. J. Stephen, Phys. Rev. B28, 6358 (1983)

[63] D. F. Escande and B. Souillard, Phys. Rev. Lett. 52, 1296 (1984)

[64] A. R. McGurn and A. A. Maradudin, Phys. Rev. B30, 3136 (1984)

[65] K. Arya, Z. B. Su, and J. L. Birman, Phys. Rev. Lett. 54, 1559 (1985)

[66] P. Sheng and Z. Q. Zhang, Phys. Rev. Lett. 57, 1879 (1986)

[67]   *B. L. Altshuler, A. G. Aronov*, and *B. Z. Spivak*, Pisma Zh. Eksp. Teor. Fiz. 33, 101 (1981) [JETP Lett. 33, 94 (1981)]

[68]   *R. B. Dingle*, Proc. Roy. Soc. **A212**, 47 (1952)

[69]   *R. D. Parks* and *W. A. Little*, Phys. Rev. **A133**, 97 (1964)

[70]   *D. Yu. Sharvin* and *Yu. V. Sharvin*, Pisma Zh. Eksp. Teor. Fiz. 34, 285 (1981) [ZETP Lett. 34, 272 (1981)]

[71]   *B. L. Altshuler, A. G. Aronov, B. Z. Spivak, D. Yu. Sharvin*, and *Yu. V. Sharvin*, Pisma Zh. Eksp. Teor. Fiz. 35, 476 (1982) [JETP Lett. 35, 588 (1982)]

[72]   *B. Pannetier, J. Chaussy, R. Rammal*, and *P. Gaudit*, Phys. Rev. Lett. 53, 718 (1984)

[73]   *D. J. Bishop, J. C. Licini*, and *G. J. Dolan*, Appl. Phys. Lett. 46, 1000 (1985)

[74]   *B. Doucot* and *R. Rammal*, Phys. Rev. Lett. 55, 1148 (1985)

[75]   *B. Doucot* and *R. Rammal*, J. Physique 47, 973 (1986)

[76]   *B. Pannetier, J. Chaussy, R. Rammal*, and *J. C. Villegier*, Phys. Rev. Lett. 53, 1845 (1984)

[77]   *V. Chandrasekhar, M. J. Rooks, S. Wind*, and *D. E. Prober*, Phys. Rev. Lett. 55, 1610 (1985)

[78]   *C. P. Umbach, C. van Haesendonck, R. B. Laibowitz, S. Washburn*, and *R. A. Webb*, Phys. Rev. Lett. 56, 386 (1986)

[79]   *Q. Li* and *C. M. Soukoulis*, Phys. Rev. Lett. 57, 3105 (1986)

[80]   *M. Büttiker, Y. Imry, R. Landauer*, and *S. Pinhas*, Phys. Rev. **B31**, 6207 (1985)

[81]   *B. L. Altshuler*, Pisma Zh. Eksp. Teor. Fiz. 41, 530 (1985) [JETP Lett. 41, 648 (1985)]

[82]   *P. A. Lee* and *A. D. Stone*, Phys. Rev. Lett. 55, 1622 (1985)

[83]   For a simple discussion, see *P. A. Lee*, Physics Today 40, 516 (1987)

[84]   *C. P. Umbach, S. Washburn, R. B. Laibowitz*, and *R. B. Webb*, Phys. Rev. **B30**, 4048 (1984)

[85]   *A. D. Stone*, Phys. Rev. Lett. 54, 2692 (1985)

[86]   *W. J. Skocpol, P. M. Mankiewich, R. E. Howard, L. D. Jackel, D. M. Tennant*, and *A. D. Stone*, Phys. Rev. Lett. 56, 2865 (1986)

[87]   *B. L. Altshuler* and *B. Z. Spivak*, Pisma Zh. Eksp. Teor. Fiz. 42, 363 (1985) [JETP Lett. 42, 447 (1985)]

[88]   *S. Feng, P. A. Lee*, and *A. D. Stone*, Phys. Rev. Lett. 56, 1960 (1985)

[89]   *B. L. Altshuler* and *D. E. Khmelnitskii*, Pisma Zh. Eksp. Teor. Fiz. 42, 291 (1985) [JETP Lett. 42, 359 (1986)]

[90]   *M. Ma* and *P. A. Lee*, Phys. Rev. **B35**, 1448 (1987)

[91]   *P. A. Lee, A. D. Stone*, and *H. Fukuyama*, Phys. Rev. **B35**, 1039 (1987)

Invar: A New Approach to an Old Problem of Magnetism

Eberhard F. Wassermann

Tieftemperaturphysik, Universität Duisburg, D-4100 Duisburg 1, Federal Republic of Germany

Summary: The anomalies in the thermal expansion of fcc FeNi-alloys are known as "Invar-effect" since a long time. Inspite of numerous experimental and theoretical efforts, the microscopic origin of the Invar-effect is todate not understood. We will show in the present paper that the Invar-effect incorporates a broad spectrum of magnetic and magnetoelastic anomalies in a wide variety of ferro- as well as antiferromagnetic alloys, which may have fcc, bcc, and hexagonal lattice structures or even be amorphous. We will also demonstrate to what extent mixed magnetic behavior or the occurrence of reentrant spin glass states contribute to the observed anomalies in Invar alloys. We will then focus on the question, how, according to new, systematic studies, the contributions from band (itinerant) magnetism and local moments can be separated. Finally, we will present results of our recent experimental investigations of the spin polarized photoemission on Fe_3Pt-Invar. Comparison of these data to band-structure calculations on fcc Fe and Fe_3Pt as a function of volume of the Wigner-Seitz cell will show that within the broad spectrum of available Invar models the "old" heuristic 2γ-states model of Weiss describes the Invar-effect in principle correct.

1 Invar Anomalies as a Function of Temperature

The "Invar-effect" originates from investigations by Ch. E.Guillaume, who detected already in 1897 that fcc FeNi alloys at concentrations around $Fe_{65}Ni_{35}$ show almost constant — "invariant" — thermal expansion as a function of temperature in a broad range around room temperature [1]. The practical importance of his detection for the construction of calibration- and seismographic devices, as well as his finding of the temperature independent elastic behavior of FeNiCr alloys — named "Elinvar" — used for decades as material for watch balances made him a Nobel Laureat in 1920.

Fig. 1a shows, how the temperature dependence of the thermal expansion coefficient $\alpha = d\,(\Delta l/l)/dT$ of an Fe 35 at% Ni alloy looks in principle. The full curve gives α_{exp} (T) as determined experimentally. The dashed curve shows the temperature dependence of a non-magnetic (nm) ("hypothetic") reference sample, α_{nm} (T), as calculated from the Grüneisen relation, using data from specific heat. (The problem of finding a reference sample from first principles will be discussed later.) Assuming that α_{exp} (T) = α_{nm} (T) + α_m (T), where α_m (T) denotes the magnetic contribution to the thermal expansion (dashed-dotted line in Fig. 1a), one can see that in this type of analysis α_m (T) is negative throughout the range from 0 K to above the Curie temperature T_c. For comparison the temperature dependence of the thermal expansion coefficient of pure Ni (dotted line) around the Curie temperature of Ni is also shown in Fig. 1a. Note that in contrast to FeNi-Invar, the anomaly in $\alpha(T)$ in Ni is sharp (and positive) at T_c, and the total length change ($\Delta l/l \approx 1.4 \cdot 10^{-4}$) is an order of magnitude smaller than for the $Fe_{65}Ni_{35}$-Invar alloy ($\Delta l/l \approx 1.6 \cdot 10^{-3}$).

Ch. E. Guillaume also determined the concentration dependence of α_{exp} at constant temperature in the FeNi-system. Fig. 1b reveals that at room temperature and a concentration of 35 at% Ni, α_{exp} reaches its smallest value. Anomalies in α, though smaller, are, however, still present in alloys containing up to 80 at% Ni. For concentrations below ≈ 33 at% Ni a structural transition from the fcc γ-phase to the bcc α-phase (which does not show the Invar-effect) occurs, so that the data are no longer reliable. We will discuss the importance of the γ-α transition with respect to the Invar-effect in chapter 2.

Turning back to Fig. 1a and the temperature dependence of the expansion coefficient, two remarks should be added. Firstly, at low temperatures $\alpha_{exp}(T)$ is negative

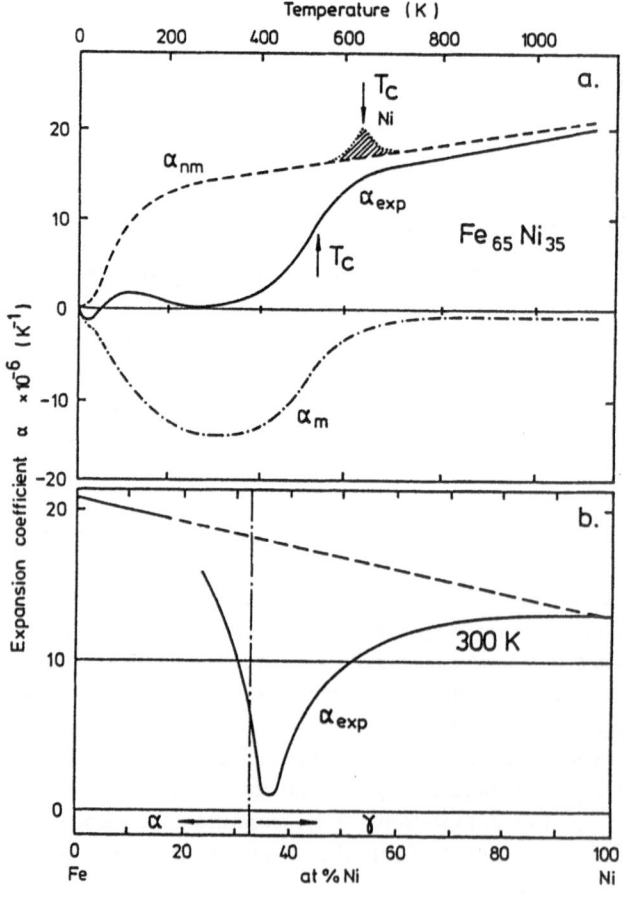

Fig. 1

a) Schematic representation of the temperature dependence of the thermal expansion coefficient α as determined experimentally, $\alpha_{exp}(T)$ (full line), for an $Fe_{65}Ni_{35}$ Invar alloy. The dashed line gives the temperature dependence of a (hypothetic) non-magnetic reference alloy as determined from the Grüneisen relation, $\alpha_{nm}(t)$. The magnetic contribution $\alpha_m(T)$, resulting from the difference $\alpha_{exp}(T) - \alpha_{nm}(T)$ is given by the dashed-dotted line. The Curie temperature is marked by an arrow. For comparison $\alpha_{exp}(T)$, as determined around T_c for pure Ni (dotted line), is also given.

b) Schematic representation of the concentration dependence of α_{exp} at room temperature in the FeNi system. The vertical line marks the structural α-γ transition in the system.

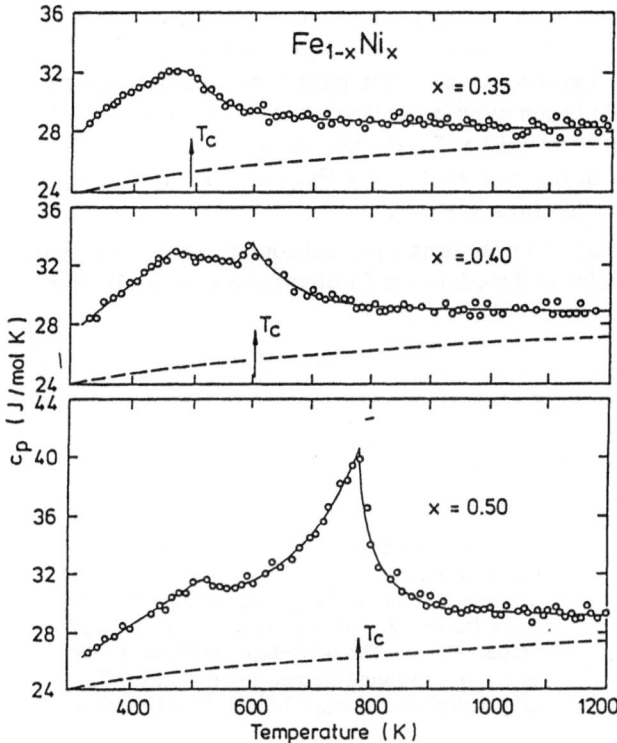

Fig. 2
Specific heat capacity c_p as a function of temperature as measured by Bendick et al. [2] on three FeNi alloys. The respective Curie temperatures are marked by arrows. The temperature dependence of c_p for respective non-magnetic reference samples are shown by the dashed lines.

and has a minimum. This means the sample first shrinks when rising the temperature from T = 0. To what extend this minimum in α_{exp} (T) is Invar typical is still debated today. Secondly, concerning the high temperature behavior, one can see that around and above the Curie temperature T_c (arrow in Fig. 1a) the expansion anomaly does not vanish (nor has any sharp "features"), a fact which excludes a Stoner type of theory for describing the Invar-effect correctly.

The importance of the high temperature contributions to the Invar-effect is revealed in Fig. 2, where the heat capacity C_p (T) for three FeNi-alloys as measured by Bendick et al. [2] in the high temperature range around T_c is shown. Above the paramagnetic background (broken line), which refers in the same way as in Fig. 1a to a non- or paramagnetic reference alloy, there is in addition to the large magnetic heat of transformation, which peaks sharply at the respective T_c-values, a considerable excess specific heat capacity due to the Invar-effect. This is revealed by the broad bumbs in c_p (T) occurring around 500 K in all three alloys. Similar effects have been seen by the same authors [2] in other Invar-alloys like, e.g., $Fe_{50}(Ni_xMn_{1-x})_{50}$ or $Fe_{80-x}Ni_xCr_{20}$ [3].

Another important Invar anomaly is the occurrence of a spontaneous volume magnetostriction. Fig. 3 shows (schematically) the volume change $\omega_s = \Delta V/V$ as a function of temperature for $Fe_{65}Ni_{35}$-Invar in comparison to $Fe_{50}Ni_{50}$ and pure Ni. $\omega(T)$ for the non-magnetic (again hypothetical) reference alloys is given by the broken lines. One can see that there exist at T = 0 a spontaneous maximum volume magnetostriction with a value of $\omega_{so} = 1.9 \cdot 10^{-2}$ for $Fe_{65}Ni_{35}$. Similar values for ω_{so} have been found in many other Invar systems.

Due to lack of space and since an extended review [4] and conference reports [5, 6] on Invar do exist, we summarize in Tab. 1 the main physical properties in which

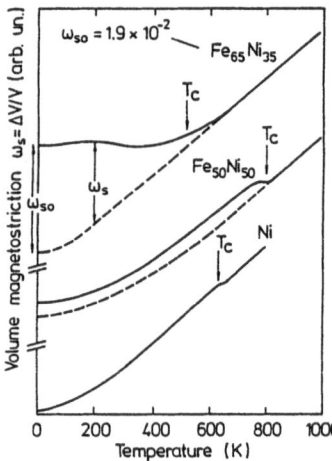

Fig. 3

Schematic representation of the spontaneous volume magnetostriction $\omega_s = \Delta V/V$ as a function of temperature for pure Ni, $Fe_{50}Ni_{50}$, and $Fe_{65}Ni_{35}$-Invar. The difference as compared to non-magnetic (hypothetic) reference alloys (dashed lines) leads to the spontaneous volume magnetostriction at T = 0, ω_{so}, with a value of $\omega_{so} = 1.9 \cdot 10^{-2}$ for $Fe_{65}Ni_{35}$.

Table 1 List of physical properties in which Invar anomalies are observed

Thermal expansion	$(\Delta l/l)(T); \alpha(T)$
Lattice constant	$a(T)$
Spontaneous volume magnetostriction	$\omega_s = (\Delta V/V)(T)$
Spontaneous volume magnetostriction at T = 0	$\omega_s(T = 0) = \omega_{so}$
Excess heat capacity at high T, linear heat capacity with large γ-values at low T	$C_V(T)$
Deviation from Brillouin-function in magnetization	$(M/M_0)(T)$
Forced volume magnetostriction	$(\delta\omega/\delta H)(T)$
Large high field susceptibility	$\chi_{HF}(T)$
Negative pressure dependence of magnetization	$-(dM/dp)_{T,H}$
Negative pressure dependence of Curie temperature	$-(dT_c/dp)$
Young- and Bulk modulus	$E(T); B(T)$
Elastic constants	$c_L(T), c_{44}(T), c'(T)$
Neutron-scattering	

Invar anomalies from low temperature to temperatures above the respective Curie temperatures have been observed. All the anomalies listed are found more or less pronounced in all Invar alloy systems known up to date (see Tab. 2).

2 Invar Anomalies as a Function of Concentration

Though the temperature dependence of the physical properties of Invar alloys as presented in the preceding chapter is characteristic and very important for the understanding of the origin of the effect, deeper insight into the problem comes from studies of the concentration dependence of the anomalies in different systems. As we have already seen in Fig. 1b, the thermal expansion coefficient α_{exp} reaches a minimum value (at room temperature) for concentrations close to the γ-α transition in FeNi. Similar plots for $(d\omega/dH)$, χ_{HF}, or other Invar typical properties (see also below Figs. 7 ... 10) reveal that these properties also reach maximum (or minimum) values on approach to the γ-α phase boundary. For a long time it was therefore believed that the origin of the Invar-effect in FeNi lies in the weakening of the magnetic moment of γ-iron, which occurs just in the same concentration range. This is shown in Fig. 4, where the magnetic phase diagram for $Fe_x Ni_{1-x}$ is given.

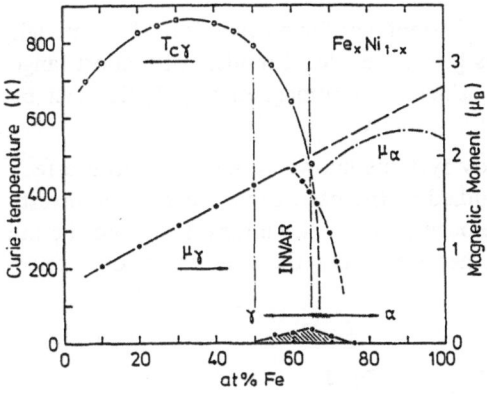

Fig. 4

Magnetic phase diagram of $Fe_x Ni_{1-x}$ as compiled from data of [7]. The Curie temperatures T_c in the γ-phase and the magnetic moment of the Fe-atoms in the γ-phase μ_{Fe} as a function of concentration are given. The Invar range close to the γ-α transition line is indicated by the dashed-dotted lines. Properties are extrapolated into the bcc α-phase. The hatched region indicates a low temperature mixed magnetic phase as observed in a.c. susceptibility measurements [8]. For details see text.

Data have originally been published by Crangle and Hallam [7] but repeatedly been verified since then. One can see that on approach to the γ-α phase transition in the Invar range (dashed-dotted lines in Fig. 4) there is a strong downward deviation of the moment μ_{Fe} from the Slater-Pauling curve, accompanied by a decrease in Curie temperature T_c. Weak itinerant ferromagnetism (WIF) was thus suspected to cause the Invar-effect.

One should mention, however, that based on, e.g., a.c. susceptibility measurements [8], neutron diffraction [9], Mößbauer effect [10], or high field susceptibility investigations [11], there is evidence that FeNi Invar alloys have an inhomogeneous magnetic ground state and that partially antiferromagnetism (AF) or even a mixed magnetic phase at low temperatures does occur. This is shown in Fig. 4 by the hatched region. We will come back to this point later in the paper (Chapt. 3).

In the approach towards an understanding of the Invar-effect a completely new situation arose, when it was found by Kussmann and v. Rittberg [12] that ordered, purely ferromagnetic (FM) FePt alloys around the concentration of the Fe_3Pt-phase (with Cu_3Au-superstructure) showed Invar properties like FeNi, yet no deviation of the moment from the Slater-Pauling curve. Later work [13] revealed that this holds for disordered FePt alloys as well. In Fig. 5 we present the magnetic phase diagram of this system. One can see that though T_c for the ordered and disordered alloys decreases on approach to the γ-α transition line, the moment does not weaken. There is also no indication for the occurrence of mixed magnetic behavior in FePt at low temperatures, neither in the ordered nor the disordered alloys, although some Invar properties might be influenced by premartensitic transformation for alloys with concentrations close to the α-γ line ($\approx$ 25 at% Pt) [14].

The detection of the Invar-effect on a hard magnetic material like Fe_3Pt, which does not show mixed magnetic behavior or signs of antiferromagnetism, excluded a series of models from the long list of theories trying to describe the Invar-effect.

It is certain to date that the model of latent antiferromagnetism [15], models emphasizing the role of inhomogeneities [16 ... 18], local models with short-range order [19, 20], and theories of weak itinerant ferromagnetism [21, 22] can be discarded.

In Tab. 2 we give a list of the main alloy systems known to show the Invar-effect. Some of these have been intensively studied in the past, some are less well investigated. It is seen from the table that the occurrence of the Invar-effect is not bound to a certain lattice structure, nor is it necessary that the systems are only ferro-

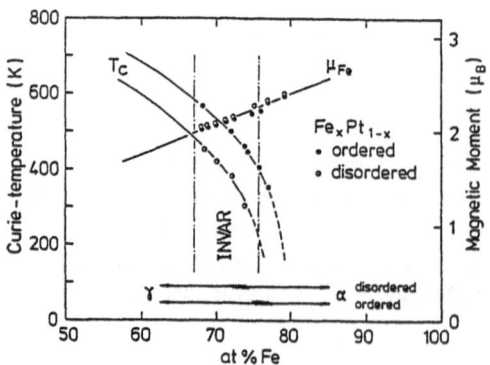

Fig. 5

Magnetic phase diagram of Fe_xPt_{1-x} as determined experimentally [13] for ordered (full dots) and disordered (open dots) alloys in the range close to the γ-α transition. Note that though the Curie temperatures decrease on approach to the γ-α phase boundary, the iron moment μ_{Fe} does not deviate from the Slater-Pauling curve.

Table 2

Invar systems	basic type of magnetic ordering	mixed magnetic or RSG-phase at low T
I) fcc structure		
$Fe_{1-x}Pt_x$ (ord., diso.)	FM	no
$Fe_{1-x}Pd_x$ (ord., diso.)	FM	no
$Fe_{1-x}Ni_x$	FM	yes
$Ni_{1-x}Cr_x$	FM, AF	yes
$Fe_{1-x}Mn_x$	AF	yes
$(Fe_{1-x}Ni_x)_yCr_{1-y}$	FM – AF	yes
$(Fe_{1-x}Ni_x)_yMn_{1-y}$	FM – AF	yes
$(Fe_{1-x}Mn_x)_yCo_{1-y}$	FM – AF	yes
$(Fe_{1-x}Cr_x)_yCo_{1-y}$	FM – AF	yes
II. bcc structure		
$Cr_{1-x}Fe_x$	FM – AF	yes
$Cr_{1-x}Mn_x$	AF	?
III. hex. structure		
$Co_{1-x}Fe_x$	FM	yes
IV. amorphous		
$Fe_{1-x}B_x$	FM	no
$Fe_{1-x}Zr_x$	FM	yes
$(Fe_{1-x}Ni_x)_{77}Si_{10}B_{13}$	FM	yes
$(Fe_{1-x}Co_x)_{77}Si_{10}B_{13}$	FM	yes
V. Laves Phase		
$Zr(Fe_{1-x}Co_x)_2$	FM	yes

magnetic (FM). Invar properties are also found in antiferromagnetic systems (AF) like $Fe_{1-x}Mn_x$ or in systems like stainless steel $(Fe_xNi_{1-x})_{1-y}Cr_y$, where the Invar-effect occurs on the FM as well as on the AF side of the phase diagram (see Chap. 4.1).

Most of the systems in Tab. 2 show mixed magnetic behavior or what is called a "reentrant" spin-glass phase (RSG) at low temperatures. There is no room to discuss the complicated properties of spin-glass systems showing the reentrant transition here in detail, and we have therefore to refer to the existing literature (see e.g. reviews [23, 24]). The microscopic details of this transition as well as the dynamics of it, e.g. the occurrence of "canting" [25] and of time effects in the remanent magnetization are also not completely understood. Yet, it is certain to date that many low temperature anomalies which have been denoted Invar typical in the past, originate from the existence of mixed magnetic behavior or RSG phases in the systems. We will discuss this in more detail in chapter 3 but mention here as an example the high γ-values found from heat capacity measurements in some Invar systems, e.g. $\gamma = 20 \dots 30$ mJ/(molK^2) for $Fe_{65}Ni_{35-x}Mn_x$ ($4 \leqslant x \leqslant 20$) [26]. This high value of γ is not Invar typical, because the heat capacity at low temperatures is enhanced due to a spin-glass contribution, which is also linear in T.

Note that according to Tab. 2 there are only three systems revealing Invar behavior, yet no sign of mixed magnetic phases. These are Fe_xPt_{1-x} (ordered and disordered) as mentioned, $Fe_{1-x}Pd_x$ $(0.3 \leqslant x \leqslant 0.45)$ [27], and amorphous $Fe_{1-x}B_x$ $(0.12 \leqslant x \leqslant 0.23)$ [28]. Comparison of the low and high temperature physical properties of these purely FM systems with those showing mixed magnetic behavior thus allows us to distinguish between those properties which are Invar relevant and those which are influenced by the presence of RSG-phases.

3 Comparison of Invar Properties in Strongly and Weakly Magnetic Systems

For comparison we choose four systems, on the one hand ordered $Fe_{1-x}Pt_x$ and a-$Fe_{1-x}B_x$, on the other hand $Fe_{1-x}Ni_x$ and a-$Fe_{1-x}Zr_x$. The magnetic phase diagrams of the two amorphous systems as determined from data in the literature are presented in Fig. 6a, b. Note that in a-FeB like in the FePt system (Fig. 5) the magnetic moment of iron does not deviate from the Slater-Pauling-curve, and the extrapolation of the Curie temperatures (dashed line in Fig. 6a) would result in $T_c = 350$ K and thus FM pure amorphous Fe [29]. In contrast, a-FeZr shows a weakening of the moment in the Invar range, and an extrapolation of T_c would give $T_c = 0$, i.e. AF or non-magnetic pure amorphous Fe [30]. This polymorphism of amorphous Fe has recently been investigated in detail [31] and finds its origin in the volume dependence of the moment, as we shall see later.

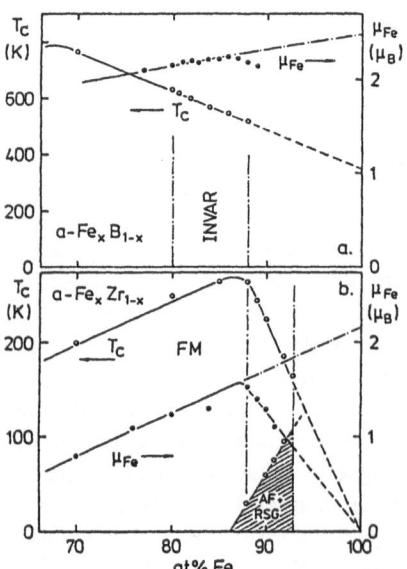

Fig. 6

a) Magnetic phase diagram of amorphous Fe_xB_{1-x} as determined experimentally [29] on melt spun ribbons. Curie temperature T_c and Fe moment μ_{Fe} as a function of concentration are given. Data are extrapolated to amorphous pure Fe, since samples with boron concentrations below ≈ 12 at% cannot be achieved.

b) Magnetic phase diagram of amorphous Fe_xZr_{1-x} from [30]. Note that both the Curie temperature and the iron moment, when extrapolated beyond the stability range of the amorphous ribbons ($x > 93$ at% Fe) result in zero values for amorphous pure Fe (dashed lines). The hatched region at low temperatures indicates the occurrence of a reentrant spin glass phase (RSG) and/or the occurrence of AF components in the system as found in TRM (t) [32], a.c. $\chi(T)$ [33], or spec. heat experiments [34].

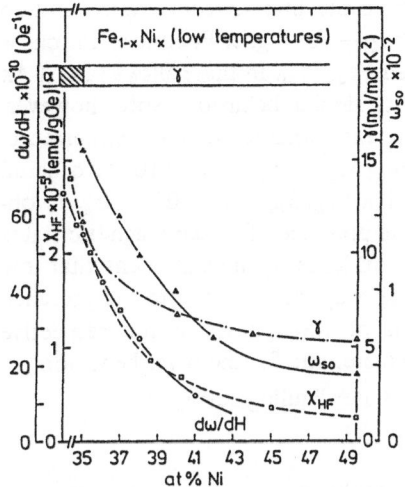

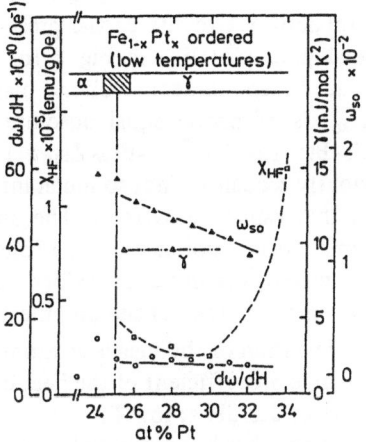

Fig. 7 Spontaneous volume magnetostriction at zero temperature ω_{so} [35], forced volume magnetostriction in fields up to 20 T ($d\omega/dH$) at 4.2 K [36], high field susceptibility χ_{HF} at 4.2 K in fields of 16 T [11], and the low temperature coefficient γ of the electronic specific heat capacity [37] as a function of the Ni-concentration for $Fe_{1-x}Ni_x$-Invar alloys. The α-γ structural phase transition occurs at 34 at% Ni. The hatched region on the top marks the concentration range, where both phases overlap.

Fig. 8 Spontaneous volume magnetostriction at zero temperature ω_{so} [38], forced volume magnetostriction $d\omega/dH$ at 4.2 K in fields up to 2.7 T [38], high field susceptibility χ_{HF} at 4.2 K in fields up to 16 T [11], and the coefficient of the low temperature specific heat capacity [39] as a function of the Pt-concentration in ordered $Fe_{1-x}Pt_x$-Invar alloys. The α-γ transition occurs at around 25 at% Pt. The hatched region marks the concentration range where both phases overlap.

Though there is still some debate about the nature of the mixed magnetic behavior in a-FeZr, we deduce from measurements of the time dependence of the thermoremanent magnetization [32], a.c. susceptibility [33], and heat capacity ($\gamma = (23.3 \pm 0.5)$ mJ/(mol K^2) for $Fe_{90}Zr_{10}$ [34]) that a RSG-phase is present in the system at low temperatures.

Concerning the low temperature behavior in a weakly magnetic system, we present in Fig. 7 the concentration dependence of the spontaneous volume magnetostriction ω_{so} [35], the forced volume magnetostriction $d\omega/dH$ [36], the high field susceptibility χ_{HF} [11], and the coefficient γ of the heat capacity [37] as measured at low temperatures by different authors in the $Fe_{1-x}Ni_x$-system. As mentioned above the rise in absolute value of these quantities on approach to the α-γ transition is obvious. A similar plot can be done for a-FeZr.

In contrast to that one can see from Fig. 8 that in the strongly magnetic system $Fe_{1-x}Pt_x$ the same quantities ω_{so} [38], $d\omega/dH$ [38], χ_{HF} [14], and γ [39] are

93

almost constant in the concentration range where the Invar-effect is found and do not increase near the α-γ line. (The increase in χ_{HF} for higher Pt-concentrations is probably due to the partial presence of ordered $Fe_{50}Pt_{50}$ in these alloys [14].) An analogous plot like in Fig. 8 for a-FeB exhibits similar behavior. Note, however, that the spontaneous volume magnetostriction extrapolated to zero temperature, ω_{so}, is of about equal absolute value for $Fe_{65}Ni_{35}$: $\omega_{so} = 1.9 \cdot 10^{-2}$, ordered Fe_3Pt: $1.4 \cdot 10^{-2}$, a-$Fe_{90}Zr_{10}$: $1.1 \cdot 10^{-2}$, and a-$Fe_{86}B_{14}$: $1.5 \cdot 10^{-2}$. ω_{so} is obviously a quantity not so much influenced by the presence of mixed magnetic states in the respective systems. The above comparison leads to the statement that low temperature anomalies in the following physical properties are due to the presence of mixed magnetic behavior, AF-components of RSG-phases in the respective systems (c.f. Tab. 2) and are, therefore, *not* characteristic for the Invar behavior:

1) deviation of the magnetic moment from the Slater-Pauling curve,
2) large γ coefficient of the heat capacity at low T,
3) large $d\omega/dH$ at low T,
4) anomalous hyperfine-field distribution in the Mößbauer spectrum at low T,
5) large χ_{HF} at low T,
6) occurrence of relaxation effects in $\Delta l/l$ and $d\omega/dH$ at low T.

Statements about the characteristic nature of the magnetic ground state of an Invar system from investigations of the physical properties as listed above have thus often been misleading in the past. We repeat: magnetic inhomogeneity has nothing to do with the Invar-effect, it is just masking it.

Evidence that the ground state of the Fe-atoms in strongly magnetic Fe_xPt_{1-x} and the weakly magnetic Fe_xNi_{1-x} is of the same nature comes from a comparison of the mean values of the hyperfine-field at the Fe-nucleus in both systems. In Fig. 9

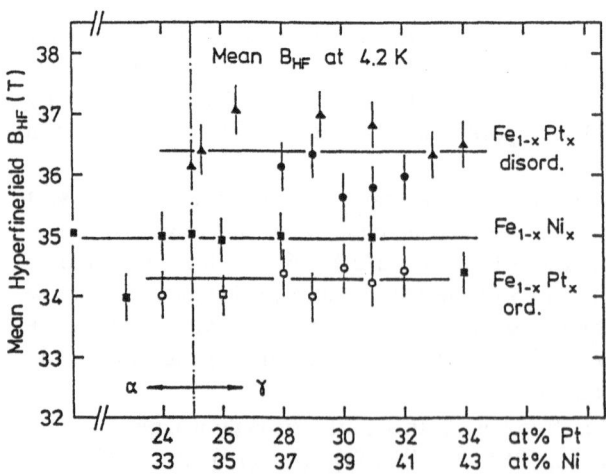

Fig. 9

Mean value of the hyperfine-field at 4.2 K as a function of the concentration for ordered and disordered $Fe_{1-x}Pt_x$-alloys [40] and $Fe_{1-x}Ni_x$ [41] in the Invar range. Dashed-dotted line indicates the α-γ structural transition.

94

we compare the concentration dependence of the main peak in the HF-distribution as determined for Fe_xPt_{1-x} (ordered and disordered) [40] and Fe_xNi_{1-x} [41] at 4.2 K. Though at 4.2 K in $Fe_{65}Ni_{35}$ a few percent of the Fe atoms are possibly flipped and thus are aligned antiferromagnetically (but become aligned ferromagnetically above $\approx$ 30 K) [42], the majority of the spins sit in a FM environment and keep their full moment like in the FM system FePt. This is important for the Invar-effect (as we shall see later, Chap. 4.2) and reflected in the spontaneous volume magnetostriction ω_s at T = 0, ω_{so}, which is – as mentioned – of the same order of magnitude in $Fe_{65}Ni_{35}$ and Fe_3Pt.

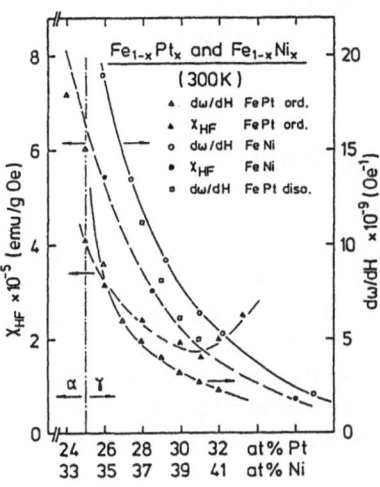

Fig. 10

Concentration dependence of the forced volume magnetostriction $d\omega/dH$ for $Fe_{1-x}Ni_x$ [36] and ordered and disordered $Fe_{1-x}Pt_x$ [43], as well as the high-field susceptibility χ_{HF} for $Fe_{1-x}Ni_x$ [44] and ordered $Fe_{1-x}Pt_x$ [14] as determined at room temperature as a function of Ni- and Pt-concentration, respectively.

Further evidence that the Invar-effect in FeNi and FePt is indeed of the same nature comes from a comparison of some physical properties of these systems at room temperature (R.T.), where the presence of mixed magnetic phases or AF is of minor importance. In Fig. 10 we present the concentration dependence of $d\omega/dH$ for FeNi [36] and ordered and disordered FePt [43], and χ_{HF} for FeNi [44] and FePt [14] as determined at R.T. The two representative quantities, which have been so different in absolute value and concentration dependence at 4.2 K (c.f. Figs. 7 and 8), exhibit now analogous concentration dependence if the α-γ transition lines are overlapping in the way shown. For disordered FePt we obtain for $d\omega/dH$ at R. T. even almost equal absolute values as compared to FeNi. Analogous behavior can be found from a similar plot comparing a-FeB and a-FeZr.

We are thus bound to say that the following physical properties are not influenced by the presence of mixed states, and thus can be considered to be characteristic for Invar alloys in general:

1) large spontaneous volume magnetostriction at low temperatures (ω_{so}),
2) small thermal expansion coefficient around R. T.,
3) large forced volume magnetostriction ($d\omega/dH$) at R. T. up to T_c,
4) large high-field susceptibility at R. T. up to T_c,
5) considerable contributions in $\alpha(T), \omega_s(T), (d\omega/dH)(T)$, and $\chi_{HF}(T)$ for $T \geqslant T_c$,
6) anomaly in the heat capacity in a wide range around T_c,
7) negative pressure dependence of the Curie temperature, dT_c/d_p.

In the following chapter we will present some new experimental data achieved by us on different Invar systems. We will show that when discussed within the frame work of modern band-structure calculations these results can provide a first step towards a general understanding of the Invar-effect.

4 Towards a New Understanding of the Invar-Effect

4.1 Magnetic Anomalies in the Thermal Expansion

As often pointed out in this paper, an intriguing problem in the field of Invar has been the search for appropriate non-magnetic or paramagnetic reference alloys, to determine the magnetic contributions to, e.g., the thermal expansion $\alpha_m(T)$, the spontaneous volume magnetostriction $\omega_s(T)$, or the heat capacity $C_m(T)$. So far, always the Grüneisen relation has been used to calculate the temperature dependence of the physical properties for a respective reference, which we have called "hypothetic", because the data achieved depend on the validity of certain theoretical models.

Quite recently we have detected for the first time that in certain Invar systems almost ideal reference samples can be found from first principles. These reference alloys are paramagnetic over a wide range of temperature, and have almost the same composition as those FM or AF alloys of the same system showing the Invar-effect. To illustrate this we present in Fig. 11 the magnetic phase diagram of $Fe_{80-x}Ni_xCr_{20}$ as determined in [45] and also verified through work of our own group. For the composition range shown ($x \geqslant 14$ at% Ni) all alloys have fcc structure at all temperatures, and martensitic transformations are not observed. For low Ni-concentrations the alloys are AF (with relatively low T_N), on the Ni-rich side the alloys are FM. RSG-phases are observed on both sides, AF as well as FM. Important is the fact that alloys with concentrations around 20 at% Ni are pure spin glasses (SG), with freezing temperatures of about 25 K. This means that for $T > T_f$ these samples are paramagnetic, and do not show the Invar-effect, which is, however, found right and left from the SG-region in the FM- as well as the AF-alloys [3]. A sample of the composition $Fe_{60}Ni_{20}Cr_{20}$ can therefore be regarded as an ideal, paramagnetic

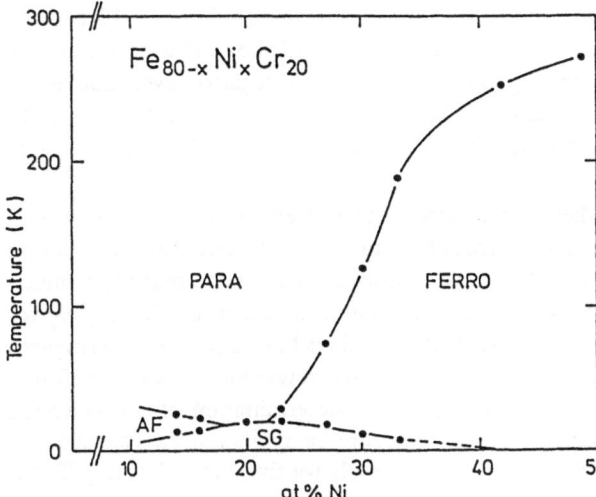

Fig. 11

Magnetic phase-diagram of the system $Fe_{80-x}Ni_xCr_{20}$ as determined in [45] (full lines) and by our group (dots).

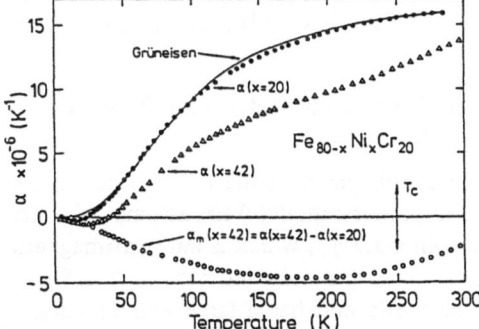

Fig. 12

Thermal expansion coefficient α as a function of temperature for two $Fe_{80-x}Ni_xCr_{20}$ samples. Full points: spin-glass alloy with $x = 20$; open triangles: FM alloy with $x = 42$. The open points give the magnetic contribution α_m (T) for the FM sample as determined from the difference of the two curves above. Full curve: Grüneisen curve as calculated from thermodynamic data using $\theta_D = 375$ K. The arrow marks the Curie temperature of $Fe_{38}Ni_{42}Cr_{20}$.

reference, since it is composed in almost the same concentration ratio of the same atoms as the respective Invar samples of the system. The phonon spectrum, the respective Debye temperatures, and the elastic properties are thus almost equal.

We have measured the thermal expansion $\Delta l/l$ as a function of temperature in the temperature range 4.2 to 300 K for different alloys in the $Fe_{80-x}Ni_xCr_{20}$ system [46]. Fig. 12 shows as an example the thermal expansion coefficient α as a function of temperature as determined experimentally for the SG reference alloy with $x = 20$ at% Ni (full dots) and a typical FM sample with $x = 42$ at% Ni (open triangles). The respective Grüneisen curve as calculated from data of the specific heat [3] for the $x = 20$ sample is also given (full curve). Disregarding the differences in the SG-range (T ⩽ 30 K), one can see from Fig. 12 that the calculated Grüneisen curve follows the α(T)-behavior of the paramagnetic sample relatively well.

For determination of the magnetic contribution α_m (T) of the FM x = 42 at % Ni alloy we use the SG sample (x = 20) as a reference. Fig. 12 shows that based on this analysis α_m (T) for the FM sample (open dots) is always negative, even above T_c, where an appreciable (negative) contribution still exists. Similar results are found for the other FM samples of the $Fe_{80-x}Ni_xCr_{20}$ system, when analyzed in the same way.

We have also carried out studies of magnetovolume effects in the ternary system $Fe_{50}Ni_xMn_{50-x}$ [47], which shows transition from AF to FM behavior on variation of the Ni concentration as well RSG-phases on both sides of the magnetic phase-diagram [48]. An analogous phase diagram has been established by Shiga [49] for $Fe_{65}(Ni_{1-x}Mn_x)_{35}$, for which the magnetovolume effects have also been investigated in detail [50]. In both ternary systems pure SG reference alloys are found ($Fe_{50}Ni_{32}Mn_{18}$; $Fe_{65}Ni_{25}Mn_{10}$) and the analysis for determination of the magnetic contribution to the thermal expansion can therefore be applied in the two systems in the same way as described for FeNiCr. As a result we find again that α_m (T) for FM alloys of the systems is negative from $T > T_f$ to $T > T_c$. In general, the magnetic contribution to the thermal expansion coefficient α_m (T) is always negative in ferromagnetic Invar alloys. This confirms again that Invar alloys are not WIF and cannot be described by the spin fluctuation theory of Moriya and Usami [51], who find a positive contribution in α_m (T) for $T > T_c$ for WIF, as, e.g., experimentally verified on MnSi [52].

4.2 Evidence for High-Spin Low-Spin State Transition in Ordered Fe_3Pt Invar Alloys

A famous Invar model, not yet discussed in this paper, is the two-γ-states model originally proposed by Weiss [53]. In this hypothetical model Weiss assumed that fcc Fe can exist in two distinct states, a low-spin state γ_1, which is antiferromagnetic and has a small volume (a = 3.57 Å) and a small magnetic moment (μ = 0.5 μ_B), and a high-spin state γ_2, which is ferromagnetic and has a large atomic volume (a = 3.64 Å) and a large moment (μ = 2.8 μ_B). For FeNi alloys, Weiss assumed the energy difference ΔE between the two states to be a function of the Ni concentration, so that in the Invar range the FM γ_2 becomes the ground state and the AF γ_1-state the excited state. Transitions from γ_2 to γ_1 with rising temperature then logically explained the Invar-effect, because the expansion of the lattice with rising T is compensated by a growing population of the low volume γ_1-state.

The Weiss model experienced some expansions and variations through the addition of a Zeeman splitting of the levels for $T < T_c$ and an energy distribution of the two levels [54, 55], but a direct proof of its validity was never given. Precise statements about the level schemes for different systems and the nature of the electronic transitions remained speculative, although the broad Schottky-like anomalies, as found in the high temperature specific heat of FeNi (see Fig. 2), FeNiMn [2], and FeNiCr [56], called for the presence of them. The only measurement giving direct evidence for the presence of levels was mostly overlooked in the literature. In an infrared

98

absorption investigation on FeNiCr, Wieting and Schriempf [57] showed that absorption bands above the Drude reference (interpreted as electronic transitions) do exist at energies which corresponded reasonably well with the energies as deduced from the specific heat [56].

The validity of the Weiss-model, however, came into doubt with the discovery of the Invar-effect on ordered Fe_3Pt, a system which – as discussed – neither shows a deviation from the Slater-Pauling-curve nor signs for the existence of antiferromagnetism.

New evidence for the correctness of the assumptions by Weiss came from band-structure calculations for pure Fe by Anderson et al. [58] and Kübler [59], showing that (at T = 0) in the fcc state a rapid increase from an AF ground state with low moment ($\mu = 0.55 \; \mu_B$) to a FM state with high moment ($\mu = 2.32 \; \mu_B$) ($\Delta E = 0.3$ eV) occurs within a narrow range, in which the radius r_{WS} of the Wigner-Seitz (WS) cell changes from $r_{WS} = 2.54$ a.u. to r_{WS} 2.65 a.u. Recent band-structure calculations by different authors [60 ... 63] confirm these results, though the transition from the high spin (HS)-state to the low spin (LS)-state is found at larger radii of the Wigner-Seitz cell. This is shown in Fig. 13a, where the iron atomic moment is plotted as a function of r_{WS} as obtained from [60] (dashed-dotted line),

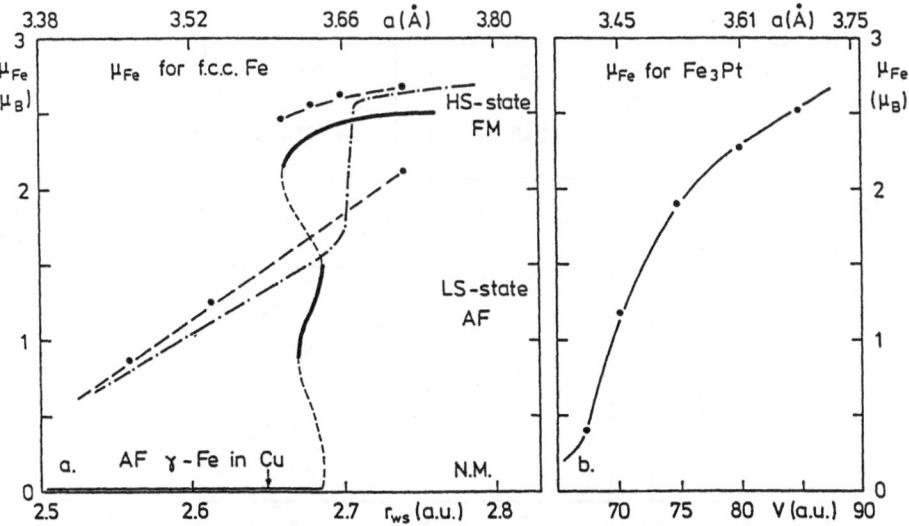

Fig. 13 a) Magnetic moment of fcc Fe as a function of the radius of the Wigner-Seitz cell r_{WS} as calculated by Bagayoko and Callaway [60] (dashed-dotted line), Wang et al. [61] (dashed lines) and Moruzzi et al. [63] (full lines for the stability ranges in zero field). The arrow marks the lattice constant as determined for AF coherent γ-Fe precipitations in Cu [64].

b) Magnetic moment of Fe in Fe_3Pt as calculated as a function of the volume of the Wigner-Seitz cell by Podgorny [65].

[61] (dotted line), and [63] (full lines and weakly dotted lines), respectively. The arrow marks the lattice constant as determined for AF γ-iron precipitations in Cu [64]. One can see that within $2.66 \leqslant r_{WS} \leqslant 2.72$ the transition from the HS-to the LS-state takes place. According to the results by Moruzzi et al. [63] there are indeed instability ranges inbetween the states, as shown by the weakly dashed lines in Fig. 13a. The ground state of fcc Fe is non-magnetic (nm) according to their calculations.

Due to the importance of the $\mu_{Fe}(r_{WS})$-dependence for the Invar-effect, we present in Fig. 13b a recent result by Podgorny [65], showing that the iron moment in ordered Fe_3Pt rises in a similar way as a function of r_{WS} as found for pure fcc Fe.

Considerable progress occurred when the step-like change in the moment of fcc Fe could be related to specific features in the band structure. In Fig. 14a, b we show part of the majority-spin $\uparrow$ band-structure of fcc Fe as calculated by Bagayoko and Callaway [60]. One can see that in particular on the transition from the high-spin state with $r_{WS} = 2.74$ in Fig. 14a (a = 3.70 Å) to the low-spin state in Fig. 14b with $r_{WS} = 2.55$ (a = 3.46 Å) a flat d-band with Z_2-symmetry in the X-W direction shifts from a binding energy of $E_B \approx 0.26$ eV from below E_F to above E_F. The near flatness of the band implies that the transition will be accompanied by a substantial change in the density of states (DOS) near E_F.

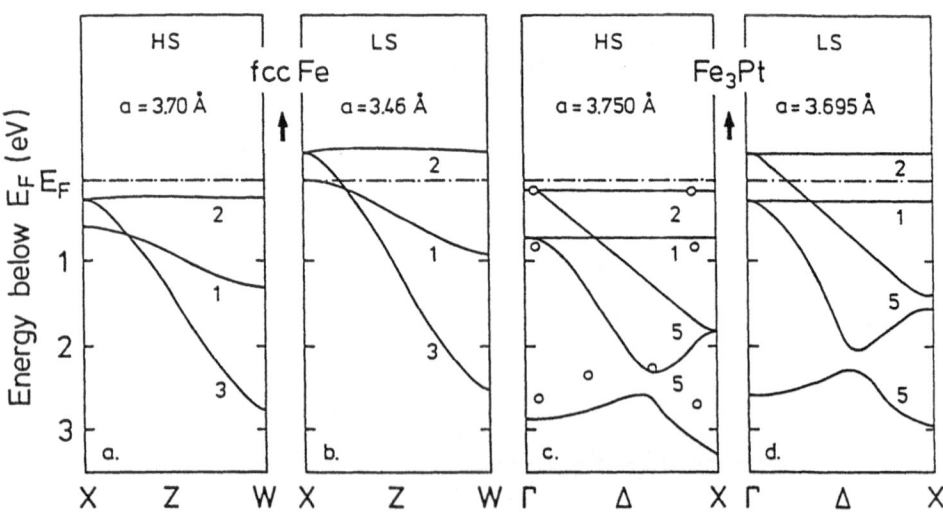

Fig. 14 Majority-spin band structure in the X-W direction for fcc Fe [60] a) in the high-spin state (a = 3.70 A), b) in the low-spin state (a = 3.46 A), and majority-spin band structure in the Γ-X direction for ordered Fe_3Pt, c) in the high-spin state (a = 3.750 A) [66] and d) in the low-spin state (a = 3.695 A) [65]. The open circles mark the energies as determined from spin polarized, angle-resolved photoemission for Fe_3Pt [67].

100

In Fig. 14c, d we show part of the majority-spin band structure in the Γ-X direction of the Brillouin zone for ordered Fe_3Pt as recently determined [65, 66]. The HS-state (a = 3.75 Å; Fig. 14c) and the LS-state (a = 3.695 Å, Fig. 15d) are given, respectively. The band structure of Podgorny [65] is almost identical to the one found by Hasegawa [66]. Some bands have been omitted for clearness in Fig. 14c, d.

In a recent investigation by spin-resolved angular dependent photoemission with synchrotron radiation on ordered Fe_3Pt [67] we have tested for the first time the electronic structure of this alloy experimentally. The agreement between our data and the calculated band structure is good. Some points determined from the photoemission spectra are shown in Fig. 14c (open dots). Moreover, we find an exchange splitting along the Δ-direction of (2.1 ± 0.2) eV, which also agrees with the calculated data (for details see Ref. [67]).

Comparing the band structures of pure fcc Fe and Fe_3Pt in Fig. 14, it is obvious that on transition from the HS- to LS-state in both systems all bands just shift upwards on reduction of the lattice constant. In particular, the flat band with d-character, which is responsible for the accompanying change in moment, mooves from a position closely below E_F to a position above E_F. How far upwards this bands shifts depends just on the lattice constant chosen for the LS-state. Moreover, since the crystal-structures of Fe_3Pt and fcc Fe are similar, the Γ-X distance in the simple cubic reciprocal lattice of Fe_3Pt is one half of that of the corresponding distance in the bcc reciprocal lattice of fcc Fe. Therefore, bands of fcc Fe in the right part of the Γ-X direction are folded back towards Γ in Fe_3Pt, and bands along X-W add to those in Γ-X. The folded band structure of fcc Fe in the high spin state (a = 3.7 Å) and Fe_3Pt (a = 3.75 Å) are then indeed similar [67]. This means that our spin resolved photoemission investigation presents so far the closest experimental proof of the electronic structure of the metastable HS-phase of fcc Fe, which is stabilized by the Pt in Fe_3Pt [67].

Decisive for the basic understanding of the Invar-effect is now that the transition from the HS- to LS-state has been detected by us in Fe_3Pt by studying the temperature dependence of the angle-resolved photoemission spectra [68]. We find that data taken at temperatures below T_c = 450 K (T/T_c = 0.6; 0.7) differ from data taken above T_c (T/T_c = 1.05; 1.22) by a significant *decrease* in intensity in the vicinity of $E_B \approx 0.4$ eV and an *increase* in intensity within an energy range of about 0.5 eV above E_F, while at E_F the intensity does not change. We attribute these features to the shift of the flat band positioned for $T < T_c$ in the HS-state slightly below E_F (for fcc Fe as well as Fe_3Pt, c.f. Fig. 14) to energies slightly above E_F in the LS-state, because the increase in intensity above E_F is an order of magnitude larger than expected from the variation of the Fermi function $F(E, T)$ with temperature alone.

The photoemission intensity above E_F at a given temperature is proportional to $F(E, T) \cdot D(E)$, where $D(E)$ is the symmetry projected DOS. Therefore, we compare in Fig. 15 the difference in the photoemission intensity (full curve and dots) as determined experimentally (normal emission, s-polarized light, 60 eV photon energy)

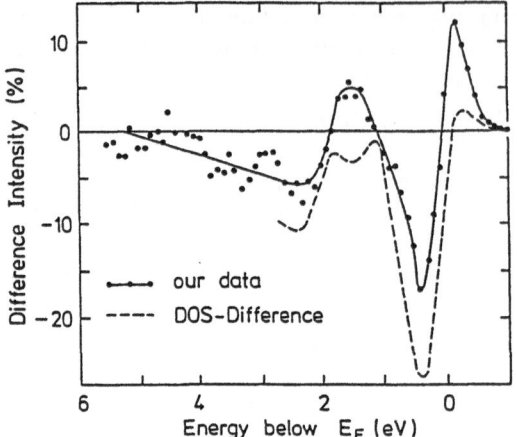

Fig. 15

Full curve and dots: difference in intensity of the angle resolved energy distribution curves of $Fe_3Pt(001)$ for normal emission and s-polarized light (60 eV photon energy) at $T/T_c = 0.6$ and $T/T_c = 1.22$ (intensities have been normalized to the photon flux prior to subtraction). The dashed curve shows the difference between the density of states for fcc Fe in the high-spin state (a = 3.70 A) and low-spin state (a = 3.46 A) [60] each convoluted with the experimental resolution function (0.4 eV FWHM Gaussian) after truncation by the Fermi function.

at $T/T_c = 0.6$ and $T/T_c = 1.22$, respectively, with the difference curve between the DOS of fcc Fe in the HS- and LS-state. We use values for fcc Fe as calculated by Bagayoko [60], since corresponding data for Fe_3Pt are not yet available. Prior to performing the subtraction of the two DOS, the effect of the Fermi function at the respective temperature has been taken into account and the resulting DOS's have been convoluted with the experimental resolution function. As can be seen in Fig. 15, both curves — the difference in the experimentally determined EDC's and the theoretical difference of the DOS's — are similar, suggesting that they are of the same physical origin. We therefore conclude that we have given the first direct experimental evidence for the transition of the high-spin to low-spin state with temperature for an Invar alloy in the sense of the Weiss-model [53]. The aforementioned flat band shifting through E_F with increasing temperature due to thermal excitation of the LS-state is the salient electronic feature, because — as discussed above — this shift is accompanied by the change in magnetic moment and volume (c.f. Fig. 13).

Finally, the amplitude of the intensity above E_F in photoemission curves should be a measure for the weight of the low-spin state if the interpretation given applies. As shown by Holden et al. [69], the fractional change of the radius of the Wigner-Seitz cell with temperature, i. e. the fractional linear expansion $(1(T_1) - 1(T = 0))/1(T = 0)$ is proportional to the change of the square of the local magnetic moment with temperature, $m^2(T_1) - m^2(T = 0)$, where T_1 is an elevated temperature $T_1 > T_c$. Therefore, the square of the intensity above E_F in the EDC's, $I^2(T)$, should be proportional to $m^2(T)$ which is proportional to the volume change $\omega_s = (\Delta V/V)(T)$. In Fig. 16 we have plotted the experimentally determined squared intensity I^2 (dashed line) as a function of temperature together with the qualitative behavior of

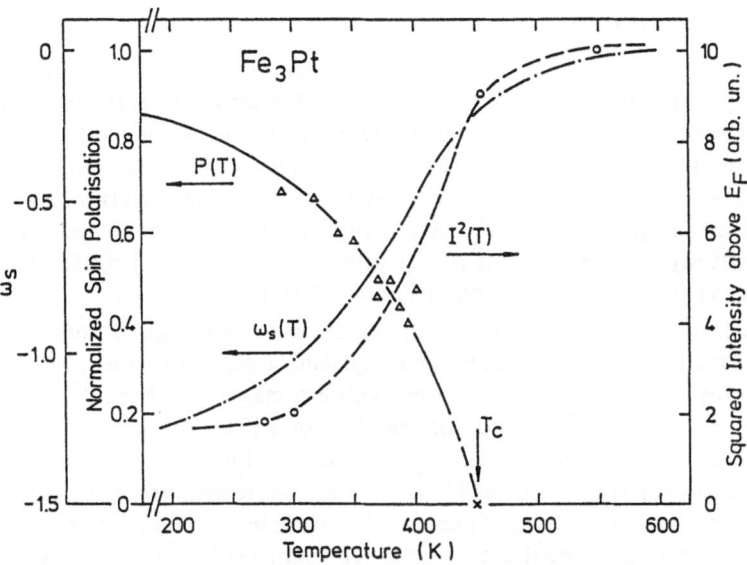

Fig. 16 Normalized spin polarization as a function of temperature as determined experimentally for Fe_3Pt (open triangles, full curve), and squared intensity above E_F of the photoemission curves at different temperatures (open circles, dashed curve). The dashed-dotted curve shows (qualitatively) the negative fractional volume change ω_s (T) for Fe_3Pt as taken from [38].

the negative spontaneous volume magnetostriction $-\omega_s$ (T) as taken from [38] for ordered Fe_3Pt. Also plotted in Fig. 16 is the spin polarization as determined by us experimentally on our Fe_3Pt sample (actual concentration $Fe_{72}Pt_{28}$) as a function of temperature, leading to a T_c value of 450 K, which corresponds to a degree of order of 60 ... 70 %. One can see that in the range where the polarization starts to decrease, both I^2 (T) and $-\omega_s$ (T) increase in the same way. This gives further support for the correctness of our analysis, and the presence of the high-spin low-spin transition in Fe_3Pt on approach of T_c.

Since temperature dependent band-structure calculations are not yet available, it is too early to conclude something about the energy difference ΔE between the two states from our results. Weiss [53] suggested that in FeNi Invar ΔE should be of the order of 0.1 eV, a value which allows access of the γ_1-state with temperature.

Support for our observations on Fe_3Pt also comes from investigations of the angle resolved photoemission of Fe_xNi_{1-x}-alloys at R.T. as a function of the Fe-concentration by Rogge et al. [70]. Their data clearly show that on approach to the Invar concentration, $Fe_{65}Ni_{35}$, the density of states near the Fermi energy increases appreciably, calling for the occurrence of a flat band closely below E_F also in this system. Indeed, density of state calculations by Podgorny [65] for a hypothetical phase Fe_5Ni_3 reveal a sharp peak in the majority DOS at $E_B \approx 0.5$ eV.

5 Conclusions and Outlook

In total, we come to the conclusion that the Invar-effect at least in systems containing iron is caused by a transition from a large volume, high moment γ_2-state to a low volume, small moment γ_1-state with temperature. In the band structure this transition is equivalent to the shift of a flat band with d-character and high density of states from values slightly below E_F in the HS-state to values slightly above E_F in the LS-state. The changes in the DOS's are accompanied by a change in moment. So far, we cannot say if the γ_1-state is AF, as proposed by Weiss [53], since small spin polarization values (c.f. Fig. 16) around T_c are difficult to determine.

Concerning the Invar-effect in systems not containing iron, there is good hope that it might be of similar physical origin as in the Fe-systems. E.g., it was shown recently that the thermal expansion behavior of bcc antiferromagnetic CrMn alloys can be explained by a two δ-state transition of the Mn atoms [71]. For bcc (and fcc) Mn analogous curves to Fig. 13 for the moment μ_{Mn} as a function of the atomic volume have been calculated by Kübler [72]. Certainly, more band-structure calculations are necessary to confirm our statement. As shown by Holden et al. [69] for a quantitative explanation of the size of the magnetovolume effect in Invar systems ($\omega_{so} = 1.2 \ldots 1.9 \cdot 10^{-2}$ in most of the alloys) and the respective loss of the moment, averaged over all atoms only 15 % of the Fe-atoms in a static model flip from $\approx 2.6\ \mu_B$ to $\approx 0.5\ \mu_B$ on approach to T_c; in a dynamical model it would be about 35 %. Though these values might be somewhat incorrect due to the crudeness of the model [69] the figures given lead us to one of the main remaining difficulties concerning the understanding of the Invar-effect: its dynamics. Obviously, one cannot say, as e.g. in the inhomogeneous models tried before, that some certain Fe-atoms in specified positions of a lattice do the transition, some not. The γ_2-γ_1 transition occurs on a certain time scale at any Fe-atom in a crystal, and the effect is "itinerant" in the real sense of the word, yet just 15 to 35 % of the atoms do it simultaneously. This also explains some of the confusion which arose from the mixed magnetic behavior presented in most of the Invar systems, as discussed extensively in this paper. We are bound to assume that those Fe (or other Invar causing) atoms sitting in the "wrong" environment, i.e. AF, RSG, or mixed magnetic regions, do not (or in a different way) participate in the γ_2-γ_1 transition, because the band structure is locally disturbed and the flat, Invar relevant band below E_F probably does not exist in these lattice regions. Concerning the transition or fluctuation time we can only guess that it should be faster than the observation time of the photoemission, since both states are seen in the spectra.

In this context we mention the long debated "hidden excitations" in Invar alloys, which have been introduced to explain the observed differences in the spin-wave stiffness constants, since D_{ns} as determined from inelastic neutron scattering is much larger than D_{mag} as determined from magnetization measurements for *all* Invar systems, magnetically weak or strong [73]. In spite of many efforts, however, these hidden excitations have so far not been detected experimentally [74, 75].

The analysis by Ishikawa et al. [73] suggested that the hidden excitations should, if they really exist, be spin-wave like excitations with a quadratic dispersion relation starting from zero energy, and should be neither Stoner-like nor optical modes with a finite energy gap at $q = 0$. Can the γ_2-γ_1-state transitions be the hidden excitations? The question mark shows that much more work on the dynamics of the Invar transition are necessary in the future.

Acknowledgements

Its my pleasure to announce that I benefited appreciably from intensive discussions with Y. Nakamura, W. Pepperhoff, W. Podgorny, P. Wohlfahrt, J. Kübler, and J. Hesse. My thanks also go to E. Kisker and his group, especially C. Carbone, for performance and interpretation of the photoemission experiments, which led to the break through in the field. Helpful discussions with people of my own group in Duisburg, M. Acet, W. Stamm, and H. Zähres, who did the thermal expansion measurements, are also gratefully acknowledged. Work was supported by Deutsche Forschungsgemeinschaft within Sonderforschungsbereich 166 Duisburg/Bochum.

References

[1]   *Ch. E. Guillaume*, Comp. Rend. Acad. Sci. **125**, 235 (1897)

[2]   *W. Bendick, H. H. Ettwig*, and *W. Pepperhoff*, J. Phys. F.: Metal Phys. **8**, 2525 (1978)

[3]   *W. Bendick* and *W. Pepperhoff*, J. Phys. F: Metal Phys. **11**, 57 (1981)

[4]   Honda Memorial Series on Material Science, No. 3: The Physics and Applications of Invar Alloys, ed. by *H. Saito* (Marunzen Company Ltd, Tokyo, Japan 1978)

[5]   The Invar Problem, Proc. Int. Symp. on the Invar Problem, Nagoya, Japan, 1978, ed. by *A. J. Freemann* and *M. Shimizu*, (North Holland Publ. Comp., Amsterdam 1979), (J. M. M. M. **10** (1979))

[6]   Proc. Int. Sympos. on Magnetoelasticity, in: Trans. Met. Alloys, Nagoya, Japan, 1982, ed. by *M. Shimizu, T. Nakamura* and *J. J. M. Franse* (North Holland Publ. Comp., Amsterdam 1983) (Reprinted from Physica 119 B + C (1983))

[7]   *J. Crangle* and *G. C. Hallam*, Proc. Roy. Soc. (London) **A272**, 119 (1963)

[8]   *T. Miyazaki, Y. Ando*, and *M. Takahashi*, J. Appl. Phys. **57**, 3456 (1985)

[9]   *S. F. Dubinin, S. G. Teplouchov, S. K. Sidorov, Y. U. Izyumov*, and *V. N. Syromyatnikov*, phys. stat. sol. (a) **61**, 159 (1980)

[10] *J. B. Müller* and *J. Hesse*, Z. Physik **B54**, 35 and 43 (1983)

[11] *H. Maruyama, R. Pauthenet, J. C. Picoche*, and *O. Yamada*, J. Phys. Soc. Japan **55**, 3218 (1986)

[12] *A. Kussmann* and *G. v. Rittberg*, Z. Metallkunde **41**, 470 (1950)

[13] *K. Sumiyama, M. Shiga, Y. Kobayashi, K. Nishi*, and *Y. Nakamura*, J. Phys. F.: Metal Phys. **8**, 1281 (1978)

[14] *H. Maruyama*, J. Phys. Soc. Japan **55**, 2834 (1986)

[15] *E. I. Kondorski* and *V. L. Sedov*, J. Appl. Phys. **31**, 331 S (1960)

[16] *S. Kachi* and *H. Asano*, J. Phys. Soc. Japan **27**, 536 (1969)

[17] *W. F. Schlosser*, J. Phys. Chem. Solids **32**, 939 (1971)

[18] *M. Shimizu*, J. Phys. Soc. Japan **45**, 1520 (1978)

[19] *S. K. Sidorov* and *A. V. Doroshenko*, Phys. Met. Metallogr. **18**, 12 (1984)

[20] *S. F. Dubinin, S. K. Sidorov*, and *E. Z. Valiev*, phys. stat. sol. *(b)* **46**, 337 (1971)

[21]   *E. P. Wohlfahrth*, J. Appl. Physics **39**, 1069 (1968)

[22]   *E. P. Wohlfahrt*, Phys. Lett. **28A**, 569 (1969)

[23]   *C. Y. Huang*, J. Mag. Magn. Mat. **51**, 1 (1985)

[24]   *H. Maletta* and *W. Zinn*, in: Handbook of the Physics and Chemistry of Rare Earth, ed. by *K. A. Gschneider* and *L. Eyring* (North Holland Publ. Comp., 1968) Vol. 12

[25]   *J. Lauer* and *W. Keune*, Phys. Rev. Lett. **48**, 1850 (1982)

[26]   *A. V. Deryabin, V. I. Rimlyand*, and *A. P. Larionov*, Sov. Phys. Solid State **25**, 1109 (1983)

[27]   *M. Matsui, T. Shimizu* and *K. Adachi*, Physica **119B**, 84 (1983)

[28]   *S. Ishio, M. Takahashi, Z. Xianyu*, and *Y. Ishikawa*, J. Mag. Magn. Mat. **31–34**, 1491 (1983)

[29]   *K. Fukamichi, M. Kikuchi, S. Arakawa*, and *T. Masumoto*, Solid State Commun. **23**, 955 (1977)

[30]   *M. Nose* and *T. Masumoto*, Suppl. Sci. Rep. RITU **A-28**, 222 (1980)

[31]   *G. Xiao* and *C. L. Chien*, preprint

[32]   *J. A. Heller, E. F. Wassermann, M. F. Braun*, and *R. A. Brand*, J. Mag. Magn. Mat. **54–57**, 307 (1986)

[33]   *N. Saito, H. Hiroyoshi, K. Fukamichi*, and *Y. Nakagawa*, J. Phys. F: Met. Phys. **16**, 911 (1986)

[34]   *Y. Obi, L. C. Wang, R. Motsay*, and *D. G. Onn*, J. Appl. Phys. **53**, 2304 (1982)

[35]   *M. Hayase, M. Shiga*, and *Y. Nakamura*, J. Phys. Soc. Japan **34**, 925 (1973)

[36]   *S. Ishio* and *M. Takahashi*, J. Mag. Magn. Mat. **50**, 271 (1985)

[37]   *R. Caudron, J. J. Meunier*, and *P. Costa*, Solid State Commun. **14**, 975 (1974)

[38]   *K. Sumiyama, M. Shiga, M. Morioka*, and *Y. Nakamura*, J. Phys. F: Metal Phys. **9**, 1665 (1979)

[39]   *K. Sumiyama, M. Shiga*, and *Y. Nakamura*, J. Phys. Soc. Japan **40**, 996 (1976)

[40]   *K. Sumiyama, M. Shiga, K. Tachi*, and *Y. Nakamura*, J. Phys. Soc. Japan **40**, 1002 (1976)

[41]   *A. Z. Menshikov* and *E. E. Yurchikov*, Sov. Phys. JETP **36**, 100 (1973)

[42]   *M. M. Abd-Elmeguid, U. Hobuß, H. Micklitz, B. Huck*, and *J. Hesse*, Phys. Rev. B, to be published

[43]   *K. Sumiyama, Y. Emoto, M. Shiga*, and *Y. Nakamura*, J. Phys. Soc. Japan **50**, 3296 (1981)

[44]   *F. Ono* and *S. Shikazumi*, J. Phys. Soc. Japan **37**, 631 (1974)

[45]   *A. K. Majumdar* and *P. v. Blanckenhagen*, Phys. Rev. **B29**, 4079 (1984)

[46]   *M. Acet, W. Stamm, E. F. Wassermann*, and *H. Zähres*, to be published

[47]   *M. Acet, W. Stamm, E. F. Wassermann*, and *H. Zähres*, in preparation

[48]   *F. Richter* and *W. Pepperhoff*, Arch. Eisenhüttenwesen **47**, 45 (1976)

[49]   *M. Shiga*, Phys. Soc. Japan **22**, 539 (1967)

[50]   *M. Shiga* and *Y. Nakamura*, J. Phys. Soc. Japan **26**, 24 (1969)

[51]   *T. Moriya* and *K. Usami*, Solid State Commun. **34**, 95 (1980)

[52]   *M. Matsunaga, Y. Ishikawa*, and *T. Nakajima*, J. Phys. Soc. Japan **51**, 1153 (1982)

[53]   *R. J. Weiss*, Proc. Roy. Phys. Soc. (London) **82**, 281 (1963)

[54]   *W. Bendick, H. H. Ettwig*, and *W. Pepperhoff*, J. Mag. Magn. Mat. **10**, 214 (1979)

[55]   *S. Chikazumi*, J. Mag. Magn. Mat. **15–18**, 1130 (1980)

[56] *W. Bendick* and *W. Pepperhoff*, J. Phys. F.: Metal Phys. **11**, 57 (1981)

[57] *T. J. Wieting* and *J. T. Schriempf*, J. Appl. Phys. **47**, 4009 (1976)

[58] *O. K. Anderson, J. Madsen, K. K. Paulsen, O. Jepsen* and *J. Kollar*, Physica **86–88 B+C**, 249 (1977)

[59] *J. Kübler*, Phys. Lett. **81A**, 81 (1981)

[60] a) *D. Bagayoko* and *J. Callaway*, Phys. Rev. **B28**, 5419 (1983)
      b) *D. Bagayoko*, Thesis (Lousiana State University, 1983)

[61] *C. S. Wang, B. M. Klein*, and *H. Krakauer*, Phys. Rev. Lett. **54**, 1852 (1985)

[62] *F. J. Pinski, J. Staunton, B. L. Gyoffry, D. D. Johnson*, and *G. M. Stocks*, Phys. Rev. Lett. **56**, 2096 (1986)

[63] *V. L. Moruzzi, P. M. Marcus, K. Schwarz*, and *P. Mohn*, Phys. Rev. **B34**, 1784 (1986)

[64] *W. Bendick, H. E. Ettwig, F. Richter*, and *W. Pepperhoff*, Z. Metallkunde **68**, 103 (1977)

[65] *W. Podgorny*, unpublished

[66] *A. Hasegawa*, J. Phys. Soc. Japan **54**, 1477 (1985)

[67] *C. Carbone, E. Kisker, K. H. Walker*, and *E. F. Wassermann*, Phys. Rev. B, to be published

[68] *E. Kisker, E. F. Wassermann*, and *C. Carbone*, Phys. Rev. Lett. **58**, 1784 (1987)

[69] *A. J. Holden, V. Heine*, and *J. H. Samson*, J. Phys. F: Metal Phys. **14**, 1005 (1984)

[70] *V. Rogge, H. Neddermeyer*, and *Th. Paul*, J. Phys. F: Metal. Phys., to be published

[71] *M. Shiga, M. Miyako*, and *Y. Nakamura*, J. Phys. Soc. Japan **55**, 2290 (1986)

[72] *J. Kübler*, Proc. of the Workshop on 3d Metallic Magnetism, Institute Laue Langevin (1983)

[73] *Y. Ishikawa, S. Onodera*, and *K. Tajima*, J. Mag. Magn. Mat. **10**, 183 (1979)

[74] *Y. Ishikawa, M. Kohgi, S. Onodera, B. H. Grier*, and *G. Shirane*, Sol. State Commun. **57**, 535 (1986)

[75] *P. Böni, G. Shirane, B. H. Grier*, and *Y. Ishikawa*, J. Phys. Soc. Japan **55**, 3596 (1986)

Festkörperprobleme 27 (1987)

Spectroscopy of Inversion Electrons on III-V Semiconductors

Ulrich Merkt

Institut für Angewandte Physik, Universität Hamburg, D-2000 Hamburg 36, Federal Republic of Germany

Summary: The quasi two-dimensional electron gases realized in inversion layers of selectively doped GaAs/Ga$_{1-x}$Al$_x$As heterojunctions and of metal-oxide-semiconductor structures on InSb are studied by far-infrared spectroscopy. Specifically, effects arising from the nonparabolic band structure of these semiconductors in the presence of the surface electric field and external magnetic fields are discussed. Taking advantage of the distinct band parameters of the two compounds GaAs and InSb, a fairly comprehensive picture of the motion of electrons in quasi two-dimensional electron systems on III-V semiconductors is obtained.

1 Introduction

Semiconductors composed of group III and V elements possess physical properties that make them perfectly suited for applications in optoelectronic and transistor devices [1]. Whereas for optical applications the direct band structure and the possibility to continuously vary band gaps and refractive indices by alloy composition is decisive, transistor devices take advantage of mobilities that are much higher than those in silicon. These devices are grown by sequential deposition of layers with sharp interfaces by advanced techniques like molecular beam epitaxy (MBE) and are commonly addressed as layered semiconductor structures [2, 3].

The technological interest has brought about considerable effort to study the fundamental physical properties of the underlying electron systems which often behave dynamically two-dimensional. This means that the electron motion is restricted in one spatial direction but is free in the two others. Such electron gases are realized in quantum wells and heterojunctions as well as in metal-oxide-semiconductor (MOS) structures [4]. In this contribution, the subband structure of inversion layers in GaAs/Ga$_{1-x}$Al$_x$As heterojunctions (see Fig. 1a, b) and in MOS-structures on InSb (see Fig. 1c, d) is studied by far-infrared spectroscopy at liquid helium temperatures. Both semiconductors, GaAs and InSb, have the same type of band structure with a nonparabolic conduction band but their band parameters are largely different (see Tab. 1) yielding different but supplementary subband characteristics in a variety of aspects.

In inversion layers on GaAs only the ground electric subband is commonly occupied, whereas we have four or more subbands occupied on InSb [5, 6]. The width of the inversion channel on InSb is much wider than on GaAs and makes possible to probe

109

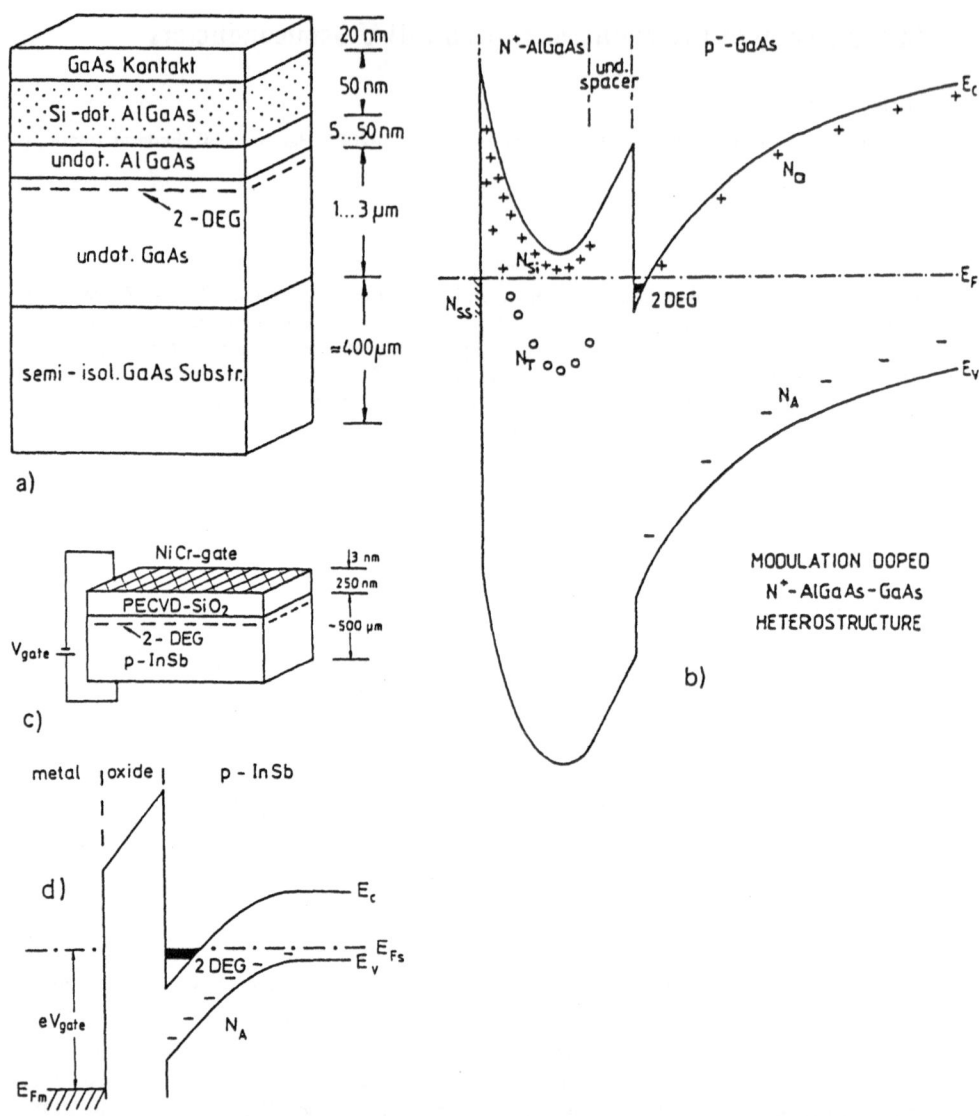

Fig. 1 a) Selectively doped GaAs/Ga$_{1-x}$Al$_x$As heterojunction grown by molecular beam epitaxy (MBE) [3].
b) Schematic band diagram with two-dimensional electron gas (2 DEG), positively and negatively charged depletion layers in the AlGaAs and GaAs layers, respectively, and undoped spacer. Deep levels N$_T$ and surface states N$_{ss}$ are indicated [3].
c) Metal-oxide-semiconductor (MOS) structure on p-type InSb. The gate contact is a thin semitransparent NiCr film, the insulating oxide is grown by plasma-enhanced-chemical-vapor deposition (PECVD).
d) Schematic band diagram of a MOS structure with 2 DEG and negatively charged depletion layer. The density of the 2 DEG is controlled by the gate voltage V$_{gate}$.

Table 1 Band parameters of GaAs and InSb (T = 0).

	E_g (eV)	Δ (eV)	m_0^*/m_e	g_0^*	α
GaAs	1.52	0.34	0.067	$-$ 0.44	0.058
InSb	0.24	0.81	0.014	-51	0.020

the transition from two-dimensional to three-dimensional behaviour of the electron motion in MOS structures on InSb [7]. The particular advantage of GaAs/Ga$_{1-x}$Al$_x$As heterojunctions is their immense electron mobility [2, 3]. In MOS structures on InSb [6] the density of electrons can be controlled readily by a gate voltage (see Fig. 1c). This is much more involved in case of GaAs/Ga$_{1-x}$Al$_x$As heterojunctions, where the density is often increased by illumination with above-band-gap radiation utilizing the persistent photoeffect [8].

We have previously studied the interaction of quasi two-dimensional electrons and optical phonos, i.e., two-dimensional polarons in both systems [9 ... 11]. This interaction is important for polar semiconductor structures and limits the electron mobility at room temperature [3]. Despite of the higher Fröhlich coupling constant α of GaAs compared to InSb (see Tab. 1), which leads to stronger polaron effects in bulk GaAs, the situation is reversed in the corresponding two-dimensional systems [12]. Since this subject has been reviewed recently [13], emphasis in this contribution is laid on the influence of band nonparabolicity on the subband structure and on the motion of electrons in external magnetic fields. In particular, on InSb we can examine the motion of semiconductor electrons in crossed electric and magnetic fields if we consider the surface electric field and apply a magnetic field parallel to the inversion layers [14]. In this configuration of external fields, the nonparabolic nature of the conduction band affects the electron motion in a rather spectacular way.

2 Theory: Electric and Magnetic Quantization in Space-Charge Layers

2.1 Perpendicular Magnetic Fields

Inversion electrons in two-dimensional systems are free to move parallel to an interface but are bound to it by the action of the surface electric field. The strength F of this field is proportional to the derivative of the conduction band energy with respect to the space variable z (see Fig. 1b, d). In the vicinity of the interface, the regime of the inversion electrons, the surface electric field may be approximated by a constant value. The resulting simplified surface potential

$$V(z) = \begin{cases} \infty & z \leqslant 0 \\ eFz & z > 0 \end{cases} \tag{1}$$

is called the triangular-well potential [4]. This model assumes an infinite potential barrier at the interface which is in fact a very good approximation for MOS structures. For GaAs/Ga$_{1-x}$Al$_x$As heterojunctions, the barrier height is called the band offset [15] and our approximation is still reasonable since the band offset is much larger than the subband energies. Although the assumptions of the triangular-well potential are rough, this model leads to intuitive pictures and simple analytical expressions that allow us to discuss the experimental results in a qualitative but instructive way.

A proper value of the electric field strength F has been derived in Ref. [16]:

$$F = \frac{\pi^2}{12} \left(n_{depl} + \frac{11}{32} n_s \right) \frac{e}{\epsilon_0 \epsilon} \tag{2}$$

with dielectric constant ϵ of semiconductor, depletion charge density n_{depl} and inversion electron density n_s. In the one-band effective-mass approximation (EMA), the eigenenergies E of inversion electrons in the triangular-well potential are given by the expression [4]:

$$E = \frac{\hbar^2 k_\parallel^2}{2m^*} + \left(\frac{9\pi^2}{8m^*} \right)^{1/3} (e\hbar F)^{2/3} \left(i + \frac{3}{4} \right)^{2/3} \tag{3}$$

for electron momentum $\hbar k_\parallel$ parallel to the interface and a semiconductor with an isotropic effective mass m^*. Thus we have quantization into discrete electric subbands i = 0, 1, ... with subband edge energies E_i (EMA) obtained from Eq. (3) for momentum $\hbar k_\parallel = 0$. If a magnetic field B is applied in the direction perpendicular to an inversion layer, the energy spectrum becomes totally discrete with a Landau ladder on each particular electric subband i:

$$E = \hbar \omega_c \left(n + \frac{1}{2} \right) \pm \frac{1}{2} g^* \mu_B B + E_i \text{ (EMA)}. \tag{4}$$

Comparing Eqs. (3) and (4), in perpendicular magnetic fields the kinetic energy becomes substituted by a spin-split Landau energy which depends on cyclotron frequency $\omega_c = eB/m^*$, Landau index n, and effective Lande' factor g^*. In cyclotron resonance experiments, the spacing between two Landau levels $E_{i,n+1}$ and $E_{i,n}$ is measured. This spacing is often expressed by the cyclotron mass which is defined by the relation [17]

$$m_c = \frac{e\hbar B}{E_{n+1,i} - E_{n,i}.} \tag{5}$$

In the parabolic EMA, the cyclotron mass is the same for all subbands i and Landau transitions $n \to n + 1$ and is identical with the effective mass m^*. Due to non-parabolic effects, this is not true for subbands on GaAs/Ga$_{1-x}$Al$_x$As heterojunctions or MOS structures on InSb. However, the definition of a cyclotron mass remains useful since it represents a very sensitive measure for deviations from the parabolic case.

The triangular-well approximation was also employed to describe nonparabolic subbands [16, 18] on semiconductors of direct gap band structure depicted in Fig.

112

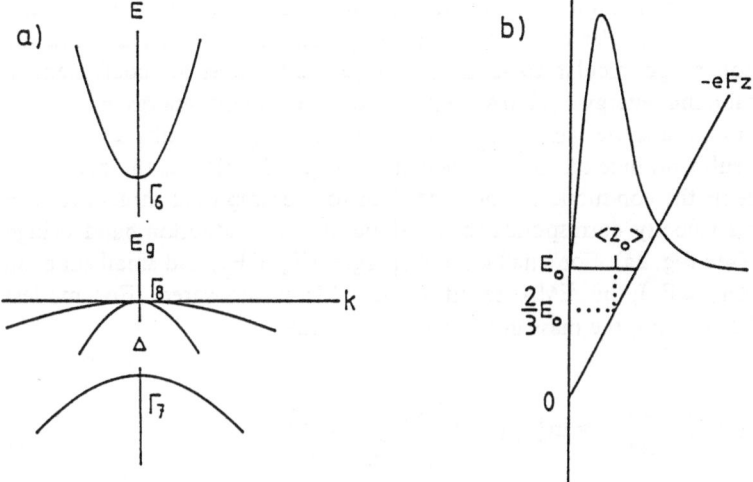

Fig. 2 a) Band structure underlying the three-level k · p model of InSb-type semiconductors. b) Triangular-well potential and wave function for the ground electric subband in the effective mass approximation [16]. The average distance of electrons away from the interface in any subband i is given by $\langle z_i \rangle = 2E_i/3eF$.

2a. Three levels at the Γ point have been taken into account in a k · p model: a Γ_6 conduction level separated by the gap energy E_g from a Γ_8 valence level, this in turn separated from a Γ_7 valence level by the spin-orbit interaction energy Δ. For subband energies smaller than the gap energy ($E \ll E_g$), a power series for the eigenenergies has been derived [16]:

$$E = \left[\left(\frac{E_g}{2}\right)^2 + E_g E_\parallel\right]^{1/2} \cdot \left[1 + 2\left(\frac{E_i^{EMA}}{E_g}\right)\left(1 + \frac{4E_\parallel}{E_g}\right)^{-2/3} - \frac{2}{5}\left(\frac{E_i^{EMA}}{E_g}\right)^2 \cdot\right.$$

$$\left.\cdot \left(1 + \frac{4E_\parallel}{E_g}\right)^{-4/3} \pm \ldots\right] - \frac{E_g}{2}. \tag{6}$$

The term $-E_g/2$ appears since the zero of the energy scale is chosen at the conduction band edge. The important parameter is the ratio of the subband energy E_i (EMA) and the gap energy E_g. The energy E_i (EMA) defined in Eq. (3) and the kinetic energy $E_\parallel = \hbar^2 k_\parallel^2 / 2m_0^*$ are calculated with the bulk effective mass m_0^* at the conduction band edge (see Tab. 1).

Equation (6) has some very attractive features: In the three-dimensional limit (F = 0) it reduces immediately to the well-known Kane formula [17]

$$E = \left[\left(\frac{E_g}{2}\right)^2 + E_g \frac{\hbar^2 k_\parallel^2}{2m_0^*}\right]^{1/2} - \frac{E_g}{2}, \tag{7}$$

113

and it describes the most prominent feature of nonparabolic subbands in a rather transparent way: The energies of the free motion parallel to the interface and the quantized motion perpendicular to it are no longer additive as the coefficient in front of the subband energy E_i(EMA) depends on the kinetic energy $E_{\parallel}$. In the limit of high kinetic energies ($E_{\parallel} \gg E_g$), the subband energy $E = \hbar h_{\parallel} u$ no longer depends on the subband index i. Here the velocity $u = (E_g/2m_0^*)^{1/2}$ is the maximum velocity possible in the conduction band according to the simplified three-level $k \cdot p$ model [18]. This velocity corresponds to the slope of the conduction band at high wave vectors k (see Fig. 2a). For small kinetic energies ($E_{\parallel} \ll E_g$) and small subband energies (E_i(EMA) $\ll E_g$), the EMA result of Eq. (3) is recovered. For modest energies $E_{\parallel}$ and E_i(EMA), the mass in the subband i reads

$$m_i^*(E) = \hbar^2 k_{\parallel} \left(\frac{\partial E}{\partial k_{\parallel}} \right)^{-1} \approx m_0^* \left(1 + 2 \frac{\frac{1}{3} E_i^{EMA} + E_{\parallel}}{E_g} \right) \qquad (8)$$

and now depends on subband edge energy E_i (EMA) and kinetic energy $E_{\parallel}$ of subband i.

Landau levels $E_{i,n}$ in a magnetic field perpendicular to the interface are also obtained from Eq. (6) when the substitution

$$E_{\parallel} = \frac{\hbar^2 k_{\parallel}^2}{2m_0^*} \longrightarrow \hbar\omega_c \left(n + \frac{1}{2} \right) \pm \frac{1}{2} g_0^* \mu_B B. \qquad (9)$$

is made. Both, the bulk Lande' factor g_0^* and the cyclotron frequency $\omega_c = eB/m_0^*$ are taken at the conduction band edge. This simple substitution is justified, provided, the spin-orbit interaction energy Δ is much larger than the gap energy ($\Delta \gg E_g$) which is approximately the case for InSb [18]. Using the definition of Eq. (5), the cyclotron mass

$$m_c \approx m_0^* \left[1 + \frac{2}{E_g} \left(\hbar\omega_c (n + 1) \pm \frac{1}{2} g_0^* \mu_B B + \frac{1}{3} E_i^{EMA} \right) \right] \qquad (10)$$

is obtained for energies $\hbar\omega_c \ll E_g$ and E_i(EMA) $\ll E_g$. Unlike in the EMA, the cyclotron mass m_c does depend on subband index i, Landau index n, and spin orientation $\pm$. Despite of its different definition, it becomes identical to the purely electric subband mass of Eq. (8) in the limit of vanishing magnetic fields. To see this, we must keep constant the Fermi energy E_F as the magnetic field goes to zero. Then, the Landau index diverges ($n \to \infty$) and the spin-split Landau energies become kinetic energies of subbands i. It is well-known from three-dimensional electron systems on narrow-gap semiconductors [17] that coupling of conduction and valence bands increases when the electron energies are further away from the conduction band edge at the Γ point. Then, the dispersion strongly deviates from a parabolic shape, and so-called nonparabolic effects come into play. In particular, the effective mass becomes energy dependent and increases with energy. If the energy is counted from the conduction band edge, the resulting mass increase is determined by this energy divided by half of the gap energy: $m^*(E) = m_0^*(1 + 2E/E_g)$. In the absence of a magnetic field, the mass increase in three-dimensional systems is determined solely by kinetic energy E (see Fig. 2a), whereas in two-dimensional systems we have kinetic energies $E_\parallel$ for the motion parallel to the interface and subband energies E_i (EMA) for the quantized motion perpendicular to it. According to Eq. (8), the nonparabolic mass increase due to subband energy is only a factor 1/3 as effective as the increase due to kinetic energy. Intuitively, this can be explained by the fact that in any electric subband the electron on an average is the energy E_i (EMA)/3 away from the conduction band edge (see Fig. 2b). This explains the factor 1/3 in Eqs. (8) and (10). In the presence of quantizing magnetic fields, we measure the mass that corresponds to a Landau transition $n^\pm \to n^\pm + 1$. In fact, the average energy of initial and final Landau states appears in Eq. (10). It should be emphasized once more that the present simplified picture only holds as long as the conduction band energies are low and the assumptions of the triangular-well potential are justified. More quantitative theories not only require to consider the interaction between more than two bands, but also to treat the subband quantization in a self-consistent way [19, 20].

2.2 Crossed Electric and Magnetic Fields

A crossed electric and magnetic field configuration for inversion electrons is created by the surface electric field directed perpendicular to the inversion layer and a magnetic field that is applied parallel to it. Again, we start with the EMA description and then treat the k · p model to account for the band structure of III-V semiconductors [14]. In the description, it is convenient to use characteristic lengths L and l that correspond to electric subband quantization ($E_0 \approx \hbar^2/m^*L^2$) and magnetic Landau quantization ($\hbar\omega_c = \hbar^2/m^*l^2$), respectively:

$$L = \left(\frac{\hbar^2}{2m^*eF}\right)^{1/3}, \quad l = \left(\frac{\hbar}{eB}\right)^{1/2}. \tag{11}$$

The electric length L describes the width of the inversion layer and is derived from the uncertainty principle ($p \cdot L \sim \hbar$) and energy conservation ($p^2/2m^* = eFL$). The magnetic length l is the well-known Landau radius. In a parallel magnetic field, electric subband energies and Landau energies are not additive as in perpendicular magnetic fields. Instead, there is complete coupling of electric and magnetic quantization and so-called hybrid electric-magnetic subbands form [14, 21]. This is a consequence of directions of Coulomb and Lorentz forces which act in the same direction in crossed fields.

Classically, the trajectories of electrons in crossed electric and magnetic fields (F ∥z, B ∥x) are cycloides along the interface in y-direction which are characterized by a center coordinate z_0 and a drift velocity $v_D = F/B$ as is discussed in [14].

Quantum mechanically, we have three quantum numbers: the subband index i, the wave vector k_x parallel to the magnetic field, and the quasi continuous center coordinate $z_0 = l^2 (k_y - k_D)$ which is related to the wave vector k_y via the constant wave vector $k_D = m^* v_D/\hbar$. The eigenenergies

$$E = \frac{\hbar^2 (k_x^2 + k_D^2)}{2m^*} + (\nu_i + \frac{1}{2}) \hbar\omega_c + eFz_0 \tag{12}$$

can be evaluated with the indices ν_i of the parabolic cylinder functions. These indices depend on the normalized center coordinate z_0/l, and approximate analytical expression $\nu_i (z_0/l)$ are given in [14, 22].

A dimensionless measure of the relative strength of electric and magnetic quantization was introduced [14] by the parameter

$$k_D l = \frac{eFl}{\hbar\omega_c} = \frac{1}{2} \left(\frac{l}{L} \right)^3. \tag{13}$$

For two values of this parameter, the surface band structure calculated from Eq. (12) is depicted in Fig. 3a, b. Note that the abszissae could equally well be given as electron wave vectors k_y. For the larger parameter ($k_D l = 5$), the resulting band structure is similar to the one of purely electric subbands, provided, we consider negative center coordinates. A magnetic field which is applied parallel to a quasi two-dimensional electron gas causes three major effects onto the subband structure [4]: The subband edge energies increase, the positions of the subband minima shift to higher center coordinates (see Fig. 3a), and the effective mass that characterizes the motion perpendicular to the magnetic field is enhanced. The corresponding subbands are addressed as diamagnetically shifted subbands and have previously been studied in MOS structures on Si [21, 23]. Within the triangular-well model, we find for the shifts of the subband minima to higher wave vectors

$$\Delta k_y = \frac{1}{l} \left(\frac{2\pi^2}{3} \right)^{1/3} \cdot \left(i + \frac{3}{4} \right)^{2/3} \left(\frac{L}{l} \right) \tag{14}$$

and for the diamagnetic shifts of the subband minima to higher energies

$$\Delta E_i = \hbar\omega_c \left(\frac{\pi^4}{2250} \right)^{1/3} \cdot \left(i + \frac{3}{4} \right)^{4/3} \cdot \left(\frac{L}{l} \right)^2. \tag{15}$$

The dispersion in the direction of the magnetic field ($B \| x$) remains unchanged and the curvature near the subband minima of the dispersion perpendicular to the magnetic field is decreased (see Fig. 3a). Equivalently, the effective mass m_y becomes higher:

$$\Delta m_y = m^* \left(\frac{16\pi^2}{81} \right)^{1/3} \cdot \left(i + \frac{3}{4} \right)^{2/3} \cdot \left(\frac{L}{l} \right)^4. \tag{16}$$

116

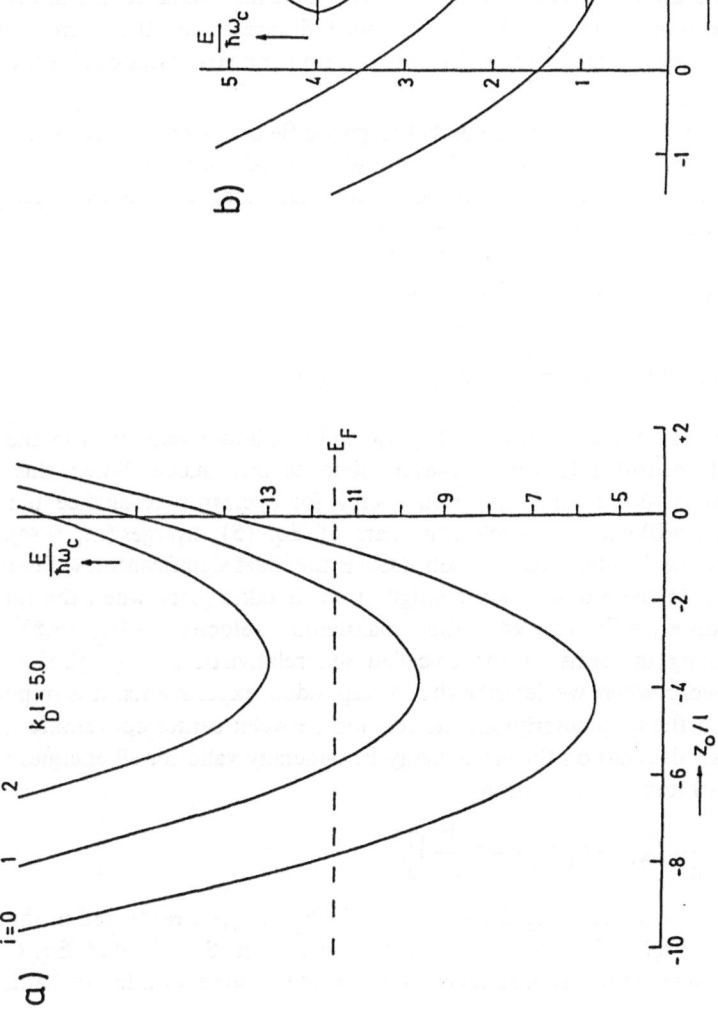

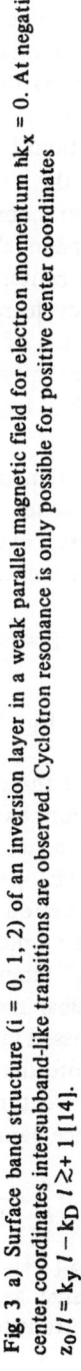

Fig. 3 a) Surface band structure ($i = 0, 1, 2$) of an inversion layer in a weak parallel magnetic field for electron momentum $\hbar k_x = 0$. At negative center coordinates intersubband-like transitions are observed. Cyclotron resonance is only possible for positive center coordinates $z_0/l = k_y l - k_D l \gtrsim + 1$ [14].
b) Surface band structure in a strong parallel magnetic field ($k_x = 0$). The inset shows a Fermi line. Cyclotron resonance is possible for most of the occupied center coordinates [14].

117

The anisotropy of the effective masses $m_y \neq m_x = m^*$ and the increase of the mass m_y in parallel magnetic fields has recently been demonstrated for $GaAs/Ga_{1-x}Al_xAs$ heterojunctions by plasmon spectroscopy [24]. Equations (14) to (16) are only valid, as long as the ratio of electric and magnetic length is small, i.e., for large parameters $k_D l$. This condition, in general, is fulfilled in $GaAs/Ga_{1-x}Al_xAs$ heterojunctions but not in MOS structures on InSb in high magnetic fields.

For the lower parameter ($k_D l = 0.1$) and a low Fermi energy E_F, only positive center coordinates are populated (see Fig. 3b). In such a situation most of the inversion electrons show cyclotron resonance since most of them occupy states with center coordinates for which the bands run parallel to each other separated by the cyclotron energy $\hbar\omega_c$. In this case, the Landau orbit of radius l fits into the width L of the inversion layer and we observe cyclotron resonance like in the bulk of an n-type semiconductor. A more detailed discussion of the corresponding classical trajectories visualizing the different surface band structures can be found in [14].

The triangular-well potential in a parallel magnetic field was also treated within the $k \cdot p$ scheme [18]. Most interestingly is the so-called magnetic case where we observe cyclotron resonance in crossed electric and magnetic fields. Again, the Landau energies can be calculated analytically:

$$E = \hbar k_y v_D + \sqrt{1 - \delta^2} \sqrt{\left(\frac{E_g}{2}\right)^2 + E_g D_i} - \frac{E_g}{2},$$

$$D_i = \hbar\omega_c \left(i + \frac{1}{2}\right)\sqrt{1 - \delta^2} + \frac{\hbar^2 k_x^2}{2m_0^*} \pm \frac{1}{2}g_0^* \mu_B B \tag{17}$$

where the ratio $\delta = v_D/u$ of drift velocity v_D and maximum velocity u in the conduction band according to the two-level model is introduced. When this ratio approaches unity ($\delta \to 1$), the transition energy for cyclotron resonance tends to zero ($E_{i+1} - E_i \to 0$) and the cyclotron mass of Eq. (5) diverges ($m_c \to \infty$). This destruction of the Landau quantization itself is the most significant new feature of the $k \cdot p$ model in the crossed field configuration. It takes place when the ratio of field strengths $v_D = F/B$ exceeds the maximum velocity $u = (E_g/2m_0^*)^{1/2}$. Its physical meaning in terms of the so-called semirelativistic analogy [25] will be discussed in Sec. 5 where we describe the corresponding experiments. It is important to note that in the $k \cdot p$ description the relation between center coordinate z_0 and wave vector k_y depends on the eigenenergy E. Generally valid for all energies in the surface band structure, the relation

$$\frac{z_0}{l} = \frac{1}{1 - \delta^2}\left[k_y l - k_D l\left(1 + 2\frac{E}{E_g}\right)\right]. \tag{18}$$

holds [14]. The corresponding EMA result $z_0/l = k_y l - k_D l$ is recovered in the limit of large gap energies ($E_g \to \infty$, $u \to \infty$). It will be seen in Sec. 5 that Eq. (18) is important to understand the basic mechanism for the strong excitation of harmonics of cyclotron resonance that accompanies the destruction of the Landau quantization in crossed fields.

2.3 Tilted Magnetic Fields

In magnetic fields that are tilted away from the surface normal by an angle Θ, the Hamiltonian no longer can be separated, and the problem must be solved numerically [23]. At present, such calculations are not available for narrow-gap semiconductors and, therefore, we rely on a perturbational approach [26, 27]. The idea is to treat the effects of the parallel magnetic field component $B_\|$ on the subband structure in perpendicular magnetic fields $B = B_\perp$ as a small perturbation [4]. In fact, in a strictly two-dimensional system the influence of component $B_\|$ vanishes since the electrons cannot react to a force directed perpendicular to their plane. Thus our procedure is justified as long as the electric length L is small compared to the magnetic length $l_\| = (\hbar/eB_\|)^{1/2}$. The resulting level diagram is visualized in Fig. 4.

In this figure, Landau ladders of two subbands $i = 0, 1$ are depicted versus magnetic field component $B_\perp$. There are diamagnetic shifts ΔE_i of the purely electric subband energies due to the parallel field component $B_\|$, the shift in the first excited subband $i = 1$ being higher according to Eq. (15). The coupling of the parallel and perpendicular motion manifests itself as a splitting of cyclotron resonance when the intersubband energy is a multiple ($p = 1, 2, \ldots$) of the cyclotron energy $\hbar\omega_c = e\hbar B_\perp/m^*$. The splittings at multiples $p = 1$ and 2 are commonly addressed as coupling at full and half field, respectively, since the strength of the magnetic field at $p = 2$ is only half the value of full field coupling at $p = 1$. The normalized gaps

$$\frac{\Delta E}{E_{10}} \approx \begin{cases} 0.855 \tan\Theta & p = 1 \\ 0.447 \tan^2\Theta & p = 2 \end{cases} \tag{19}$$

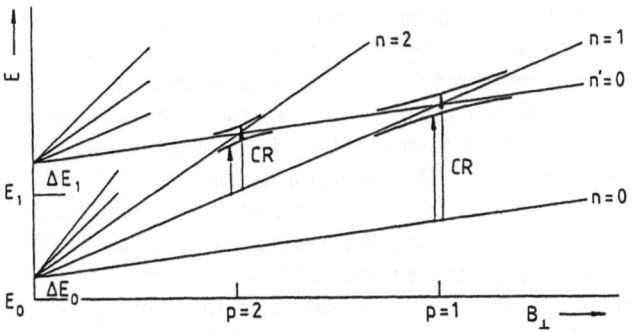

Fig. 4 Schematic level diagram of an inversion layer in tilted magnetic fields. Landau levels of two electric subbands E_0 and E_1 are depicted vs magnetic field component $B_\perp$ perpendicular to the layer. The diamagnetic shifts ΔE_0 and ΔE_1 due to the parallel magnetic field component are indicated. Coupling of electron motion parallel and perpendicular to the layer causes splitting of cyclotron resonance (CR) at full field coupling ($n = 0 \rightarrow 1$, $p = 1$) and half field coupling ($n = 1 \rightarrow 2$, $p = 2$).

119

for p = 1 and p = 2 at resonance $\hbar\omega_c = E_{10}/p$ are obtained by first and second order perturbation theory [27], respectively. The splitting vanishes in strictly perpendicular fields ($\Theta = 0$) and decreases with order p.

3 Subband Spectroscopy in GaAs/Ga$_{1-x}$Al$_x$As Heterojunctions

The determination of subband spacings in GaAs/Ga$_{1-x}$Al$_x$As heterojunctions is of great interest to characterize their interface potential (see Fig. 1b) and its dependence on inversion electron density. Originally, intersubband transitions were detected in Raman scattering [28, 29], but a drawback of Raman scattering is the strong band-gap radiation that causes quasi-accumulation conditions [4]. More recently, intersubband resonances were studied with infrared laser [26] and Fourier spectroscopy [30, 31] in tilted magnetic fields. These experiments make use of the splitting of cyclotron resonance that was described in Sec. 2.3. This spectroscopic method has proved most powerful to measure subband spacings in GaAs/Ga$_{1-x}$Al$_x$As heterojunctions.

Experimentally, it is much more convenient to study coupling at half field than coupling at full field since smaller magnetic fields are necessary. On the other hand, a precondition for the study of half field coupling is the population of Landau level n = 1 of the ground electric subband i = 0 (see Fig. 4) which is only possible in samples with sufficiently high electron densities. We have studied both couplings [31] but here we restrict the discussion to the half field coupling. In this case, we generally observe three resonances a, b, and c as shown in the left hand inset of Fig. 5 for a constant magnetic field component $B_\perp = 7.53$ T. The origin of these resonances is explained in the right hand inset. When the tilt angle is increased from $\Theta = 30°$ to $48°$, much of the oscillator strength is transferred from resonance b into a. This is a consequence of the diamagnetic shift and is discussed in more detail in [30]. Figure 5 itself shows resonance positions versus magnetic field component $B_\perp$ for the tilt angle $\Theta = 39°$. Whereas in magnetic fields $B_\perp$ up to 6T usual cyclotron resonance is detected, a pronounced splitting into resonances a, b, and c occurs in the range from 6.5 to 9 T. A negative slope of the energy versus magnetic field relation is present at magnetic field component $B_\perp \leqslant 7$ T for resonance b and at $B_\perp \geqslant 7$T for resonance a, respectively. This behavior is easily understood by considering the lengths of the arrows in the right hand inset and is characteristic for anticrossing at p = 2 but does not occur at p = 1 (see Fig. 4). At magnetic fields $B_\perp > 12.6$T the Landau level n = 1 becomes depopulated ($\nu \leqslant 2$) and only resonance c is left. At still higher fields, resonance c becomes subjected to coupling at p = 1, and its transition energy strongly deviates from the straight line that was determined in purely perpendicular magnetic fields. The value $E_{10}/2$ of the intersubband spacing we determine from the center of the energy gap.

The diamagnetic shift of subband energies due to the parallel magnetic field component given in Eq. (15) is observable in intersubband transitions since it is higher in excited subbands and, hence, causes an increase of the subband spacing. This is directly apparent in Fig. 6: As the tilt is increased, the center of the gap shifts to higher energies. Simultaneously, the splitting of the two branches *a* and *b* is increased. The resonance position of resonance *c* is not much affected, but, as a

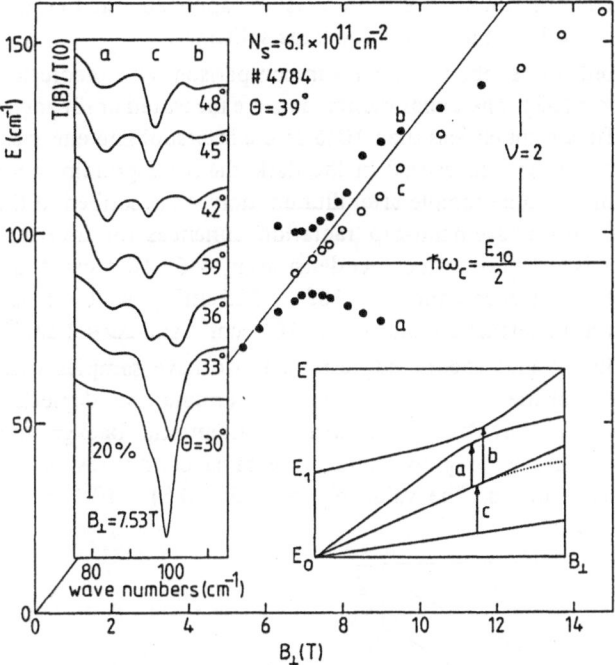

Fig. 5 Resonance positions of coupled cyclotron-intersubband transitions at half field coupling. The left hand inset shows spectra at a fixed magnetic field component $B_\perp$ for different tilt angles. Three resonances *a*, *b*, and *c* indicated in the right hand inset are observed. From the position of the center of the energy gap of resonances *a* and *b* the intersubband energy $2\hbar\omega_c$ is determined [30].

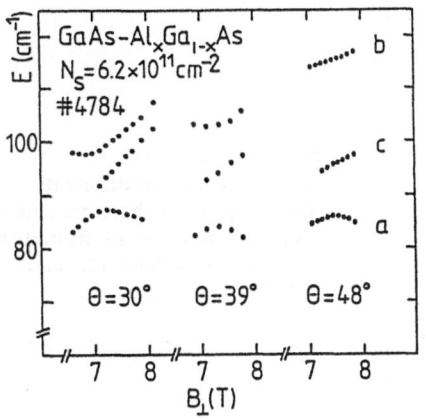

Fig. 6

Tilt angle dependence of coupled cyclotron-intersubband resonances at half field coupling. Enhanced splittings and diamagnetic shifts of the gap center between resonances *a* and *b* are evident [27].

121

consequence of coupling at full field, the slope of the experimental points is decreased. For the particular sample and angle $\Theta = 48°$, we obtain a diamagnetic shift which is 12% of the subband spacing of 180 cm^{-1}. This is well accounted for by Eq. (15) which gives 14% for a depletion charge density $n_{depl} = 5 \cdot 10^{10} \text{ cm}^{-2}$. Equation (18) for the splitting of resonances a and b is not applicable at the high tilt angles of the particular experiment and gives too large splittings.

Figure 7 shows the measured intersubband energies in comparison with theoretical results of Stern and DasSarma [32]. The experimental data are measured at tilt angles where the diamagnetic shifts are much less than 10% of the intersubband energies. The closed symbols indicate energies measured in the dark, the corresponding open symbols indicate energies for the same sample after illumination. As described in the introduction, illumination with above-band-gap radiation enhances the inversion electron density via the persistent photoeffect. At densities $n_s \approx 6 \cdot 10^{11} \text{ cm}^{-2}$, the data of three samples differ from each other ($\nu = 165 \dots 225 \text{ cm}^{-1}$), well outside the range of the experimental uncertainties $\Delta n_s \approx 0.1 \cdot 10^{11} \text{ cm}^{-2}$ and $\Delta\nu \approx 1 \text{ cm}^{-1}$.

The comparison of experiment and theory suggests that we have samples with different depletion charge densities n_{depl}. In order to determine the depletion charge, we measured by the van der Pauw method the unintentional background doping of a thick epitaxial GaAs layer grown under the same conditions as the spectroscopically investigated samples. The value $N_A = (1.0 \pm 0.5) \cdot 10^{14} \text{ cm}^{-3}$

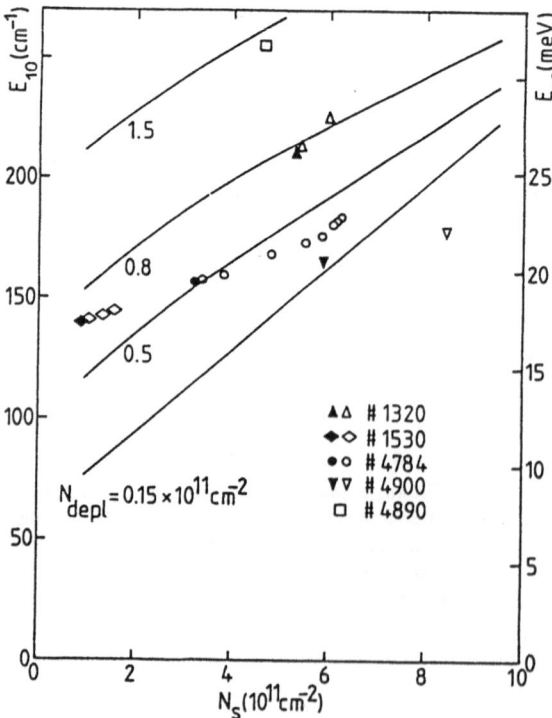

Fig. 7

Intersubband energies vs electron density for different GaAs/Ga$_{1-x}$Al$_x$As heterojunctions [30]. The solid lines are theoretical results [32] calculated for various depletion charge densities.

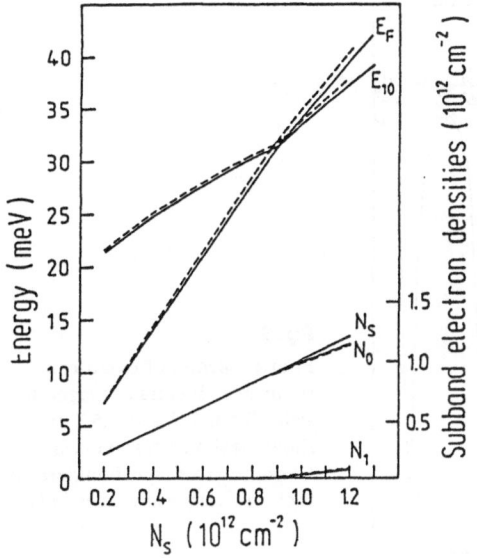

Fig. 8

Theoretical intersubband energies. Solid and dashed lines correspond to calculations with and without nonparabolic corrections to the density of states [20].

of background doping of these layers corresponds to a depletion charge $n_{depl} = (0.5 \pm 0.2) \cdot 10^{11}$ cm^{-2}. The agreement between experiment and theory is fairly good if one considers this uncertainty of the experimentally determined depletion charge. The intersubband energies increase with electron density n_s and roughly follow the theoretically predicted curves. The influence of the non-parabolic band structure of GaAs does not seem to influence the subband energies strongly. In fact, nonparabolic effects were taken into account by Malcher, Lommer, and Rössler [20] in self-consistent subband calculations and were found to be small (see Fig. 8). However, nonparabolic effects are clearly observed in cyclotron resonance experiments which we discuss in the next section.

4 Cyclotron Resonance

4.1 GaAs/Ga$_{1-x}$Al$_x$As-Heterojunctions

In this section we discuss cyclotron resonance in GaAs/Ga$_{1-x}$Al$_x$As heterojunctions in perpendicular magnetic fields. Typical Fourier spectra taken at various magnetic fields at liquid helium temperatures are depicted in Fig. 9. From the minima of normalized transmittances, cyclotron masses $m^* = eB/\omega_c$ can be read and are given in Fig. 10 together with masses of a sample of lower density [33]. The filling factors $\nu = hn_s/eB$, which give the number of filled spin-split Landau levels [4], are marked by dashed and solid arrows, respectively. The filling factor is also related to Landau transitions $n^{\pm} \to (n+1)^{\pm}$ which can contribute to the experimental signal. In the so-called spin polarized state ($\nu < 1$) only transitions $0^+ \to 1^+$ are possible, whereas in general up to three spin-split transitions are involved.

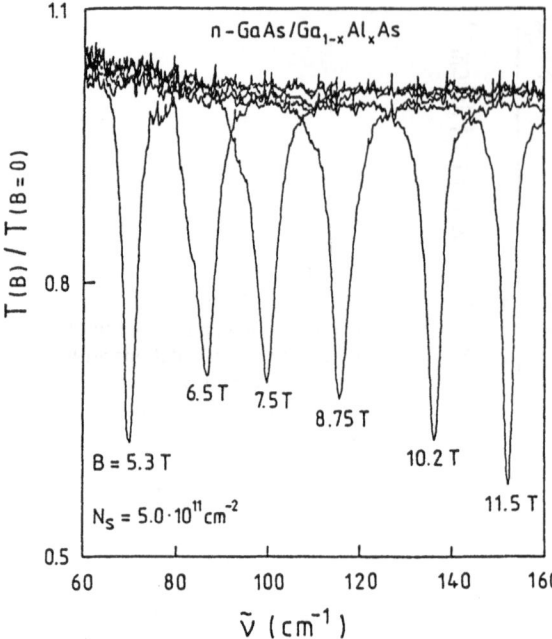

Fig. 9

Fourier spectra of cyclotron resonance. Spectra at magnetic fields B = 6.5 T and 7.5 T ($\nu \approx 3$) show maximum normalized transmittances and shoulders on their low frequency sides [33].

In the experiments, we cannot resolve the effect of electron spin even in the best samples due to the small effective Lande' factor $g_0^* = -0.44$ of GaAs [17]. However, the contribution of different Landau transitions $n \rightarrow n + 1$ are observable in Fig. 9. There is a maximum of the normalized transmittance at resonance and a corresponding maximum in the linewidths of traces B = 6.5 and 7.5 T ($\nu \approx 3$). Simultaneously, these traces show shoulders on their low frequency sides. We interpret these observations as being caused by a superposition of Landau transitions $0 \rightarrow 1$ and $1 \rightarrow 2$ with slightly different transition energies, i. e., nonparabolic cyclotron masses.

At all magnetic field strengths, the masses of the sample with the higher electron density are higher than the masses of the sample with the lower density (see Fig. 10). This is due to the fact that with increasing electron concentration the Fermi energy is increasingly separated from the bottom of the electric subband i = 0 and, hence, the electrons are subject to stronger influence of nonparabolicity. In the magnetic quantum limits ($\nu < 2$) above magnetic fields B = 2.2 T and 10.4 T, respectively, the experimental and theoretical masses steadily increase. Here the Landau levels shift away from the subband minimum and, again, become subject to increased nonparabolicity. The anomaly at magnetic fields B $\approx$ 9T for the low density sample ($n_s = 1.1 \cdot 10^{11}$ cm^{-2}) is a consequence of the intersubband-cyclotron resonance coupling. As described in Sec. 2.3 and Sec. 3, this coupling is caused by a tilt angle between surface normal and magnetic field direction. It may be as small

124

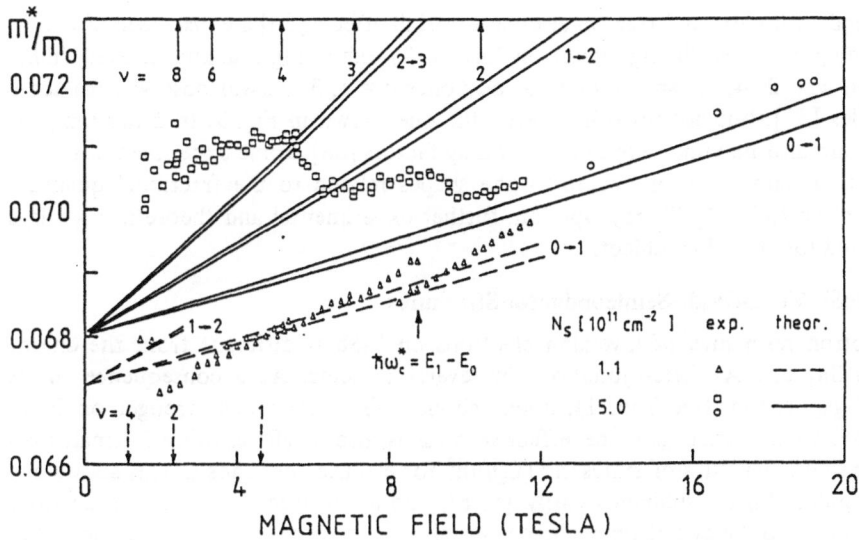

Fig. 10 Experimental and theoretical cyclotron masses in two GaAs/Ga$_{1-x}$Al$_x$As hetero-junctions (x $\approx$ 0.3) of different electron densities 1.1 X 10^{11} cm^{-2} and 5.0 X 10^{11} cm^{-2}. The corresponding filling factors ν are indicated at the bottom and the top, respectively. Theoretical values are for spin-split Landau transitions n$^{\pm}$ $\rightarrow$ n$^{\pm}$ + 1 assuming a depletion charge density n$_{depl}$ = 5 X 10^{10} cm^{-2} [33].

as 0.5° and can hardly be avoided experimentally. At magnetic fields B $\geqslant$ 15T, the resonant polaron effect becomes important which has not yet been included into the self-consistent calculations [33]. This explains the discrepancy between experimental and theoretical masses at high magnetic field strengths [11, 12].

At magnetic fields below the magnetic quantum regimes ($\nu > 2$), we observe mass oscillations which we attribute to a transfer of oscillator strength between different Landau transitions as the filling factor is changed. In the limit of vanishing magnetic fields, continually higher transitions become important, and we can extrapolate (B $\rightarrow$ 0) a cyclotron mass which no longer depends on Landau index but is solely determined by the Fermi energy. This is the electric subband mass defined in Eq. (8), and there is excellent agreement with the theoretical values of Lommer, Malcher, and Rössler over a wide range of electron densities [33].

Whereas the energetic subband structure of GaAs/Ga$_{1-x}$Al$_x$As heterojunctions is now fairly well understood, this is not the case for the cyclotron linewidths in quantizing magnetic fields. At higher temperatures scattering by longitudinal optical phonons becomes the prominent scattering process, and the mobility drops from values 10^6 cm^2 V^{-1} s^{-1} at liquid helium temperatures to about 10^4 cm^2 V^{-1} s^{-1} at room temperature [3]. At low temperatures, however, the situation is more complicated and, in particular, the role of filling factor dependent screening [34,

35] and of integer and fractional quantum Hall effects [36] are discussed. Electron screening with oscillating cyclotron linewidths exhibiting maxima at even filling factors ν = 2, 4, ... and minima at odd ones ν = 1, 3, ... was observed for some samples [37], but not for others, e.g., the one shown in Fig. 9. In other samples, there are also maxima between even filling factors [38]. Some experimental studies report on anomalies of cyclotron line shapes related to the fractional quantum Hall effect [36, 39]. To my opinion, further experimental and theoretical work is required to settle this subject.

4.2 InSb Metal-Oxide-Semiconductor Structures

Cyclotron resonance of inversion electrons on InSb is different from the one of GaAs/Ga$_{1-x}$Al$_x$As heterojunctions in several respects. As a consequence of its small gap energy (see Tab. 1), nonparabolic effects are much stronger on InSb. Related to the small gap, the effective mass is also small resulting in small two-dimensional densities of states m* (E$_i$)/$\pi\hbar^2$ for electric subbands i. This allows one to populate higher subbands easily [5, 6], and a multitude of Landau transitions contributes to the cyclotron resonance signals. On the other hand, typical mobilities at liquid helium temperature are 2 to $5 \cdot 10^4$ cm^2 V^{-1} s^{-1} and much less than in GaAs/Ga$_{1-x}$Al$_x$As heterojunctions and, in general, the various Landau transitions are not clearly resolved on InSb despite of their larger nonparabolic energy shifts. Thus, the general situation requires careful analysis of the experimental spectra as was done in [6, 40]. There are two situations where direct interpretation of experimental spectra is possible, namely, the quasi classical limit ($\hbar\omega_c \ll E_F$) and the magnetic quantum regime ($\hbar\omega_c \approx E_F$). We first consider the classical limit. In Fig. 11a laser spectra are displayed that were taken at a relatively low laser energy ($\hbar\omega$ = 17.6 meV) in correspondingly low magnetic fields (B $\approx$ 2 to 4T). From these spectra cyclotron masses were extracted for three electric subbands i = 0, 1, 2 shown in Fig. 11b. These masses can be considered as electric subband masses defined in Eq. (8). In all subbands the masses increase with electron density showing some deviations from monotonic increase by the influence of quantum oscillations [6, 40]. The increase of masses is different in different subbands and is strongest in the ground subband i = 0.

Subband masses of InSb have been calculated with self-consistent models in the absence of magnetic fields and agree qualitatively with the experiments [6, 19]. The most important features, however, already can be understood within the framework of the simple theory given in Sec. 2.1. The masses in all subbands increase with electron density since subband edge energies and occupations, i.e., kinetic energies of subbands increase. The masses are less in higher subbands since their subband edge energies are larger yielding lower kinetic energies. This immediately follows from Eq. (8) if the nominator $\frac{1}{3}$E$_i$ (EMA) +E$_{\parallel}$ is replaced by E$_F$ − $\frac{2}{3}$E$_i$ (EMA). In the limit of zero electron density, the subband energies and the kinetic energies vanish, and we observe the bulk mass m$_0^*$ of InSb at conduction band edge (see Tab. 1).

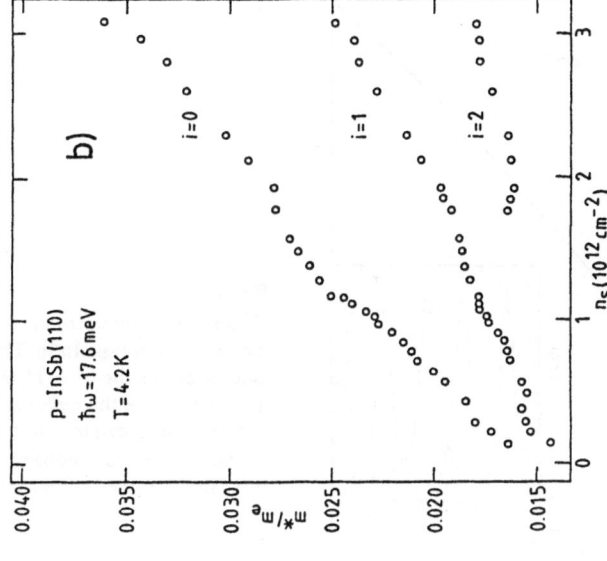

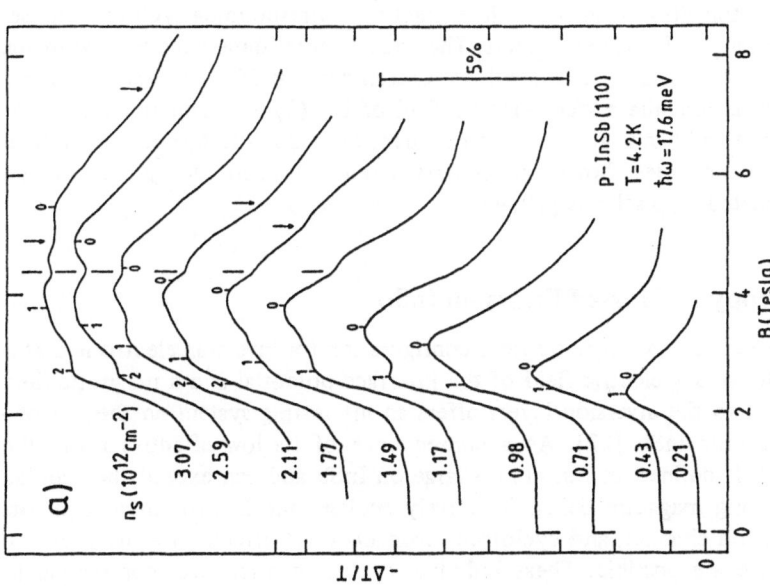

Fig. 11 a) Cyclotron spectra of inversion electrons on p-type InSb taken at the laser energy $\hbar\omega = 17.6$ meV and various electron densities n_s. The resonance positions of subband cyclotron resonances are obtained from theoretical fits ($i = 0, 1, 2$). The arrows mark more pronounced quantum oscillations, the dashes cyclotron resonance of bound holes in the p-type substrate [6]. b) Subband cyclotron masses vs electron density for three electric subbands $i = 0, 1, 2$. The masses are determined from spectra taken at a fixed laser energy $\hbar\omega = 17.6$ meV, i.e., modest magnetic fields [6].

127

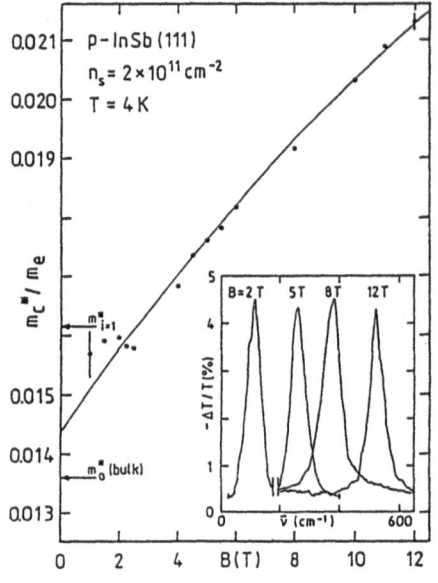

Fig. 12

Experimental cyclotron masses of inversion electrons on p-type InSb. The solid line is calculated for the $0^+ \rightarrow 1^+$ transition. The arrows indicate the bulk band-edge mass m_0^* and the calculated electric subband mass of the first excited subband $m_{i=1}^*$. The inset shows some Fourier spectra [16].

We also studied cyclotron masses in the spin polarized state ($\nu < 1$) when only Landau transitions $0^+ \rightarrow 1^+$ of the ground electric subband $i = 0$ are possible [16]. At the low electron density $n_s = 2 \cdot 10^{11}$ cm^{-2} chosen in Fig. 12, this is the case in magnetic fields $B \geqslant 8.3$ T. However, even in lower fields the spectra are dominated by the $0^+ \rightarrow 1^+$ transition, and we could extract the corresponding cyclotron masses over a wider range of magnetic fields. The experimental data agree well with the solid line calculated for $0^+ \rightarrow 1^+$ transitions according to Eqs. (5), (6) and (9). In particular, this means that the surface electric field of Eq. (2) is a good approximation for the ground subband at low electron densities. Quantitative description at higher electron densities, however, requires a self-consistent theory, including a more sophisticated k · p scheme [19, 41].

5 Spectroscopy in Crossed Fields on InSb

The crossed electric and magnetic field configuration for inversion electrons that is created by the strong electric field of the interface potential and a magnetic field applied parallel to the inversion layers offers an interesting system on the narrow-gap semiconductor InSb [22]. As a consequence of its low effective mass, the electric length L defined in Eq. (11) is large on InSb and can exceed the Landau radius l of strong magnetic fields. Intuitively spoken, the Landau circle then fits into the inversion channel, and cyclotron resonances not affected by the presence of the interface are possible. These bulk-like cyclotron resonances correspond to

128

transitions at positive center coordinates of the hybrid electric-magnetic band structure depicted in Fig. 3b. However, there are also electrons that do feel the influence of the potential barrier at the interface, namely those at negative center coordinates. Their optical excitations can be regarded as diamagnetically shifted intersubband resonances [14, 42].

Typical Fourier spectra shown in Fig. 13 were taken with radiation polarized parallel to the surface for various electron densities and magnetic fields B. The spectra show sharp cyclotron resonances and relatively weak and broad diamagnetically shifted intersubband-like resonances superimposed on the Drude background [42]. From the spectra for the density $n_s = 2 \cdot 10^{12}$ cm^{-2}, it is seen that the excitation strength of the $1 \rightarrow 2$ intersubband transitions increases with magnetic field $B = 4$ to 6 T. At magnetic fields $B \leqslant 2T$ we could not clearly observe the relatively weak intersubband transitions. This indicates that the parallel magnetic field plays an important role in the excitation mechanism [14]. Reflecting the diamagnetic shifts of the electric subbands, the intersubband energies increase with magnetic field strength as it is shown in Fig. 14. A weak splitting of the intersubband-like resonances in high magnetic fields is observed. This splitting is interpreted as spin splitting of hybrid electric-magnetic subbands [18]. From an extrapolation ($B \rightarrow 0$), the purely electric intersubband energies can be obtained. They have previously been measured directly in the absence of magnetic fields as a doublet pair of peaks for the transition $0 \rightarrow 1$ [43]. The large diamagnetic shifts revealed in Fig. 14 no

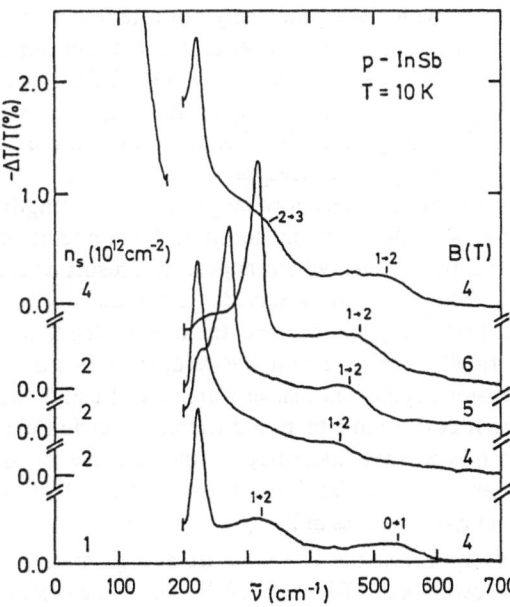

Fig. 13

Fourier spectra of inversion electrons in magnetic fields applied parallel to the inversion layers. The prominent maxima are cyclotron resonances, the relatively weak structures are diamagnetically shifted intersubband resonances $0 \rightarrow 1$, $1 \rightarrow 2$, and $2 \rightarrow 3$ [42].

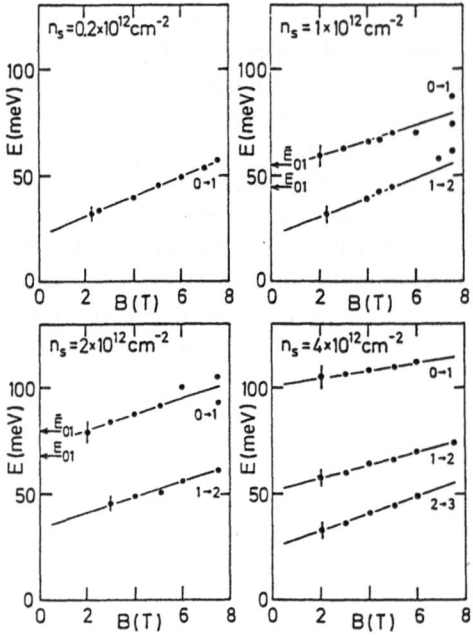

Fig. 14

Diamagnetic shifts of intersubband resonances at various electron densities. The arrows indicate the values of purely electric intersubband resonances (B = 0) that were found in [43] as a doublet E_{01} and $\tilde{E}_{01}$ separated by the depolarization energy. Spin-splitting is observed at magnetic fields B ≥ 7 T. The straight lines only serve as guides to the eye [42].

longer can be described by perturbation theory. A qualitative description within the k · p scheme and the triangular-well approximation was given in [18].

We now discuss the cyclotron resonances which are practically not affected by the presence of the interface and, therefore, can be regarded as three-dimensional cyclotron resonances in crossed electric and magnetic fields (see Sec. 2.2). It was predicted more than twenty years ago [44] that the Landau quantization in crossed fields is destroyed at high field ratios F/B. Only recently, this prediction could be verified with use of inversion layers on InSb [18]. Corresponding laser spectra for a fixed energy $\hbar\omega = 10.4$ meV, which yields relatively low magnetic field strengths at resonance, are shown in Fig. 15a. At low electron densities $0 \rightarrow 1$ cyclotron transitions and at higher densities $1 \rightarrow 2$ transitions, which appear as a result of the inhomogeneity of the surface electric field [18], are observed. As the density and the related electric field strength described in Eq. (2) increase, the $0 \rightarrow 1$ resonances shift to higher magnetic fields and finally vanish. From the resonance positions measured with various laser lines, apparent cyclotron masses were read. In Fig. 15b masses for two electron densities which correspond to two different electric fields $F \neq 0$ and cyclotron masses of bulk n-type InSb measured in the absence of an electric field are shown. The solid lines were calculated from Eq. (17) for $0^+ \rightarrow 1^+$ transitions using the definition of the cyclotron mass in Eq. (5).

The cyclotron masses in crossed fields approach the F = 0 values at high magnetic field strengths but strongly differ at lower ones: Whereas the F = 0 masses extra-

130

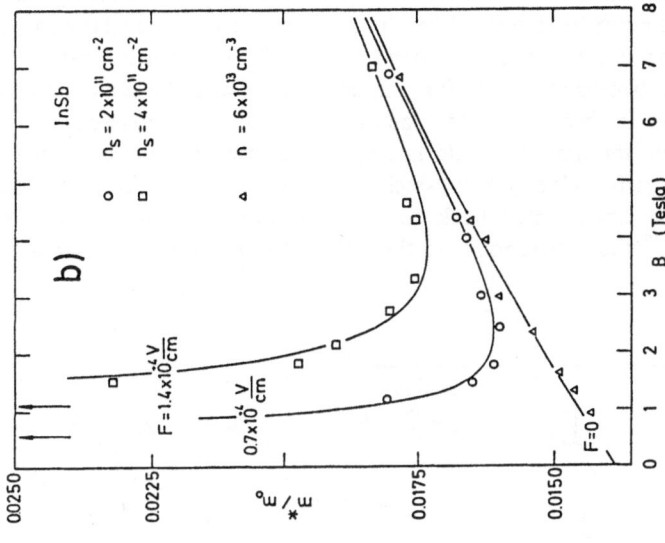

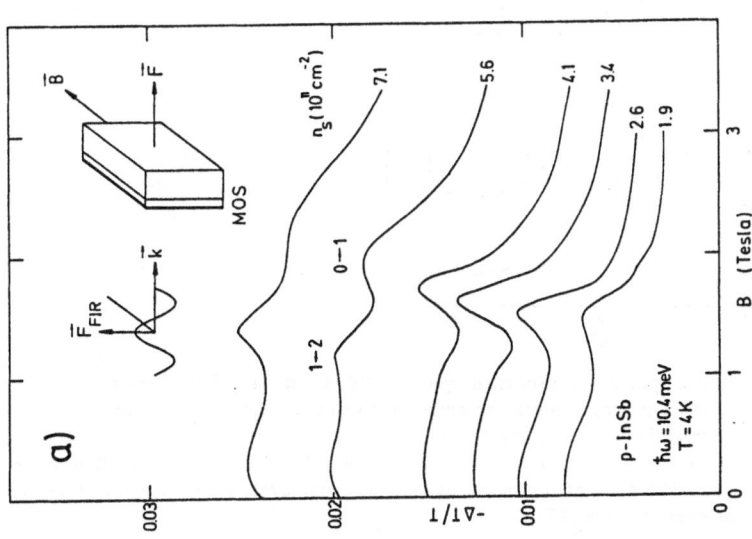

Fig. 15 a) Cyclotron resonance spectra of inversion electrons on InSb taken at fixed laser energy ℏω in parallel magnetic fields. The inset shows the geometrical configuration of crossed fields and incident far-infrared radiation [18]. b) Cyclotron masses in InSb in crossed electric and magnetic fields for two inversion electron densities n_s or electric field strengths F. Masses of n-type InSb (F = 0) are included for comparison [18].

polate to the band edge mass (see Tab. 1), the masses in crossed fields exhibit a steep increase. Theoretically, they diverge as described by the relation $m_c^* \approx m_0^*/(1 - \delta^2)$ which is obtained from Eq. (17) in the limit $\delta \to 1$. This divergence of the cyclotron mass cannot be described by the EMA and is a spectacular nonparabolic effect. Within the so-called semirelativistic analogy [25], which utilizes the formal similarity of the two-band model for semiconductors and the relativistic description of electrons in free space, the factor $(1 - \delta^2)^{-1}$ is a consequence of two relativistic effects: the magnetic field has to be Lorentz transformed to the system moving with the drift velocity F/B which gives one factor $(1 - \delta^2)^{-1/2}$. Since the energy has to be transformed back to the laboratory system, we have another factor $(1 - \delta^2)^{-1/2}$ that corresponds to the relativistic transverse Doppler shift.

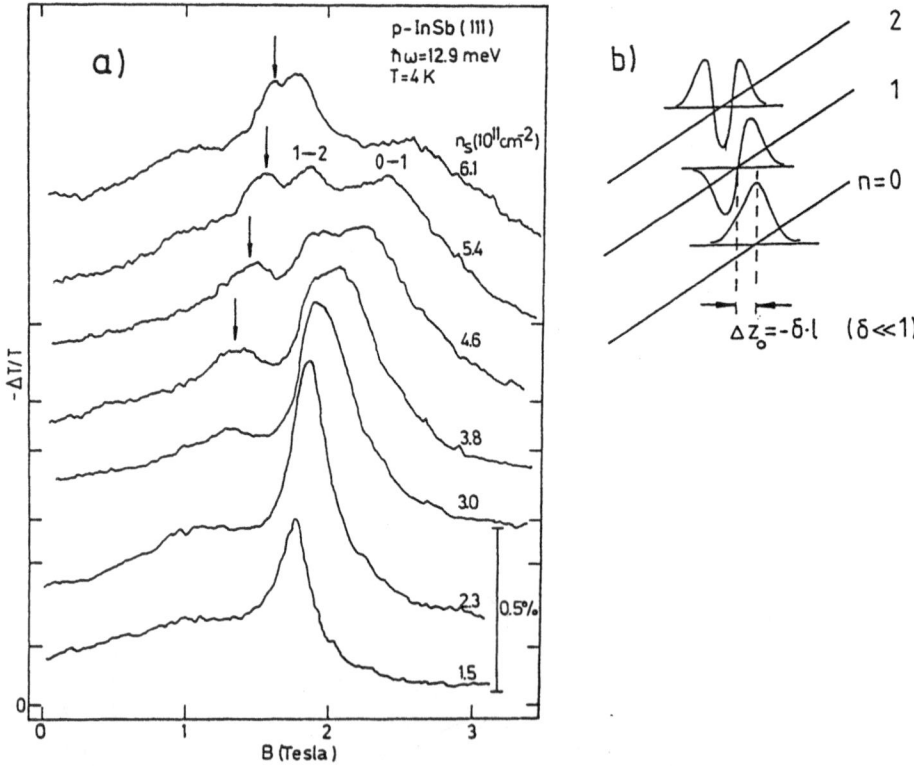

Fig. 16 a) Transmittance spectra in parallel magnetic fields taken at a fixed laser energy and various electron densities. The arrows mark harmonic cyclotron resonances $0 \to 2$ [46].
b) Intuitive picture explaining harmonic cyclotron resonance in crossed electric and magnetic fields. Oscillator wave functions related to the conduction band and Landau levels tilted in the electric field are shown. The shift of the center coordinate Δz_0 is responsible for the violation of the usual oscillator selection rule [47].

The destruction of the Landau quantization in a transverse electric field is accompanied by the breakdown of the well-known oscillator selection rules, as has also been predicted many years ago [45]. In bulk n-type InSb harmonic transitions $0 \rightarrow 2$ are rather weak with oscillator strengths that are only about 10^{-3} to 10^{-4} of the one for the fundamental resonance, and harmonics are induced by band warping, inversion asymmetry, or impurities [17]. The experimental situation in crossed fields is depicted in Fig. 16a. At low densities, the predominant maxima at magnetic fields $B \approx 1.8$ T are due to $0 \rightarrow 1$ cyclotron transitions. At higher densities these resonances shift to higher magnetic fields and vanish as was described above. At densities $n_s \approx 3 \cdot 10^{11}$ cm^{-2}, in addition to the $1 \rightarrow 2$ transitions, new resonances occur that are marked by arrows. When their resonance position is extrapolated to zero density, it agrees with the one of harmonic $0 \rightarrow 2$ transitions in bulk n-type InSb in the absence of an electric field [46]. The oscillator strength of these harmonic transitions in crossed fields is comparable to the one of fundamental cyclotron resonance.

Resonance positions and oscillator strengths of harmonic cyclotron resonance can be described qualitatively within the $k \cdot p$ model [47] which also provides an intuitive picture for the excitation mechanism depicted in Fig. 16b. There is a shift Δz_0 of the center coordinate which can be calculated from Eq. (18). In other words, momentum and center coordinate are no longer proportional as they are in the EMA, and conservation of momentum $\hbar k_y$ does no longer mean that the center coordinate z_0 is conserved too. Intuitively spoken, the electron takes a step into the direction of the electric field, and the common selection rules for the harmonic oscillator no longer are valid. However, this simplified picture does not make use of the full four-component wave function of the two-level $k \cdot p$ model and, thus, cannot account quantitatively for the theoretical result of [47].

6 Summary and Conclusions

Electrons of quasi two-dimensional inversion layers in heterojunctions and MOS structures on semiconductors GaAs and InSb, respectively, have been examined by far-infrared spectroscopy in order to determine their subband structure. As a consequence of the reduced dimensionality of these electron systems, magnetic fields applied in various directions were found to be a most versatile and valuable tool [21]. The most decisive parameter for the behaviour of inversion electrons in such experiments is the ratio of Landau radius l and width of inversion layer L. This ratio is much larger on GaAs than on InSb [14] and, hence, different spectroscopic possibilities are offered.

The situation in various magnetic field directions is much more involved in inversion layers than in bulk electron systems of these nearly isotropic semiconductors. In the bulk, quantizing magnetic fields create Landau levels with one-dimensional continua, and one can probe the slight anisotropy of the bulk band structure [48].

In quasi two-dimensional inversion layers, a totally discrete, i.e., zero-dimensional system of Landau ladders on each electric subband is created in perpendicular magnetic fields ($\Theta = 0$). In parallel magnetic fields ($\Theta = 90°$), complete coupling of electric and magnetic quantization occurs and two-dimensional electric-magnetic subbands form. In this configuration, on InSb there are electrons which are largely unaffected by the presence of the interface and, thus, behave bulk-like. This enabled us to study cyclotron resonance in crossed fields under equilibrium conditions making use of the strong internal surface electric field. Pronounced consequences of the narrow-gap nature of InSb were observed in this configuration, such as the destruction of Landau quantization [18] and strong excitation of harmonics of cyclotron resonance [46].

At intermediate tilt angles ($0 < \Theta < 90°$), the Landau ladders of distinct electric subbands essentially are preserved on GaAs since the width of its inversion channel is narrow enough. However, there is coupling of the electron motion perpendicular and parallel to the interface resulting in a splitting of cyclotron resonances, i.e., coupled intersubband-cyclotron resonances. This effect enabled us to study subband spacings in $GaAs/Ga_{1-x}Al_xAs$ heterojunctions [30].

The polar nature and the nonparabolic conduction band are decisive for the properties of bulk and two-dimensional electron systems in III-V semiconductors. The present contribution focused on nonparabolic effects of inversion layers which are more pronounced on InSb than on GaAs since the energy gap is smaller on InSb. The energetic subband structure of n-$GaAs/Ga_{1-x}Al_xAs$ heterojunctions is presently already well understood, including the effects of nonparabolicity that were treated by perturbation theory [33]. For InSb with its stronger nonparabolicity, quantitative description of eigenenergies in inversion layers in the presence of quantizing magnetic fields is still missing. In particular, this holds for the crossed-field configuration. Generally, nonparabolic effects are important in all layered semiconductor structures based on narrow-gap materials. As only one further example, I like to mention the experimental work on space-charge layers on the ternary compound $Hg_{1-x}Cd_xTe$ which gives supplementary insight into the consequences of nonparabolicity for two-dimensional electron systems [49, 50].

Acknowledgements

This review is based on a collaborative effort as it is evidenced by the list of references. The $GaAs/Ga_{1-x}Al_xAs$ heterojunctions were provided by K. Ploog and G. Weimann. Part of the spectroscopic work in high magnetic fields was carried out at the Hochfeld-Magnetlabor of the Max-Planck-Institut für Festkörperforschung, Grenoble. Financial support of the Deutsche Forschungsgemeinschaft also is gratefully acknowledged.

References

[1] For a short comprehensive review see: *J. C. Dyment*, Can. J. Phys. **63**, 651 (1985)

[2] *K. Ploog*, J. Cryst. Growth **79**, 887 (1986)

[3] *G. Weimann*, in: Festkörperprobleme/Advances in Solid State Physics, ed. by *P. Grosse* (Vieweg, Braunschweig 1986), Vol. 26, p. 231

[4] *T. Ando, A. B. Fowler*, and *F. Stern*, Rev. Mod. Phys. **54**, 437 (1982)

[5] *A. Därr, J. P. Kotthaus*, and *F. Koch*, Solid State Commun. **17**, 455 (1975)

[6] *U. Merkt, M. Horst, T. Evelbauer*, and *J. P. Kotthaus*, Phys. Rev. **B34**, 7234 (1986)

[7] *J. H. Crasemann, U. Merkt*, and *J. P. Kotthaus*, Phys. Rev. **B28**, 2271 (1983)

[8] *H. L. Störmer, R. Dingle, A. C. Gossard, W. Wiegmann*, and *M. D. Sturge*, Solid State Commun. **29**, 705 (1979)

[9] *M. Horst, U. Merkt*, and *J. P. Kotthaus*, Phys. Rev. Lett. **50**, 754 (1983)

[10] *M. Horst* and *U. Merkt*, Solid State Commun. **54**, 559 (1985)

[11] *M. Horst, U. Merkt, W. Zawadzki, J. C. Maan*, and *K. Ploog*, Solid State Commun. **53**, 403 (1985)

[12] *W. Xiaoguang, F. M. Peeters*, and *J. T. Devreese*, Phys. Rev. **B34**, 8800 (1986)

[13] *U. Merkt, M. Horst*, and *J. P. Kotthaus*, Physica Scripta **T13**, 272 (1986)

[14] *U. Merkt*, Phys. Rev. **B32**, 6699 (1985)

[15] *H. Heinrich* and *J. M. Langer*, in: Festkörperprobleme/Advances in Solid State Physics, ed. by *P. Grosse* (Vieweg, Braunschweig 1986), Vol. 26, p. 251

[16] *U. Merkt* and *S. Oelting*, Phys. Rev. **B35**, 2460 (1987)

[17] *C. R. Pidgeon*, in: Handbook on Semiconductors, ed. by *M. Balkanski* (North-Holland, Amsterdam 1980), Vol. 2, p. 233

[18] a) *W. Zawadzki, S. Klahn*, and *U. Merkt*, Phys. Rev. Lett. **55**, 983 (1985)

b) *W. Zawadzki, S. Klahn*, and *U. Merkt*, Phys. Rev. **B33**, 6916 (1986)

[19] a) *Y. Takada, K. Arai, N. Uchimura*, and *Y. Uemura*, J. Phys. Soc. Jpn. **49**, 1851 (1980)

b) *Y. Takada*, J. Phys. Soc. Jpn. **50**, 1998 (1981)

[20] a) *F. Malcher, G. Lommer*, and *U. Rössler*, Superlattices and Microstructures **2**, 267 (1986)

b) *G. Lommer, F. Malcher*, and *U. Rössler*, Superlattices and Microstructures, **2**, 273 (1986)

[21] *F. Koch*, in: Physics in High Magnetic Fields, Springer Series in Solid-State Sciences, ed. by *S. Chikazumi* and *N. Miura* (Springer, Heidelberg 1981), Vol. 24, p. 262

[22] *U. Merkt*, in: Physics of the Two-dimensional Electron Gas, ed. by *J. T. Devreese* (Plenum), in press

[23] *T. Ando*, Phys. Rev. **B19**, 2106 (1979)

[24] *E. Batke* and *C. W. Tu*, Phys. Rev. **B34**, 3027 (1986)

[25] *W. Zawadzki*, in: Optical Properties of Solids, ed. by *E. D. Haidemenakis* (Gordon and Breach, New York 1970), p. 179

[26] *G. L. J. A. Rikken, H. Sigg, C. J. G. M. Langerak, H. W. Myron, J. A. A. J. Perenboom*, and *G. Weimann*, Phys. Rev. **B34**, 5590 (1986)

[27] *A. D. Wieck* and *U. Merkt*, to be published

[28] *G. Abstreiter* and *K. Ploog*, Phys. Rev. Lett. **42**, 1308 (1979)

[29] *A. Pinczuk, J. M. Worlock, H. L. Störmer, R. Dingle, W. Wiegmann*, and *A. C. Gossard*, Solid State Commun. **36**, 43 (1980)

[30] *A. D. Wieck, J. C. Maan, U. Merkt, J. P. Kotthaus, K. Ploog*, and *G. Weimann*, Phys. Rev. **B35**, 4145 (1987)

[31] *A. D. Wieck, J. C. Maan, U. Merkt, J. P. Kotthaus*, and *K. Ploog*, in: Proceedings of the 18th International Conference on the Physics of Semiconductors, ed. by *O. Engström* (World Scientific, Singapore 1987), Vol. 1, p. 601

[32] *F. Stern* and *S. DasSarma*, Phys. Rev. **B30**, 840 (1984)

[33] *F. Thiele, U. Merkt, J. P. Kotthaus, G. Lommer, F. Malcher, U. Rössler*, and *G. Weimann*, Solid State Commun. **62**, 841 (1987)

[34] *R. Lassnig* and *E. Gornik*, Solid State Commun. **47**, 959 (1983)

[35] *T. Ando* and *Y. Murayama*, J. Phys. Soc. Jpn. **54**, 1519 (1985)

[36] *G. L. J. A. Rikken, H. W. Myron, P. Wyder, G. Weimann, W. Schlapp, R. E. Horstman*, and *J. Wolter*, J. Phys. **C18**, L 175 (1985)

[37] *Th. Englert, J. C. Maan, Ch. Uihlein, D. C. Tsui, and A. C. Gossard*, Solid State Commun. **46**, 545 (1983)

[38] *E. Batke*, private communication

[39] *E. Gornik*, private communication

[40] *M. Horst, U. Merkt*, and *K. G. Germanova*, J. Phys. **C18**, 1025 (1985)

[41] *F. Malcher, G. Lommer*, and *U. Rössler*, in: Proceedings of the 18th International Conference on the Physics of Semiconductors, ed. by *O. Engström* (World Scientific, Singapore 1987), Vol. 1, p. 493

[42] *S. Oelting, U. Merkt*, and *J. P. Kotthaus*, Surf. Sci. **170**, 402 (1986)

[43] *K. Wiesinger, H. Reisinger*, and *F. Koch*, Surf. Sci. **113**, 102 (1982)

[44] *W. Zawadzki* and *B. Lax*, Phys. Rev. Lett. **16**, 1001 (1966)

[45] *W. Zawadzki*, in: Proceedings of the 9th International Conference on the Physics of Semiconductors (Nauka, Leningrad 1968), Vol. 1, p. 312

[46] *S. Klahn, M. Horst*, and *U. Merkt*, in: Proceedings of the 18th International Conference on the Physics of Semiconductors, ed. by *O. Engström* (World Scientific, Singapore 1987), Vol. 2, p. 1161

[47] *U. Merkt*, in: Application of High Magnetic Fields in Semiconductors Physics, Springer Series in Solid-State Sciences, ed. by *G. Landwehr* (Springer, Heidelberg 1987), Vol. 71, p. 256

[48] *H. Sigg, J. A. A. J. Perenboom, P. Pfeffer*, and *W. Zawadzki*, Solid State Commun. **61**, 685 (1987)

[49] *F. Koch*, in: Two-Dimensional Systems, Heterostructures, and Superlattices, Springer Series in Solid-State Sciences, ed. by *G. Bauer, F. Kuchar,* and *H. Heinrich* (Springer, Heidelberg 1984), Vol. 53, p. 20

[50] *J. Singleton, F. Nasir*, and *R. J. Nicholas*, Surf. Sci. **170**, 409 (1986)

Festkörperprobleme 27 (1987)

Magnetic Quantization in Superlattices

Jan-Kees Maan

Max-Planck-Institut für Festkörperforschung, Hochfeld-Magnetlabor 166X, F-38042 Grenoble Cédex, France

Summary: The energy-level structure of superlattices in a magnetic field parallel and perpendicular to the layers, when the periodicities are comparable to the cyclotron orbit radius, is studied; experimentally with interband excitation spectroscopy and theoretically with the quasi-classical quantization of the superlattice bandstructure in a magnetic field as well as with the direct solution of the Schrödinger equation of a particle moving in a field and in a periodic potential. A critical analysis, both qualitative and quantitative, of the experimental results using these theoretical descriptions is presented.

1 Introduction

The word superlattice was introduced in 1970 by Esaki and Tsu [1] to describe a structure which consists of periodic alternating thin layers (≈ 10 nm) of different semiconducting materials and/or doping. The main point of their proposal was to show that such structures would have a new bandstructure which would be determined by the artificial (super)periodicity rather than by the properties of the individual layers. In this sense superlattices can be considered like solids made from solids, which is why they are called that way. An example would be a structure of periodic alternating layers of GaAs and GaAlAs, where the GaAs bandgap is entirely contained within the larger GaAlAs bandgap, resulting in a rectangular potential (see Fig. 1) for both the conduction and the valence band, with potential wells in the GaAs and barriers in the GaAlAs for both electrons and holes. For very thick or very high barriers, there is no coupling between successive wells and the system is just a set of isolated quantum wells. For thin barriers, the coupling leads to a dispersion relation of a finite width in the superlattice direction (minibands) separated by gaps (minigaps). To behave as a real superlattice the penetration depth of the wavefunction in the barriers must be much less than the periodicity. This length is characterized by $(2m^*V/\hbar^2)^{1/2}$, with m^* the effective mass, $\hbar$ Planck's constant, and V the barrier height. For GaAs/GaAlAs superlattices it can be found that layer thicknesses < 10 nm are needed to see "real superlattice effects". In simple cases, the bandstructure can be obtained from the solution of the Kronig-Penney-model. Fig. 2 shows how the original bandstructure is changed by the superlattice periodicity and how the minibands and gaps are formed.

The central theme of this paper will be to show both experimentally and theoretically an interesting aspect of this new type of solids, namely, the quantization of

137

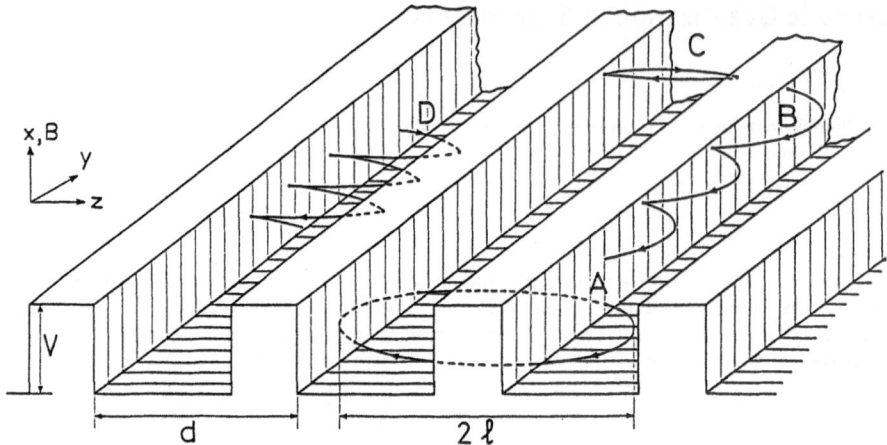

Fig. 1 Schematic representation of different classical electron orbits in a periodic one-dimensional potential V(z) and a magnetic field (‖ x) with barrier heights higher than the cyclotron energy and the orbit size (ℓ) comparable to the periodicity (d). The relative length scales are representative for a 5 nm period superlattice and a field of 20 T. Depending on the energy, tunnelling orbits (A) and orbits where carriers are reflected at the barriers, like skipping (B), lenslike (C), and doubly reflecting orbits (D) can be expected.

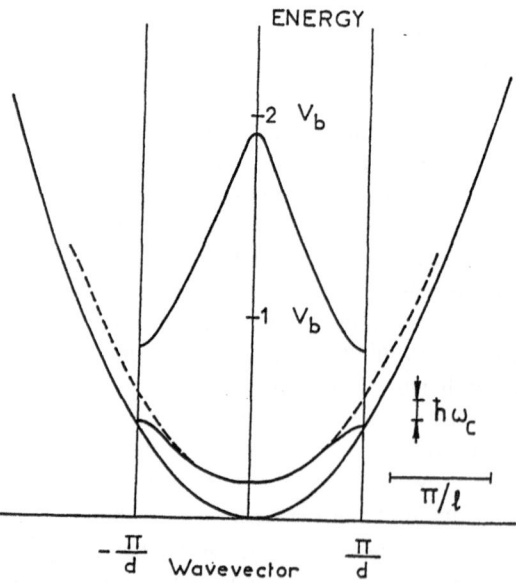

Fig. 2

Schematic diagram of the minibands in a superlattice and the original free electron-like dispersion relation (drawn parabola). Note the similarity between the curvature of the bottom of the lowest miniband and that of the free electron-like dispersion (dashed line). With the appropriate values for d, V_b, $\hbar \omega_c$, and ℓ at 20 T the scales of the figure correspond to those in the experiment.

its bandstructure by a magnetic field parallel to the layers. In a magnetic field, charged carriers describe circular orbits in the plane perpendicular to the field with a cyclotron orbit radius R_N due to the Lorentz force. This cyclotron orbit radius depends on the magnetic field B and Landau quantum number N as:

$$R_N^2 = (2N + 1) \, \ell^2 \qquad (1)$$

with $\ell^2 = \hbar/eB$, the magnetic length. At a field of 20 T this length is ≈ 5 nm which is comparable to the superlattice periodicity, d, and in a naive manner several types of orbits can be imagined, as is shown schematically in Fig. 1. This is a rather unique situation because for normal solids with a lattice period of typically 0.5 nm such conditions could only be attained with unrealistical high magnetic fields (of the order of 2000 T). As can be guessed from Fig. 1, experiments on superlattices in parallel field are a way to study transport through the barriers (now known as "vertical transport") which has recently become an important research topic. From a fundamental point of view, it is also clear that the conventional methods to treat the effect of a magnetic field on carriers in a periodic system may then have to be reconsidered.

Together with the orbital quantization, the energy is also quantized by a field, in this case in Landau levels which for simple bands are given by:

$$E_N = \left(N + \frac{1}{2} \right) \hbar \omega_c \qquad (2)$$

with ω_c the cyclotron frequency, eB/m^*. In a field of 20 T in GaAs $\hbar \omega_c \approx 30$ meV which is of the order of the width of the minibands in a GaAs/GaAlAs super-lattice. To illustrate these points in Fig. 2 is indicated how the typical scales for the energy and length compare with those of a superlattice. The scales of energy and lengths in Fig. 2 correspond to a GaAs/GaAlAs sample (x = 0.44) with 3.92 nm well and 1.12 nm barrier width in a magnetic field of 20 T. These parameters are those of a sample studied experimentally and theoretically in the rest of this paper.

In the original proposal by Esaki and Tsu [1] it was argued that the superperiodicity leads to a smaller Brillouin zone and to narrower (sub)bands. The narrower subbands have an inflection point (a change of the sign of the curvature of the band) in the E-k relation at relatively low energies above the bottom of the band (see Fig. 2). Such a peculiar bandstructure was shown to lead to interesting transport properties, perpendicular to the layers, like, for instance, negative differential conductivity. Despite the fact that in 1970 the creation of such structures would "require considerable effort with the use of the most advanced technologies" [1] the initial progress was impressively rapid. In early 1974 Chang, Esaki, and Tsu [2] measured resonant tunnelling through double barriers, and subsequently Esaki and Chang [3] measured quantum transport perpendicular to the layers. Simultaneously Dingle et al. showed quantum confinement in isolated wells [4] and tunnel-splitting between coupled double wells [5,6] using optical experiments.

However, in some sense the experimental results by Esaki and Chang [3] on vertical transport in superlattices were somewhat disappointing, because their results showed a reproducible but very irregular behaviour of the conductance through the layers as a function of the voltage applied over the superlattice. This behaviour was explained by the formation of domains of high electric field where tunnelling between neighbouring layers took place, and the spiky structure they observed, was explained as the successive breakthrough of the individual layers. In other words instead of conduction in a superlattice miniband, rather tunnelling through multiple barriers was observed. This experimental result on one hand and on the other hand the new possibilities which had now become accessible through the improved growth technique (by Molecular Beam Epitaxy initially) permitting the growth of high purity layered systems with almost atomically sharp interfaces, leads to a temporary loss of interest in "real superlattice effects". In particular between 1974 and 1983 the main interest shifted to optical properties of quantum wells ("Quantum well laser") [7], electron transport along the interface after the invention of the GaAs-GaAlAs modulation doped heterojunction [8], the Quantum Hall effect [9], and later on to the Fractional Quantum Hall effect [10]. This development is demonstrated by the review articles on these subjects in this series [11] and also in the historical overview given by Esaki [12]. As far as "real super-lattice effects" are concerned, during this period Shubnikov-de Haas-experiments on GaAs [13] and later on InAs-GaSb [14] superlattices had shown an influence of the component of the magnetic field parallel to the layer on the layers on the quantum oscillations. This behaviour indicates the possibility of transport through the layers [15]. Furthermore optical transitions at the superlattice Brillouin-zone edge, thereby measuring the width of the superlattice miniband, had been observed [16].

Since 1983 a renewed interest in transport perpendicular to the layers can be noted. Sollner et al. [17] have shown clearly negative differential conductivity in transport through a double barrier. As this type of structure is easier to realize than super-lattices, it is probably more important from a device point of view [18]. Chomette et al. [19] and Deveaud et al. [20] have demonstrated transport through a super-lattice with optical techniques. In their experiment they show that electrons and holes excited optically in the superlattice, give rise to luminescence from a single wider well which is buried in the sample below the superlattice. Calecki et al. [21] and Palmier et al. [22] have measured mobilities in doped superlattices, treating the superlattice as an effective medium. Yoshino et al. [23] have studied the angular dependence of Shubnikov-de Haas-oscillations in superlattices and have interpreted their result in terms of a quasi three-dimensional Fermi surface. Davies et al. [29] have shown for the first time negative differential resistivity in a double superlattice separated by a wider barrier. However, although their experiment definitely shows transport through the superlattice, the negative differential resistivity is due to the wider barrier only. Recently, Capasso et al. [25] showed a large photocurrent gain caused by the insertion of a superlattice between the p

140

and the n regions of a pn junction, and their experiment may be the first practical application of a superlattice.

However, although most of these experiments show the possibility of transport through the layers they do not show that a superlattice may be considered as a new material with a new bandstructure, which has properties that can be described by its bandstructure only, without referring explicitly to the layered nature of the sample. In other words most of these experiments show vertical transport, in the sense that a carrier put in the first layer of the superlattice may eventually end up in the last layer, which is, however, not the same as demonstrating the existence of a new bandstructure.

In this paper the results of magneto-absorption experiments [26, 27] will be reviewed with the purpose of showing that these results can indeed be understood as a direct manifestation of the new superlattice bandstructure. The main emphasis will be on the interband magneto-optical adsorption experiment by Belle et al. [26]. In Sect. 2 these experimental results will be described. In Sect. 3 the quasi-classical magnetic quantization of the superlattice bandstructure is discussed, i.e. the level structure in a magnetic field is derived from the band structure at zero field, without any explicit reference to the layered nature of the sample.

In Sect. 4 a complementary analysis will be given, i.e. the magnetic levels will be calculated from the solution of the Schrödinger equation of a particle moving simultaneously in a magnetic field and in a periodic potential (as was done previously in [28]). These two approaches give a rather complete qualitative understanding of the results. Finally, in Sect. 5, a more quantitative analysis of the experimental results is presented, and possible future uses of this technique are suggested.

2 Experimental Results

Two experimental results are most relevant to the questions addressed above. Belle et al. [26] have measured interband magneto-absorption in a superlattice in a field parallel to the layers using excitation spectroscopy. They have observed more or less regular Landau levels in the energy region of the electron and hole subbands and a smearing out of these levels beyond this energy range. Duffield et al. [27] have observed a slight anisotropy in cyclotron resonance with a field parallel and with a field perpendicular to the layers. Most relevant for the present paper are the data reported by Belle et al. [26], which will be described in more detail.

In this experiment, the conduction to valence band luminescence intensity is measured at fixed energy, while the exciting radiation is scanned (excitation spectroscopy). The experiment is performed at several fixed values of the magnetic field which is either perpendicular or parallel to the layers of a superlattice. The mechanism of this technique is that in an absorption maximum more carriers will be excited and therefore more carriers will in general relax to the ground state giving rise to an increase in the luminescence intensity. Hence the luminescence intensity as a function of the exciting energy is a measure of the energy dependence of the absorption coefficient if relaxation rate of photo excited carriers is assumed to be independent of energy and/or magnetic field. In an interband absorption-experiment a free electron and hole are created which interact with each other through their Coulomb attraction to form an exciton. Therefore this type of experiment does not, in principle, measure the free electron and hole states but rather excitonic states. However, if the binding energy is small enough, these excitonic effects are only a small correction to the free particle states. For the majority of the experimental results this simplication is justified, and excitonic effects will therefore be neglected. The sample consisted of 40 periods of alternating 4.48 nm GaAs and 1.12 nm GaAlAs (x = 0.44), (values determined from the growth parameters). The excitation spectra were measured at low temperatures (1.5 K) at different fixed values of the magnetic field (fields up to 23 T were generated with a 10 MW, 50 mm inner bore polyhelix magnet). In order to cover a sufficiently large range of energies two different dyes were used, DCM pumped with an Ar laser and LD700 pumped with a Kr laser.

Fig. 3 shows spectra of a superlattice with a magnetic field perpendicular and parallel to the layers at 19 T, in the upper half and of a single quantum well, in the lower half. The main difference between these results is that in the case of the superlattice in both field configurations a clear, more or less regular, Landau-level structure can be observed while for the quantum well this is only the case for the perpendicular field. A plot of the field dependence of the peaks for the superlattice is shown in Fig. 4 which basically shows equidistant, linearly field dependent transitions in both $B_\perp$ and also $B_\parallel$ with a slightly lesser slope than in $B_\perp$, however. A notable feature of the $B_\parallel$ spectra is that above a certain energy (≈ 1.8 eV) no transitions are observed anymore; as the field is increased transitions shift to higher energy and the transitions close to this cut-off energy broaden and finally disappear. This cut-off energy is illustrated more clearly in Fig. 5 where the high energy part of the $B_\parallel$ spectra is shown in more detail.

The general behaviour can be understood qualitatively in the following way. In bulk materials in a magnetic field the valence and the conduction band continua are split into discrete Landau levels. For simple parabolic bands, like the conduction band, these levels are given by Eq. (2) and are therefore equidistant and linearly field dependent. In the perpendicular configuration for both the superlattice and the quantum well electrons orbit in the plane of the layers, and the bands are split by the field like in a bulk material as indeed is observed. The hole Landau levels are more complicated, but since their splitting is much less than for the conduction-band levels the regular behaviour of the latter levels dominates the spectra. In a parallel field for the quantum well the orbit radius is larger than the well width, and therefore no Landau levels can be observed. The real interesting fact is, however, that one does indeed observe these levels in the superlattice although, there too, the width of the wells is much less than the orbit radius; the

142

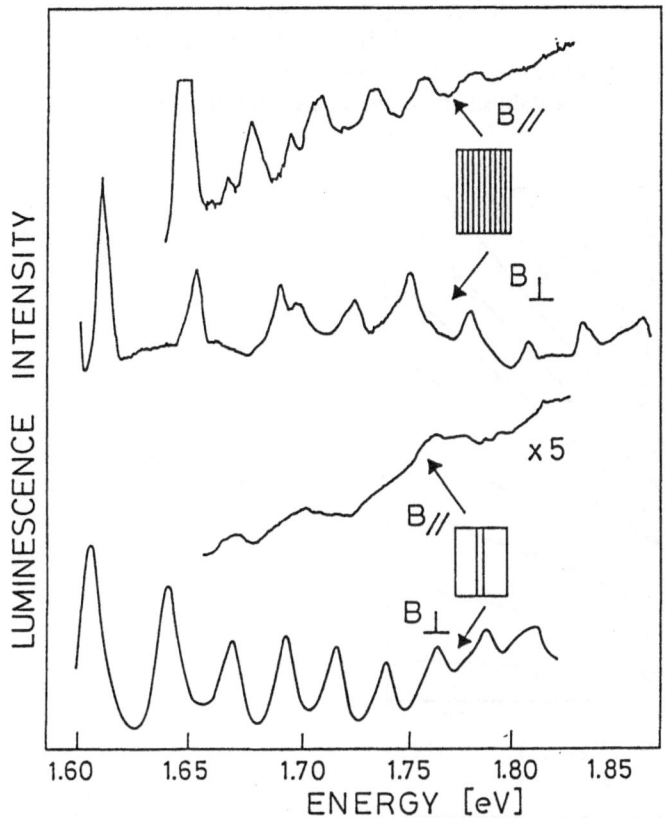

Fig. 3 Excitation spectra at 19 T of a superlattice (top) and a quantum well (bottom) for $B_\parallel$ and $B_\perp$ to the layers.

reason of this difference being of course that in a superlattice there are many wells and that they are coupled. An important feature of the results is that at a certain energy (≈ 1.8 eV) in the case of the superlattice no peaks can be observed anymore. This cut-off energy is, although not really abrupt, rather well defined. Transitions can be followed up to this energy, where they start broadening, and at a slightly higher magnetic field they cannot be observed anymore (see Fig. 5). One further aspect of the parallel spectra can be observed, namely that the E-B relation tends to flatten near the cut-off energy (Fig. 4b). Qualitatively, most of these results are in concordance with the cyclotron resonance measurements reported by Duffield et al. [27] who observed cyclotron resonance (transitions between the lowest two Landau levels in their case) both in the perpendicular and in the parallel field configuration. They observed a slightly heavier mass in the parallel configuration, which corresponds to the smaller slope of the Landau levels in $B_\parallel$ in the present case.

A word should be said about the use of high magnetic fields in the experiments. As it will become clear from the next sections, high fields are in no fundamental way essential to observe the behaviour described above. Practically, however, it was necessary to use high fields to establish unambiguously that the transitions dis-

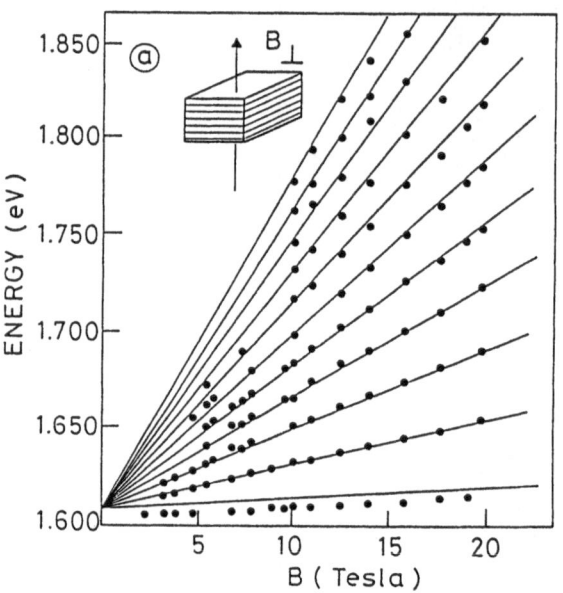

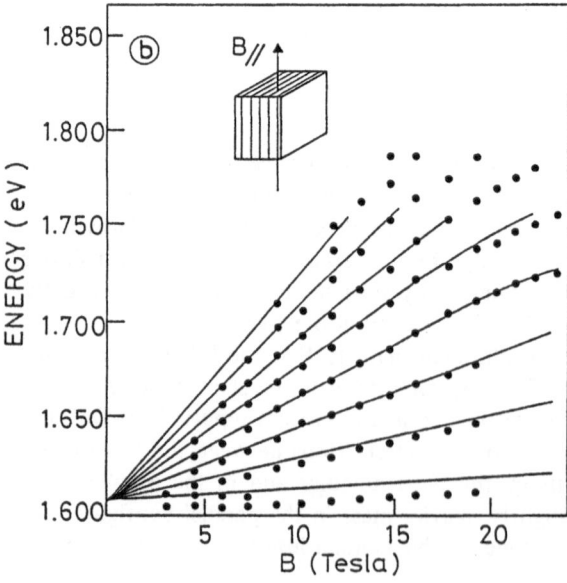

Fig. 4

Magnetic field dependence of the maxima in the excitation spectra of the superlattice in a perpendicular magnetic field (a) and in a parallel magnetic field (b). The drawn lines in (a) are a simple Landau-level fan with a reduced mass of $\mu = 0.069\ m_0$, in (b) the lines represent interband transitions calculated from the magnetic levels in a superlattice.

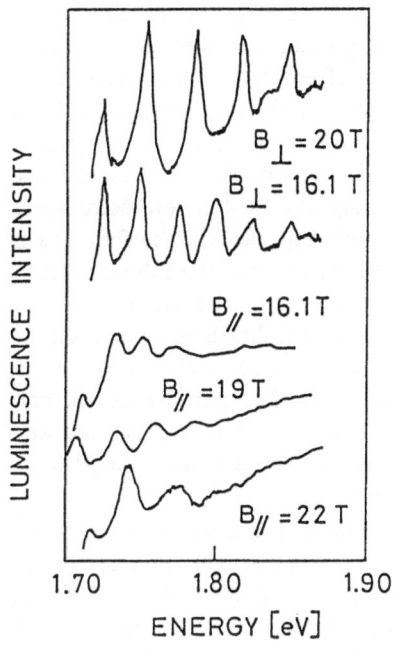

Fig. 5
Excitation spectra of a superlattice at different values of the magnetic field which is either parallel or perpendicular to the layers. Note in particular the disappearance of the peaks in the spectra beyond ≈ 1.8 eV.

appear beyond a certain energy. Interband Landau-level transitions in both quantum wells and superlattices tend to smear out at energies far from the bandedge. This effect is probably due to the fact that the more the energy increases from the bandgap the more initial and final states separated by the right energy become available. In order to observe well separated transitions at more than 200 meV from the ground-state luminescence, the Landau level separation has to be made as large as possible. The reason why so much importance is given to the disappearance of the Landau levels beyond a certain energy is that this cut-off energy will be shown to correspond to the edge of the superlattice miniband, and therefore the observation of this disappearance is a direct manifestation of the superlattice bandstructure. This point will be discussed more thoroughly in the next sections.

3 Quasi-Classical Quantization of the Electronic Motion in a Magnetic Field in a One-Dimensional Periodic Potential

The purpose of this section is to obtain the magnetic levels of the superlattice from its bandstructure. The basic idea behind this approach is to calculate the superlattice bandstructure without a magnetic field first and then to quantize this bandstructure using more or less standard text-book models. The theoretical models used in this analysis are more familiar in metal physics and were developed in particular to determine complex shaped Fermi surfaces from measurements of the de Haas-van Alphen-effect. In the next section a different approach will be used, namely, to solve numerically the Schrödinger equation of a particle in a one-dimensional periodic potential and in a magnetic field. This method is more used to describe the effect of a parallel or tilted magnetic field in heterostructures and quantum wells [29, 30]. It is useful to use both approaches. The first one is more intuitive and elegant, but since here $\hbar\omega_c$ is of the order of the subband width and π/ℓ of the order π/d (see Fig. 2) its validity is not a-priori warranted; i.e. it is not clear whether the magnetic field does not alter the bandstructure itself or whether it only quantizes an existing bandstructure. The second approach is more exact, but the results are less transparent and intuitive. The purpose of this and the next section is to describe these theoretical analysis and to explain the key features of the experimental results, i.e. why Landau levels are observed with $B_{\parallel}$ in superlattices and why they disappear at a certain energy.

The general problem of a particle in a periodic potential in a magnetic field, for cases where ℓ is much larger than the lattice constant, is well understood and quasi-classical quantization as formulated by Onsager is treated in many text books [31 ... 33]. For very high magnetic fields in a one-dimensional periodic potential an analysis based on this method has been given by, among others, Pippard [34] and Harper [35]. In the explicit context of superlattices Tsu and Janak [36] and also Ando [15] have studied related problems. A very thorough treatment for more general cases has been given by Zil'berman [37]. In the following more or less the standard text-book method will be followed with emphasis on the steps which may need further examination considering the extreme conditions here.

As was briefly mentioned before, the GaAs bandgap is entirely contained within that of GaAlAs and the difference of the bandgaps is distributed between the valence and the conduction band. For simple bands the electrons move therefore in a square-wave potential with eigenvalues:

$$\frac{\hbar^2}{2\,m_A^*}\,(k_A^2 + k_{\parallel}^2) \qquad\qquad \text{in material A,}$$

$$\frac{\hbar^2}{2\,m_B^*}\,(k_B^2 + k_{\parallel}^2) + V_b \qquad\qquad \text{in material B.}$$

(3)

Here m_A^* and m_B^* are the (generally) different masses in GaAs (A) and GaAlAs (B), k_A, k_B the z component, $k_{\parallel}$ the in-plane component of the wavevectors, and V_b the barrier height. The bandgap of GaAlAs is taken to be $1.25 \cdot x + E_{G\text{-}GaAs}$ and in the following a 75/25 rule for the distribution of this difference between the conduction- and the valence-band offset will be used. The effect of the band offsets on the quantitative results will be discussed in more detail in Chapt. 5. The eigen-

146

values in a superlattice can be obtained from the boundary conditions for the continuity of the wavefunctions and of the expectation value of the current operator at the interface [38, 39]. This latter criterium simplifies to the continuity of the first derivative of the wavefunction in cases where the masses in the two materials are equal. In that case of course the system reduces to the well-known Kronig-Penney-problem [40]. Taking plane waves for the wavefunctions in the layers and solving the boundary conditions leads to the following equation [38]:

$$\cos k_z d = \cos k_A d_A \cdot \cos k_B d_B - \frac{1}{2} \left[x + \frac{1}{x} \right] \sin k_A d_A \cdot \sin k_B d_B,$$

(4)

$$\text{with} \quad x = \frac{m_A k_A}{m_B k_B} .$$

Here k_z is the new superlattice wavevector, k_A and k_B are related to the energy through Eq. (3), and the superlattice bands are of course obtained for energies for which the right hand side of Eq. (4) is between -1 and $+1$. The superlattice sub-bands $E_z (k_z)$ in Fig. 2 are obtained this way. In fact for $V_b = 410$ meV, d = 5.04 nm, using the same masses in the barriers and in the wells, the figure corresponds to the conduction subbands in a GaAs/GaAlAs superlattice with 1.12 nm barriers of GaAlAs (x = 0.44) and 3.92 nm of GaAs wells, parameters which are a best fit of the experimentally studied sample.

The next step is to calculate the magnetic levels from the bandstructure. First it is assumed that the quasi-classical equation of motion of a particle having a dispersion relation $E (k)$ in a field B may be given by:

$$\frac{d}{dt} (\hbar \vec{k}) = -e \vec{v} \times \vec{B},$$

(5)

where $\vec{v}$ is the speed of the particle, e its charge and $e\vec{v} \times \vec{B}$ expresses the Lorentz force. Note that this relation is only proven to be valid when $\hbar \omega_c$ is much less than the width of the band [31, 32] and may in principle not be applicable in the present case. The crucial assumption in the derivation of Eq. (5) is that the magnetic field does not alter the bandstructure but only allows certain energies of the existing bandstructure, and for high fields this may not be true anymore. The velocity $\vec{v}$ is related to the dispersion relation $E (k)$ by:

$$\vec{v} = \frac{d\vec{r}}{dt} = \frac{1}{\hbar} \nabla_k E (k),$$

(6)

with $\nabla_k E (k)$ the gradient of a surface of constant energy with respect to k. It can be seen that since the "velocity" in k-space is perpendicular to $\vec{v}$ and to $\vec{B}$, Eq. (5), while $\vec{v}$ according to Eq. (6) is always normal to a constant energy surface the particle motion in k-space is along contours of equal energy. It depends of course on the energy and on the dispersion relation whether these contours are closed or open. By integrating Eq. (5) once with respect to time one obtains a relation between the motion in k and that in real space. I.e.

$$\vec{r} \times \frac{e}{\hbar} \vec{B} = \vec{k}. \tag{7}$$

Thus the motion in r-space is the same as that in k-space scaled by a factor $1/\ell^2$, the magnetic length squared, and rotated through $\pi/2$. Therefore closed or open contours in k-space can directly be translated to closed or open orbits in real space. Since no motion is involved in the direction of the field, the wavevector in this direction is irrelevant and is in the following neglected making the calculation two-dimensional.

If the orbits are closed, the quasi-classical quantization rule of Bohr-Sommerfeld can be used to obtain the stationary states (eigenvalues):

$$\oint \vec{p} \, d\vec{q} = \oint (\hbar \vec{k} - e \vec{A}) \, d\vec{r} = (N + \gamma) \, 2\pi \hbar \tag{8}$$

with N an integer and γ a phase factor at present undefined, $\vec{q}$ the position coordinate along the path and $\vec{p}$ the canonical momentum given by $(\hbar \vec{k} - e \vec{A})$, $\vec{A}$ being the vector potential defined by $\mathrm{rot}(\vec{A}) = \vec{B}$. (Strictly speaking the Bohr-Sommerfeld-Quantization is only valid for high N [33].) After some manipulation Eq. (8) can be transformed into:

$$\Phi = BS = (N + \gamma) \, 2\pi \hbar/e \tag{9}$$

with Φ the magnetic flux threading the projected real space orbit with an area S and this flux is, apart from a phase factor, an integral multiple of the fundamental flux quantum $\hbar/e$. The areas of allowed orbits in real space are therefore given by $(N + \gamma) \, 2\pi\ell^2$ and this formula can be transformed into an area of allowed contours in k-space using Eq. (7). Finally one obtains for the quantum condition of the areas in k-space:

$$A_k(E) = (N + \gamma)/\ell^2, \tag{10}$$

where $A_k(E)$ is the area in k-space of a surface of equal energy E. Eq. (10) is very useful because it gives a prescription of how the dispersion relation for the superlattice can be quantized in a magnetic field. The phase factor is as yet undefined, however, for simple bands with a parabolic dispersion relation it can easily be shown that taking $\gamma = 1/2$ results in Landau levels given by Eq. (2) which would also be obtained from the exact solution. In the case of the superlattice the dispersion in the z direction is not parabolic, and then the full bandstructure is given by:

$$E(k) = E_z(k_z) + \frac{\hbar k_y^2}{2m^*} \tag{11}$$

148

with $E(k)$ obtained from Eq. (4). Since the motion in the field (x) direction is neglected, k_x is taken to be zero. From Eq. (11) the area in k-space is calculated as a function of energy, and using Eq. (10) the energy levels for different N and B are calculated.

Fig. 6 shows several equal energy contours for different energies. The parameters used are those described in the beginning of this chapter. As one can see from Eq. (11), closed contours can only be obtained for energies less than the width of the subband,, namely, $E(k) < E_z(\pi/d)$. Therefore only for these energies quantized stationary states can be calculated using Eq. (10). The closed orbits in Fig. 6 correspond to the magnetic levels with $N = 0, 1, 2, 3$, and 4 for $B = 20$ T and the open orbits to arbitrary energies higher than the width of the subbands. The Landau levels obtained with this procedure from the closed orbits are shown in Fig. 7 and it can be seen that for energies below the subband edge (indicated by the dashed line in Fig. 7) almost normal equidistant linearly field dependent Landau

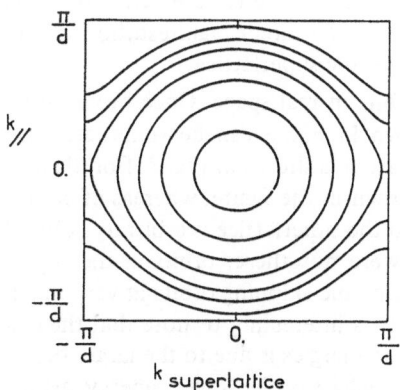

Fig. 6
Equal energy contours of Eq. (11) for the conduction (mini) band with the parameters of the experimental sample. The closed contours satisfy the quasi-classical quantization (Eq. (10)) of energy levels in a field of 20 T. The open contours correspond to energies of 20 and 50 meV, respectively, above the subband edge.

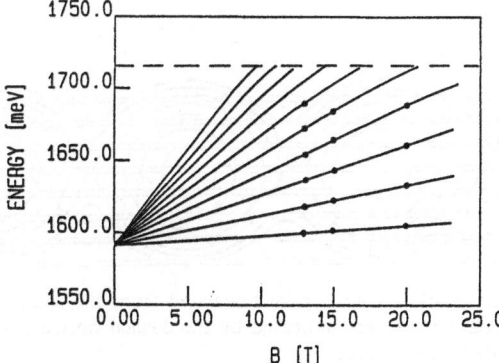

Fig. 7
Quasi-classical calculated magnetic levels of the conduction (mini) band of the experimental sample (drawn lines). The dashed horizontal line is the upper miniband edge above which no quantized energy levels can be calculated with Eq. (10). The dots are the energies of a few representative flat Landau levels calculated from the numerical solution Eq. (12).

149

levels are obtained. Close to the subband edge these Landau levels are somewhat flattened, and at higher energies no stationary states can be obtained anymore using Eq. (10). This flattening of the Landau levels can be seen to correspond to an elongation of the orbits in Fig. 6, as a consequence of the increasing non-parabolicity of the $E-k_z$ dispersion relation at π/d (Fig. 2). Zil'berman [37] has extended the present argument to energies above the subbands and shown that then Landau levels are broadened into bands with gaps at certain high symmetry points; an effect also shown by Harper [35].

Qualitatively, the results of this analysis provide a simple explanation for the experimental findings. At energies within the subband width closed orbits can be constructed and these are observable as sharp absorption peaks. The field dependence of these calculated stationary states is simply linear and successive states are equidistant. Open orbits lead to broadened levels as shown by Zil'berman [37] and Harper [35] and these become unobservable. Higher energy levels which approach the subband edge, have a shallower magnetic field dependence and finally become open orbits. The theory explains therefore the experimental findings and establishes that magnetic quantization in a superlattice is observed. From the preceding analysis it is clear that this observation is a direct manifestation of the quasi three-dimensional, artificial bandstructure of the superlattice.

It is interesting to have a closer look at the orbits in real space. Once more, using Eq. (7), the contours of equal energy as shown in Fig. 6 can be translated into orbits in real space. The result of this transformation is shown in Fig. 8. For illustration purposes the position of the barriers is shown in the figure, whereas of course the position of the orbit center with respect to the superlattice is arbitrary. Closed orbits tunnel through the barriers as if they were not there, orbits in the super-lattice bandgap are reflected at the walls and resemble skipping orbits in very much the same way as intuitively expected (Fig. 1). It is interesting to note that the disappearance of a pronounced absorption at higher energies is due to the larger orbits which sample a larger area of the superlattice, whereas the lowest energy orbits,

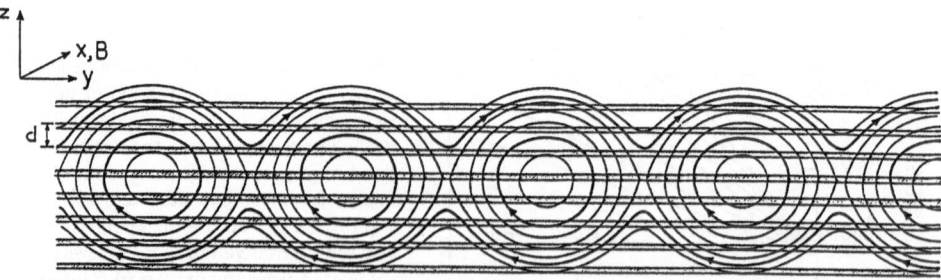

Fig. 8 Quasi-classical trajectories of electron motion in real space calculated from the energy contours in Fig. 6 using Eq. (7). The potential barriers are indicated by the dashed horizontal lines. For clarity one open and one closed orbit are indicated.

which almost "fit in the well", behave as if the barriers were not there. This observation confirms what has been said earlier, namely, that the disappearance of the Landau levels at the subband edge is the critical point of the experimental observations. It is clear now that the periodicity in the superlattice is responsible for this disappearance. It is also obvious from Fig. 8 that this experiment indeed measures "vertical transport" through the superlattice, which was one of its purposes.

Although a satisfactory qualitative explanation of the experimental results has been given, it should be borne in mind that this has been done using theoretical tools which are, strictly speaking, only valid for high quantum numbers, for energy ranges where the cyclotron energy is much less than the width of the bands involved, and for magnetic lengths longer than the periodicity of the system, etc., [31 ... 33]; these conditions are not full-filled here. Therefore in the next section a different, more direct, approach will be adopted, namely, to solve directly the Schrödinger equation for this problem. It will be seen that this procedure justifies and extends the results derived in this section.

4 Quantum Mechanical Calculation of Magnetic Levels in a Superlattice

The Hamiltonian of a particle moving in a potential $V(z)$ and in a vector potential A is given by:

$$(\vec{p} - e\vec{A})^2 + V(z) = H.$$

In the asymmetric gauge a magnetic field in the x direction can be described by the vector potential $\vec{A} = (0, -zB, 0)$. Making use of the translational symmetry in the x and y direction, the wavefunction can be chosen as $\exp(ik_x x) \cdot \exp(ik_y y) \cdot \varphi(z)$. Substituting and making a further simplification $z = z' + \hbar k_y/eB$ the following one-dimensional equation results for the motion in the z-direction.

$$\left[-\frac{\hbar^2}{2m^*} \frac{d^2}{dz'^2} + \frac{e^2 B^2 z'^2}{2m^*} + V(z) \right] \varphi(z') = E\varphi(z'). \tag{12}$$

Note that in Eq. (12) $V(z)$ is shifted with respect to the new coordinates z'. For $V(z) = 0$ Eq. (12) is the normal Schrödinger equation for a charged particle in a magnetic field with the eigenvalues $E_N = \hbar\omega_c \cdot (N + 1/2)$. The total wavefunctions are harmonic oscillator eigenfunctions centered around $\hbar k_y/eB$ in the z-direction and plane waves with momentum $\hbar k_y$ in the y direction. This means that for a certain value of $\hbar k_y/eB$ (proportional to the momentum in the y-direction) they are displaced with respect to the origin of the coordinate system in the z-direction. Due to translational symmetry in the z-direction for $V(z) = 0$ the choice of the origin is arbitrary, and all eigenvalues for different k_y are degenerate; the usual Landau-level degeneracy. Formally, a k_y dispersion relation of the Landau levels can be imagined as a set of flat lines distant $\hbar\omega_c$ from each other with the same energy for all k_y. It is customary to associate $\hbar k_y/eB$ with the center of the cyclotron orbit, although strictly speaking this is misleading because the eigenfunctions in the asymmetric gauge adopted here, do not describe circular orbits. Nevertheless it will be seen later that the notion of the orbit center, as a loose description of the location of the center of the wavefunction in the z-direction has a precise meaning once $V(z) \neq 0$ breaks the translational invariance along z. In fact, if $V(z) \neq 0$, no Landau-level degeneracy exists anymore a-priori because in general the eigenvalues will depend on where the cyclotron orbit is with respect to the potential or on $\hbar k_y/eB$. Consequently, a non-flat dispersion relation in the y direction develops. As a simple example one can imagine a potential which is very weak and very slowly (on the scale of the magnetic length) varying: in this case the eigenvalues of Eq. (12) can be obtained by treating $V(z)$ as a perturbation. Then, the Landau levels are given by $(N + 1/2)\hbar\omega_c + \langle V(z) \rangle$ where $\langle V(z) \rangle$ is the expectation value of the potential. As z and k_y are interrelated the Landau levels will develop a k_y dispersion, and in this particular case a set of equidistant energy levels spaced by $\hbar\omega_c$ added to the potential variation is found. Therefore instead of degenerate Landau levels a Landau band develops. For strongly varying $V(z)$ the perturbation approach cannot be used anymore, and the effect of the potential is a mixing between different Landau levels, and only a full calculation can give the energy-level structure.

To clarify this point Fig. 9 shows the solutions of Eq. (12) in the vicinity of a potential step of height V_b. The results can be understood in the following way: far away from the step the potential is constant and normal equidistant degenerate Landau levels are obtained; on the left side E_N and on the right side $E_N + \langle V_b \rangle$. In

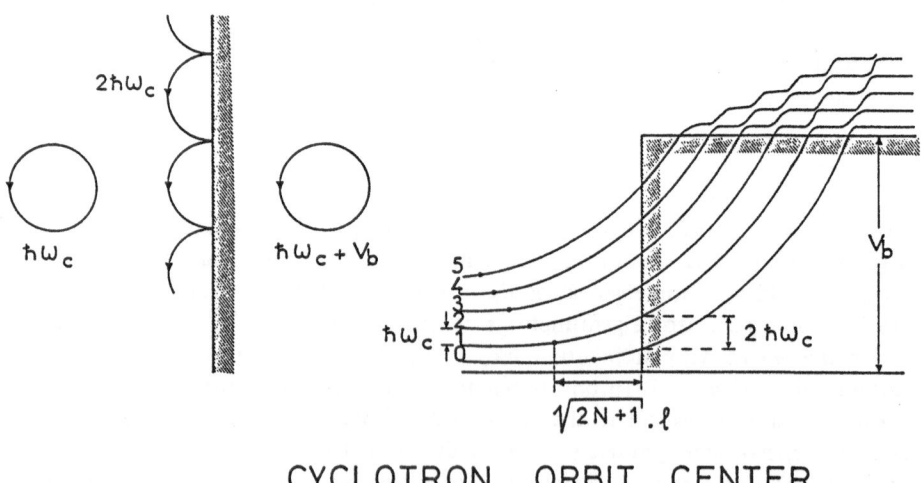

CYCLOTRON ORBIT CENTER

Fig. 9 Quasi-classical electron motion in real space at a potential step (left) and the dependence of the energy levels on the cyclotron orbit center position. Note that on the left side of the barrier Landau level N develops a dispersion for orbit center positions $(2N + 1)^{1/2} \varrho$, from the barrier.

the neighbourhood of the interface the Landau levels develop a dispersion and are not equidistant anymore. It is instructive to see that it is indeed at the length scale of the cyclotron radius that deviations from unperturbed behaviour occur. The Larmor radius of a normal cyclotron orbit with Landau-level quantum number N being given by Eq. (1) it can be seen from the figure that indeed each Landau level N develops a dispersion at a position $(2N + 1)^{1/2} \cdot \ell$ from the barrier. This observation corresponds to the intuitive classical image, namely that the energy levels will change "when the cyclotron orbit touches the barrier". One can carry this picture even further as is shown in the left part of the figure. As long as the cyclotron orbit radius is less than the distance from the center coordinate to the barrier, normal circular orbits can be expected. In the particular case that the center coordinate coincides with the barrier, in one cyclotron period ω_c the electron is reflected twice to the barrier, and for that center coordinate the spacing between Landau levels is $2 \hbar \omega_c$. This example of magnetic levels near a potential step has been studied in the literature [40, 41] quite extensively, and in metallic samples transitions between these so-called magnetic surface states have been observed [41]. One final aspect of Fig. 9 should be mentioned, which has no simple classical analog, namely that one sees that the flat Landau levels "at the top of the barriers" to the right, tend to intersect the continuation of the surface states which come from the left. Where these levels come close in energy they interact, leading to a small gap between the two sets of Landau levels, i.e., the system behaves as a superposition of two types of Landau levels which interact with each other only at their intersections.

In Figs. 10 and 11 the evolution of the Landau levels when going towards a superlattice in two different ways are shown. In Fig. 10 the well width is kept constant, and the barrier width is gradually reduced while in Fig. 11 the barrier and the well width are kept the same but the number of wells is increased. The superlattice bandwidth and the position of the quantized levels without magnetic field, corresponding to the cases shown in Fig. 10 and 11, respectively, are given in Fig. 12a and b. Going from the single interface to the single well (Fig. 10 top left and right) one observes that the size quantization shifts the lowest energy upwards and that the orbit center dispersion relation becomes parabolic-like. The origin of this parabolic dispersion relation can be most easily understood when the confinement energy (the shift of the subband edge) is much larger than the cyclotron energy. In this case the magnetic field can be treated as a perturbation. Following Ando [43] and Beinvogl et al. [29] the eigenvalues of Eq. (12) in first order perturbation are given by:

$$E = E_n + \frac{e^2 B^2}{2 m^*} (\langle z^2 \rangle_n - \langle z \rangle_n^2) + \frac{1}{2 m^*} (\hbar k_y + e B \langle z \rangle_n)^2, \tag{13}$$

where $\langle z \rangle_n$ and $\langle z^2 \rangle_n$ are the expectation values of z and z^2 for the unperturbed wavefunctions of subband n. For a symmetric well these wavefunctions are symmetric or antisymmetric, therefore $\langle z \rangle_n$ is always zero. The first term in

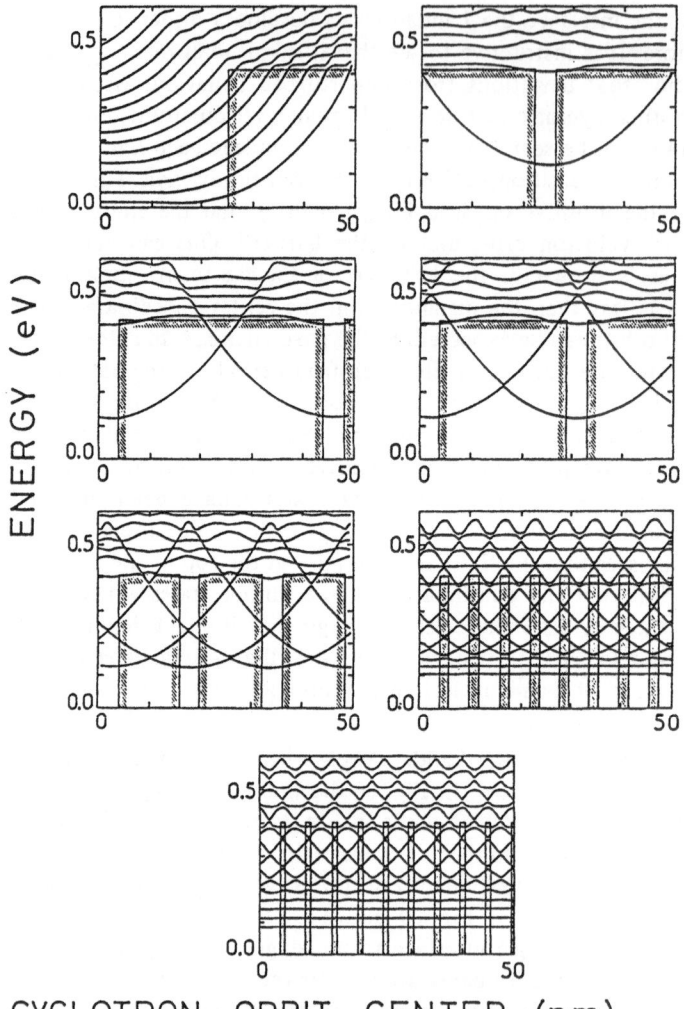

ENERGY (eV)

CYCLOTRON ORBIT CENTER (nm)

Fig. 10 The magnetic energy levels at a field of 20 T, of a potential step of 0.41 eV (top left), of a single well of 3.92 nm between barriers (top right), and of various superlattices with the same well width and barrier height but with decreasing barrier width, as indicated in the figure. The lowest figure corresponds to the experimental superlattice, and the level structure can directly be compared with the quasi-classical contour and trajectories in Figs. 6 and 8.

154

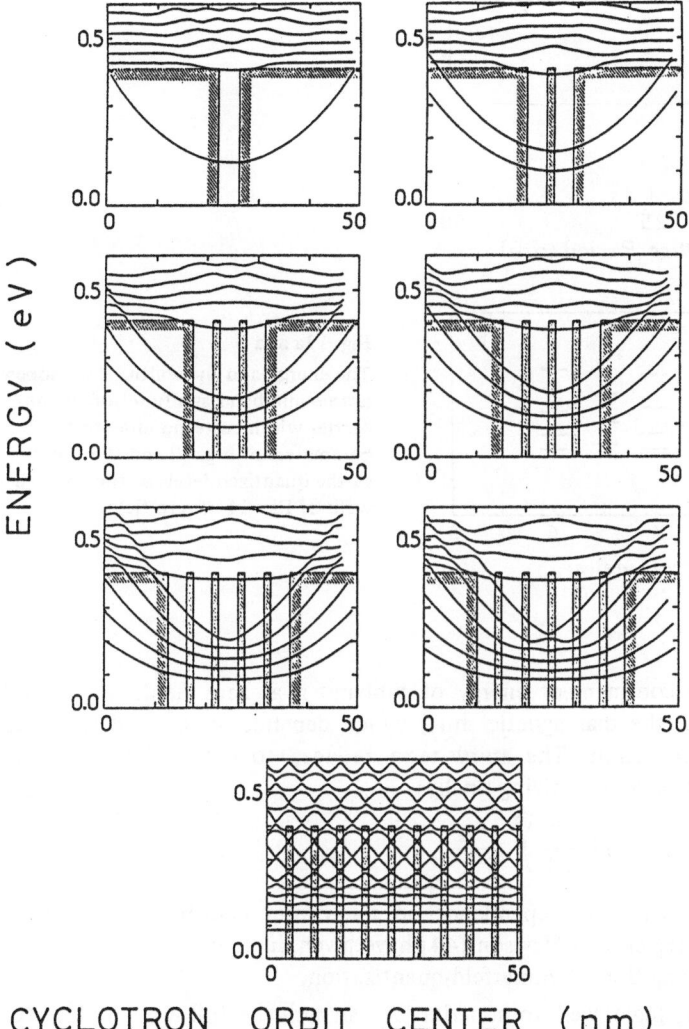

ENERGY (eV)

CYCLOTRON ORBIT CENTER (nm)

Fig. 11 The magnetic levels of resp. one, two, up to six (coupled) quantum wells each 3.92 nm wide with 0.41 eV barrier height and 1.12 nm barrier width at a field of 20 T and a superlattice with the same parameters. Note correspondence between the number of wells and the number of levels, and in particular that already for six wells the lowest levels in the centre are very similar to those in the superlattice.

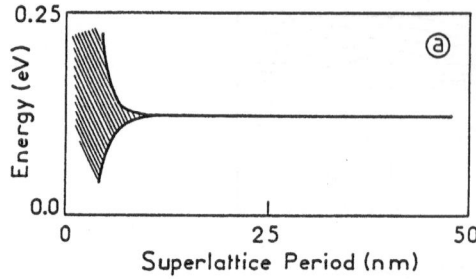

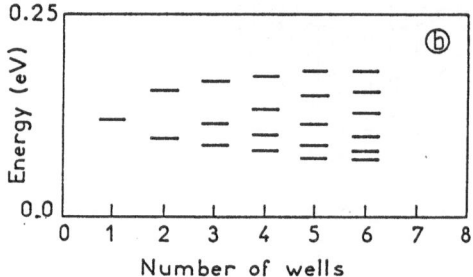

Fig. 12a and b

The energy and the width of the super-lattice miniband as a function of the barrier width corresponding to the parameters of Fig. 10 and the energy of the quantized levels of the coupled wells of Fig. 11 at zero field.

Eq. (13) is just the confinement energy of subband n at zero field, the second term is a rigid shift (the diamagnetic shift) which depends on B^2 and $\langle z^2 \rangle$, the spread of the wavefunction. The third term reduces to $(\hbar k_y)^2/2\,m^*$, the k_y dispersion relation at zero field, giving finally:

$$E = E_n + \frac{e^2 B^2}{2\,m^*}\langle z^2\rangle_n + \left(\frac{\hbar k_y}{eB}\right)^2 \frac{e^2 B^2}{2\,m^*}, \tag{14}$$

which explains the parabolic dependence on the center coordinate $\hbar k_y/eB$ in Fig. 10. Recently, Bergren and Newson [44] have given an analytic expression for more general cases using Bohr-Sommerfeld-quantization.

The evolution from single well to superlattice can now be followed in Fig. 10. As long as the barriers are thick and the wells do not interact with each other, every well has its own set of magnetic levels and the levels from different wells are just superposed, without anticrossing at the intersections. As soon as they start to interact, which without magnetic field is indicated by a development of a sub-band in k_z which has a finite width (see Fig. 12a), the parabolas from different wells show an anticrossing at their intersections. However, this process occurs more strongly at energies where at zero field the subband exists (note the analogy between Fig. 10 and 12a) and is much weaker in the superlattice minigaps. For sufficiently strong interacting wells, this anticrossing is so important that the Landau levels become flat (dispersionless) within the miniband width. As can be seen from Eq. (14) the parabolas become flatter at lower and steeper at higher

156

magnetic fields. Therefore at lower fields more parabolas from more distant wells interact within the subband width increasing the number of flat Landau levels. With increasing magnetic field parabolas interact at a higher energy, and at sufficiently high magnetic fields this energy is outside the width of the miniband. Therefore the energy-level separation increases with field and those levels which move out of the subband width become dispersive. Finally, an interesting observation in this series of pictures is that the results can be summarized in the following way: without potential variation degenerate flat Landau levels are found; when the potential varies on the scale of the CR orbit size the levels become very dispersive and finally, when the potential varies very rapidly and periodically, once again flat Landau levels are found. Somewhat intuitively one can say that since in the absence of a potential variation Landau levels are flat due to translation symmetry, in the case of a very rapidly periodically varying potential, in some way translation symmetry is recovered again.

A similar description can be given from Fig. 11 where the evolution from one to more quantum wells is shown. Each subband at zero magnetic field (see Fig. 11 and 12b) experiences a diamagnetic shift and has a parabolic dispersion relation according to Eq. (14). However, since the barriers are thin each of these parabolas interacts strongly with the others and the evolution to flat Landau levels is in this case brought about by the fact that a sufficient number of wells is needed to obtain flat bands. From Figs. 11 and 12b it can be seen that there is a one to one correspondence between the number of Landau levels in the wells and the number of interacting wells. Also it can be seen that a finite number of wells, as is the case for a real sample, leads to a broadening of the higher energy levels, while the lowest ones are less affected. Relating these results to the experiments it is clear that the number of observed peaks is a lower bound of the number of interacting wells. Furthermore since a finite number of wells is equivalent to the absence of long-range order, it can be inferred that a poor periodicity in a sample will lead to a broadening of the higher Landau levels first, whereas the lower ones will be less sensitive to sample quality. The observation of many peaks in the spectra up to the subband edge is therefore a demonstration of sample quality and homogeneity.

There is a close analogy between the results reported here and those of the previous section. Fig. 10 bottom and Fig. 8 are calculated for the same parameters and can be compared directly realizing that the smallest orbit corresponds to the lowest Landau level, the next larger orbit to the second level, etc. Fig. 10 demonstrates that the Landau levels at the subband edge result from the interaction between wells which are far apart, whereas for the lowest Landau levels only neighbouring wells are important. Analoguously, the quasi-classical result shows that the closed orbits at the upper subband edge extend over many periods while the low energy orbits are much smaller. It is interesting to compare the theoretical results obtained in this section with those of the preceding paragraph more quantitatively. For this purpose for a few fields the energy position of the flat Landau levels is compared with the quasi-classically quantized subband dispersion in Fig. 7, and it can be seen that in the lower part of the subband width the agreement is perfect, which a posteriori supports the results of Chapt. 3. However, close to the subband edge the comparison is more difficult because the exact solutions show already a substantial broadening, even at energies within the width of the subband, and it is difficult to decide which energy must be taken. In this connection, comparing Fig. 10 and Fig. 8, an analogy between the behaviour of the higher Landau levels in the two different descriptions can be observed. Quasi-classically the orbits close to the subband edge (like, e.g., the fifth orbit in Fig. 8) are elongated and similarly Landau levels in the exact calculation (the fifth Landau level in Fig. 10 bottom) develop a dispersion in the orbit center position.

It is important to note that the discussion given so far in this chapter is only true for the lowest subband. As it can be seen from Fig. 2 a second subband starts slightly below the top of the barrier and extends almost to energies twice the barrier height. Indeed, a slightly more regular Landau-level structure with partially flat dispersion relations can be observed in Fig. 10 at these energies, but clearly no simple behaviour as for the lowest subband is found. Similarly, in cases where two subbands are involved, the quasi-classical prescription of the preceding section cannot be used anymore because the contours of equal energy involve two subbands, and it is not clear which surface has to be calculated with the quantization condition of Eq. (10).

At this stage it should be mentioned that this chapter contains basically only a description of numerical results which are somewhat made plausible but not explained in detail. Concretely, for instance, it has not been made clear what exactly determines the gap between the anticrossing Landau levels from different wells. There is still ample room for further study on the subject both experimentally and theoretically, and it is hoped that the results reported here will contribute to an increase of interest in the subject.

5 Discussion

In the two preceding sections the magnetic levels in a superlattice are analyzed, and it has been shown that the experimental results reported in Sect. 2 can qualitatively be understood. In this chapter the data are analyzed more quantitatively and critically, and in the process of this analysis possible further developments will implicitly become clear.

In the experiment transitions are measured between excitonic states in a magnetic field and not directly between free electron and hole Landau levels. This excitonic nature can be recognized by the fact that the lowest energy transition (which corresponds to the exciton ground state) extrapolates to zero field to a slightly lower energy than the higher energy transitions (Fig. 4a and b). The conventional [45] interpretation is that in principle all transitions are exciton-like but that the higher energy transitions are weakly bound and behave basically as free electron-hole transitions, whereas the ground state, which is much deeper in energy, is clearly distinct from the others. The difference in energy between the extrapolation of the higher energy transitions and the ground state can then be used as an estimate of the exciton binding energy. The exciton binding energy in bulk GaAs is 4.2 meV. In quantum wells the binding energy is enhanced with respect to the bulk [45 ... 47] because of the confinement of the exciton in the well. Typically for thin wells (≈ 5 nm) binding energies of 12 ... 13 meV [45, 46, 48] are measured. From the extrapolation to zero field in Fig. 4a and b, a binding energy of 6 meV can be obtained. This value is substantially lower than that of quantum wells of similar thicknesses, and this observation in itself is already an indication of the quasi 3D character of the superlattice However, since the binding energy is not too high, the excitonic effect will be neglected in the rest of this chapter.

The two preceding theoretical sections have dealt with energy levels in super-lattices for simple parabolic bands. However, the valence subbands in quantum wells and superlattices are anything but simple [49 ... 53]; a fact that is also experimentally well established [45, 51]. At the present state of knowledge no directly applicable theory exists which calculates the hole Landau levels in a super-lattice in parallel field, although recently Kriechbaum [54] has published theoretical results which start to address this problem. Due to the lack of a better theory the hole will therefore in the following be described by one simple parabolic band. Fortunately the holes, although complicated, have on average effective masses which are much heavier than that of the electrons and therefore the main contribution to the experimental results comes from the electrons.

Despite these two additional difficulties, in comparison with cyclotron resonance, the technique of interband absorption was preferred above that of intraband absorption because with interband absorption magnetic levels over a wide energy range can be observed, and not only those in the vicinity of the Fermi energy as in the case of cyclotron resonance. In particular the disappearance of Landau levels at the subband edge which is the most spectacular manifestation of the superlattice bandstructure can be observed this way. However, the price one pays is that it becomes much more complex to analyse the data.

Bearing all these reservations in mind, the analysis proceeds in the following way. Within a simple scheme of parabolic hole and electron bands (with masses m_e^* and m_h^*), the perpendicular magnetic field Landau levels are just a simple Landau level ladder as in Eq. (2) but with a reduced mass μ.

$$\mu = \frac{m_e^* \, m_h^*}{m_e^* + m_h^*}. \tag{15}$$

Therefore the transitions in Fig. 4a can be calculated using μ as a fitting parameter. The drawn lines in Fig. 4a show the best fit of Eq. (2) to the data with a reduced mass of 0.069 m_0. It can be seen that this fit is not bad except for the lowest energy transition which is shifted to lower energies due to excitonic effects, neglected in this scheme. Generally, a reduced mass is always lighter than the two individual masses, but here the reduced mass is already heavier than the conduction band edge mass in GaAs (0.0665 m_0), which is due to non-parabolicity. Transitions are observed between 0.1 and 0.3 eV above the band edge of GaAs and a correspondingly substantial mass enhancement may be expected. Using standard non-parabolicity formula, a mass in the middle of the electronic subband is calculated as 0.078 m_0, and using Eq. (15) the "hole-mass" is then found to be 0.61 m_0. However, considering the oversimplification of the hole subbands this number should be seen as a fit parameter. In any case the "hole mass" is much heavier than the electron mass and its influence on the results is secondary.

The preceding theoretical sections have demonstrated that closed orbits, respectively flat Landau levels can be found in the energy region of the superlattice miniband. Experimentally, interband transitions are observed between 1.605 and ≈ 1.80 eV and, it is therefore plausible to identify the edges of the energy range with the superlattice bandgap at $k_z = 0$ and at $k_z = \pi/d$, in which case this range corresponds to the combined width of the superlattice electron and hole minibands. It should be pointed out that the experimental information is rather abundant because simultaneously the effective masses, the energy difference between the hole and electron minibands at $k_z = 0$ and at $k_z = \pi/d$ are measured, and therefore a theoretical fit has to be consistent with all these experimental facts. Furthermore the type of samples showing this superlattice effects have necessarily very thin layers, and small variations in the layer thicknesses lead to substantial variations in the energies. Finally, since the superlattice bandwidth is related to the coupling between successive wells it depends on the barrier heights (which are determined by the band offsets) and the masses in the GaAlAs layers, and all these quantities are not known very precisely. To exemplify this point in Fig. 13 the calculated bandwidth (using Eq. (4)) is shown for the electrons for the base of 1.12 nm GaAlAs barriers (x = 0.44) and 4.48, 3.92, and 3.38 nm GaAs well. In the calculation the 60/40 rule (1 and 2 in Fig. 13) and the 85/15 rule for the band offset are used (3 and 4). Furthermore the same masses for GaAs and GaAlAs are used in 1 and 3 and a GaAlAs mass which is higher proportional to the increase in the bandgap (30 %) in 2 and 4. This latter choice was based on the analysis by Duffield et al. [31] who have analysed their parallel field cyclotron resonance this way. In all cases with the sample thicknesses derived from the growth conditions (4.4 nm GaAs and 1.12 nm GaAlAs) the experimental results do not agree with the calculations, since both the subband width comes out too small and the band gap at $k_z = 0$ too low. (A similar calculation is of course performed for the hole to obtain the gaps.) For a layer thickness of 3.92 nm (one lattice constant less) the band edge at $k_z = 0$ is at the right energy, but the subband width is still too small (145 instead of 190 meV) for even thinner layers of GaAs the subband width agrees, but the transition at $k_z = 0$ is too high in energy. Within the simple scheme adopted here no really detailed agreement can be found with none of the theoretical models. However, a reasonably accurate description can be obtained using 75/25 for the band offset, the same masses in the GaAs and the GaAlAs, and 3.92, 1.12 nm as well and barrier width. These values are only one lattice constant less than that determined from the growth parameters

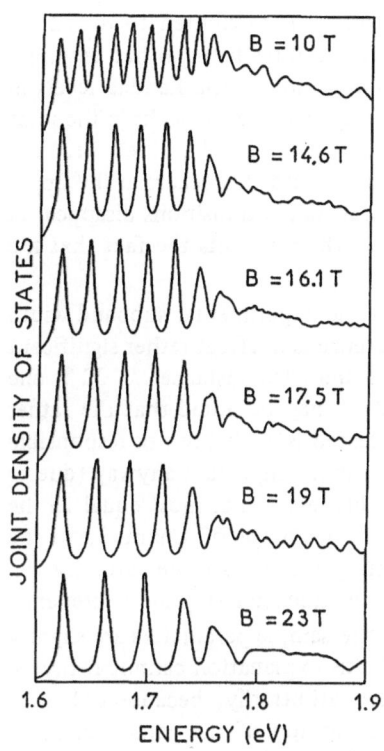

JOINT DENSITY OF STATES

B = 10 T

B = 14.6 T

B = 16.1 T

B = 17.5 T

B = 19 T

B = 23 T

1.6   1.7   1.8   1.9

ENERGY (eV)

Fig. 13

Joint density of states calculated from the magnetic levels in the experimentally studied superlattice (Fig. 10 and 11 bottom). The magnetic field dependence of the maxima are shown as the drawn lines in Fig. 4b.

The Landau levels in Fig. 10 bottom for 20 T are calculated for these parameters, and a similar calculation was performed for the holes. To compare with the experiment it is assumed that transitions take place only between the first hole and the first electron level, the second hole to the second electron, etc., whereby the cyclotron orbit center will be conserved (vertical transitions). Furthermore the matrix elements are assumed to be the same for all possible transitions. This set of assumptions are in fact the selection rules for normal Landau levels described by harmonic oscillator wavefunctions, which need not a priori to be valid in the present case. In this framework the spectra are entirely determined by the joint density of states which can be obtained from the calculated Landau levels. The result of such a calculation for a few different values of the magnetic field is shown in Fig. 14. Qualitatively, the agreement with the experimental observations is striking. More quantitatively, the magnetic field dependence of the peaks in the joint density of states is compared with the observed maxima in the excitation spectra in Fig. 4b (drawn lines). It can be seen that even quantitatively the agreement is rather good, since the correct slopes of all the transitions is reproduced by the calculations. Both experimentally and theoretically the slopes at $B_\parallel$ are somewhat shallower on average than those at $B_\perp$ even when the same mass is used for the barriers and the wells.

161

Therefore the slightly higher "parallel effective mass" is attributed to the super-lattice bandstructure. That at higher energy no peaks can be distinguished anymore in the calculated spectra of course reflects the broadening of the Landau levels in the superlattice minigap. Roughly speaking the disappearance of peaks in the joint density of states takes place when the broadening of the Landau levels becomes equal to the average energy separation between them, and this occurs at the upper edge of the superlattice miniband. However, the calculated transitions disappear at too low energies compared with the experiment, which reflects the fact that the calculated width is too small.

The main remaining discrepancy is therefore that the experimental subband width is too large compared with the theory. The difference is in effect rather significant because as can be seen from Fig. 14 it implies that, for instance, at 22 T one Landau level more is observed than is calculated. It may be noted that the rather sloppy way in which the hole states have been treated is probably not responsible for this fact, because, since the electron states are more important anyway (due to their much lighter mass), the number of observable levels is at most equal to the number of flat electron Landau levels, or in the language of Chapt. 3, to the number of closed orbits for the electrons. Treating the holes more carefully can therefore only reduce the number of calculated transitions marking the discrepancy worse. An alternative explanation might be that the sample parameters are slightly different from those used in the calculation and this explanation cannot really be excluded. However, thicknesses cannot be varied aribitrarily, because only full lattice periods can be added, and within the exploration of the parameter space for different thicknesses, for mass barrier heights discussed before no satisfactorily agreement could be found. Many more fundamental arguments may be advanced to explain this discrepancy, however, as will be shown they usually tend to make the subband width less rather than larger. For instance, it might be argued that the GaAs CB non-parabolicity has been treated too simply. However, magneto optical

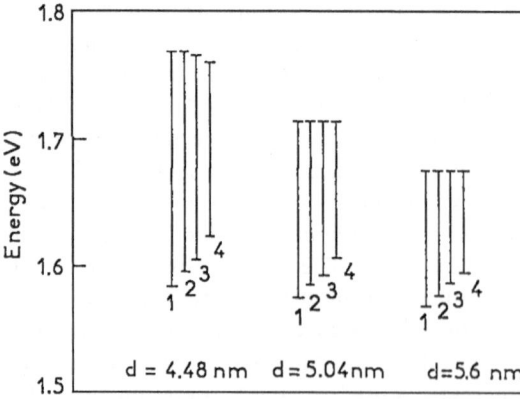

Fig. 14

The position and the width of the electron subband for a GaAs/GaAlAs superlattice with 1.12 nm barrier width and periodicity indicated in the figure. 1 and 2 are calculated using the 60/40 and 3 and 4 using the 85/15 rule for the distributing of the bandgap difference between the conduction and the valence band. 1 and 3 are calculated using the same mass in the barriers and in the wells and 2 and 4 with a 30 % higher mass in the barriers.

experiments on quantum wells [53, 55], which equally measure effective masses at energies high in the band, show that the two band model used here underestimates the non-parabolicity, and a heavier mass for GaAs would lead to an even smaller subband. Furthermore one can consider the effect of thickness variations. However, as it was argued before, the lack of long-range order in the superlattice affects the higher Landau levels first, because they are due to the interaction of wells which are far apart, contrary to experiments. For the same reason the finite number of layers of the sample (40 in this case) cannot be responsible for the discrepancy as was demonstrated in the preceding chapter. In addition, it should be noted that the Landau-level peaks disappear at the same energy independent of the magnetic field. If the finite size of the sample was responsible for the disappearance the cut-off energy should be field dependent because at lower fields the orbits are larger and they would be affected by the sample dimensions at lower energy already. In this context it is interesting to come back on the remark at the end of Chapt. 4 that there is a one to one correspondence between the number of Landau levels and the number of interacting wells. Concretely, in this experiment typically eight well resolved peaks are observed which therefore means that at least eight wells are interacting in coherent way. In the theory of the de Haas-van Alphen-effect a well-known phenomenon is magnetic breakdown. This effect is that electrons instead of following open orbit in k-space, for instance the outer orbit in Fig. 6, make a jump from some $k_{\parallel}$ to $-k_{\parallel}$ at $k_z = \pm \pi/d$ in order to follow a closed path. This effect would indeed explain the observations, namely that at energies slightly beyond the subband edge still adsorption peaks are observed. In the quasi-classical description this phenomenon is well established, although the equivalent phenomenon in the more exact description of Chapt. 4 is not directly apparent. This question too is best left for further theoretical and experimental studies.

There could be eventually a very interesting although also very speculative explanation for the results. The entire theoretical analysis of this paper is based on the validity of the envelope function approximation (effective mass theory) in this case. Concretely, a layer of two lattice constants thickness of GaAlAs (x = 0.44) is treated as some bulk material A entirely described by bulk properties, which is matched to a layer B of GaAs of seven lattice constants, equally treated as a bulk layer. First of all it may be asked whether such a very thin layer of GaAlAs can be treated in the virtual crystal approximation. Secondly, the basis of effective mass theory is the assumption that the dynamical properties of a bulk material can be described by envelope wavefunctions which are assumed to be slowly varying on the scale of the lattice periodicity, and this condition is not fullfilled here. Both theoretically and experimentally the subject alloy versus superlattice behaviour is still under discussion [56 ... 58]. For the sake of illustration in Fig. 15 schematically a superlattice with thicknesses and Al content comparable to the actual sample is drawn. Each dot represents a unit cell of GaAs or AlAs, and in the GaAlAs layers 40 % of the cells are AlAs. Looking at this figure the question

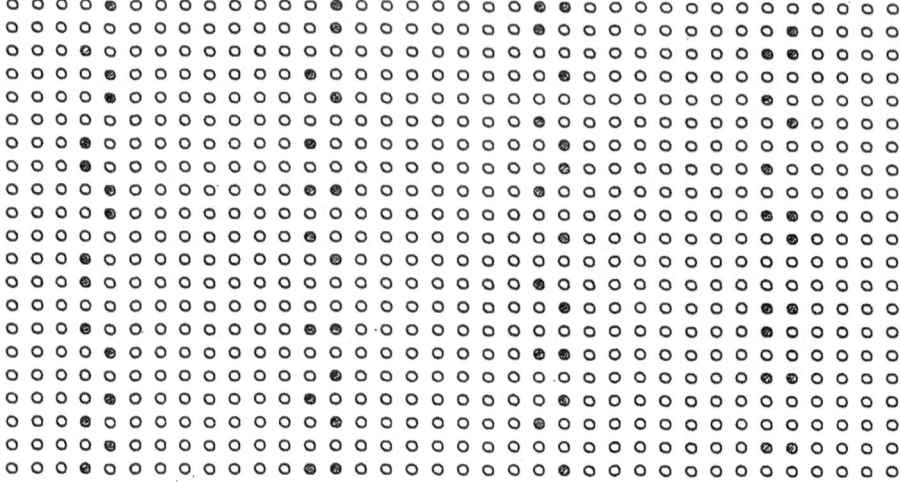

Fig. 15 Schematic representation of a superlattice with two lattice parameters of GaAlAs ($x = 0.4$) and 7 of GaAs. Each open circle represents one unit cell of GaAs and each dashed circle one of AlAs.

whether such a structure behaves like a superlattice or like an GaAlAs alloy with some average Al content (in this case 0.088) comes to the mind. Concretely, for the present sample it can naively be calculated that such an alloy would show a gap of ≈ 1.61 eV, have a mass which would be some 10 % higher than GaAs, would have an exciton binding energy of about 5 meV, would show Landau levels both with perpendicular and parallel magnetic field, in fact would behave almost exactly like what is seen experimentally. The important qualitative difference between the alloy and the superlattice is of course that a superlattice has minigaps due to the superperiodicity and that an alloy has not. Once more the observation of the disappearing Landau levels rules out such an explanation. Nevertheless it is very well possible that the theoretical description of a system with such thin layers, using the envelope-function approximation, is not accurate anymore. The residual discrepancy between experiment and theory might eventually be explained in such a way.

It was mentioned in the introduction that the application of a magnetic field parallel to the layers can be seen as a technique to measure "vertical transport", and the preceding sections can be seen as an illustration of this statement. To be more concrete, it is possible to estimate a relaxation time or a mobility for transport through the barriers from the data in a simple manner. The linewidth of the peaks is typically 5 meV, corresponding to a scattering time of 10^{-13} s which leads to a "mobility for vertical transport" of roughly 5000 cm/Vs. Furthermore it is interesting to note that the width of the peaks is only slightly wider in the $B_{\parallel}$ case compared to the $B_{\perp}$, showing that the scattering time is almost isotropic, and

164

this despite the fact that carriers cross several barriers in one direction. In fact, the mobility obtained is high compared to that found in dc transport measurements through the layers as reported by Palmier [27]. This is probably due to the fact that in this case no intentional doping is needed to study transport through the barriers.

Acknowledgement

It is a pleasure to acknowledge the pleasant collaboration with G. Belle during the experimental part of this work. Furthermore I would like to thank M. Altarelli for many very illuminating discussions on the theory of the subject. Finally I wish to express my gratitude to G. Weimann for the growth of the excellent superlattice samples which were indispensable for the success of the experiments.

References

[1] L. Esaki and R. Tsu, IBM J. Res. Dev. **14**, 61 (1970)

[2] L. L. Chang, L. Esaki, and R. Tsu, Appl. Phys. Lett. **24**, 593 (1974)

[3] L. Esaki and L. L. Chang, Phys. Rev. Lett. **33**, 495 (1974)

[4] R. Dingle, W. Wiesmann, and C. H. Henry, Phys. Rev. Lett. **33**, 827 (1974)

[5] R. Dingle, A. C. Gossard, and W. Wiegmann, Phys. Rev. Lett. **34**, 1327 (1975)

[6] R. Dingle, in: Festkörperproblem/Advances in Solid State Physics, ed. by H. J. Queisser (Vieweg, Braunschweig 1975) Vol. XV, p. 21

[7] N. Holonyak, R. M. Kolbas, W. D. Laidig, B. A. Vojak, R. D. Dupuis, and P. D. Dapkus, Appl. Phys. Lett. **33**, 737 (1978)

[8] R. Dingle, H. L. Störmer, A. C. Gossard, and W. Wiegmann, Appl. Phys. Lett. **33**, 665 (1978)

[9] K. von Klitzing, G. Dorda, and M. Pepper, Phys. Rev. Lett. **45**, 494 (1980)

[10] D. C. Tsui, H. L. Störmer, and A. C. Gossard, Phys. Rev. Lett. **48**, 1562 (1982)

[11] a) K. von Klitzing, in: Festkörperprobleme/Advances in Solid State Physics, ed. by J. Treusch (Vieweg, Braunschweig 1981) Vol. XXI, p. 1
      b) G. Döhler, in: Festkörperprobleme/Advances in Solid State Physcis, ed. by P. Grosse (Vieweg, Braunschweig 1983) Vol. XXIII, p. 207
      c) N. T. Linh, in: Festkörperprobleme/Advances in Solid State Physics, ed. by P. Grosse (Vieweg, Braunschweig 1983) Vol. XXIII, p. 227
      d) H. L. Störmer, in: Festkörperprobleme/Advances in Solid State Physics, ed. by P. Grosse (Vieweg, Braunschweig 1984) Vol. XXIV, p. 25
      e) G. Abstreiter, in: Festkörperprobleme/Advances in Solid State Physics, ed. by P. Grosse (Vieweg, Braunschweig 1984) Vol. XXIV, p. 291
      f) G. Wiemann, in: Festkörperprobleme/Advances in Solid State Physics, ed. by P. Grosse (Vieweg, Braunschweig 1986) Vol. 26, p. 231

[12] L. Esaki, in: Heterojunctions and Semiconductor Superlattices, ed. by G. Allal, G. Bastard, N. Boccara, M. Lannoo, and M. Voos (Springer, Heidelberg 1986)

[13] L. L. Chang, H. Sakaki, C. A. Chang, and L. Esaki, Phys. Rev. Lett. **38**, 1489 (1977)

[14] L. L. Chang, E. E. Mendez, N. J. Kawai, and L. Esaki, Surf. Sci. **113**, 306 (1982)

[15] T. Ando, J. Phys. Soc. Japan **50**, 2978 (1981)

[16] J. C. Maan, Y. Guldner, J. P. Vieren, P. Voisin, M. Voos, L. L. Chang, and L. Esaki, Solid State Commun. **39**, 683 (1981)

[17]   T. C. L. G. Sollner, W. D. Goodhue, P. E. Tannenwald, C. D. Parker, and D. D. Peck, Appl. Phys. Lett. **43**, 588 (1983)

[18]   T. Nakagawa, H. Imamoto, T. Kojima, and K. Ohta, Appl. Phys. Lett. **49**, 73 (1986) and also reference cited therein

[19]   A. Chomette, B. Deveaud, J. Y. Emery, A. Regreny, and B. Lambert, Solid State Commun. **54**, 75 (1985)

[20]   B. Deveaud, A. Chomette, B. Lambert, A. Regreny, R. Romestain, and P. Edel, Solid State Commun. **57**, 885 (1986)

[21]   D. Calecki, J. F. Palmier, and A. Chomette, J. Phys. C (Solid State Physics) **17**, 5017 (1984)

[22]   J. F. Palmier, M. Leperson, C. Minot, A. Chomette, A. Regreny, and D. Calecki, Superlattices and Microstructures **1**, 76 (1985)

[23]   J. Yoshino, H. Sakaki, and T. Furuta, p. 519, 49, Proc. of the 17th Int. Conf. on the Physics of Semiconductors, San Francisco 1984 (Springer, New York 1985)

[24]   R. A. Davies, M. J. Kelly, and T. M. Kerr, Phys. Rev. Lett. **55**, 1114 (1985)

[25]   F. Capasso, K. Mohammed, A. Y. Cho, R. Hull, and A. L. Hutchinson, Phys. Rev. Lett. **55**, 1152 (1985)

[26]   a) G. Belle, J. C. Maan, and G. Weimann, Solid State Commun. **56**, 65 (1985)
       b) G. Belle, J. C. Maan, and G. Weimann, Surface Sci. **170**, 611 (1986)

[27]   T. Duffield, R. Bhat, M. Koza, D. M. Hwang, P. Grabbe, and S. J. Allen Jr., Phys. Rev. Lett. **56**, 2724 (1986)

[28]   J. C. Maan, Springer Series in Solid State Sciences, Vol. 53, ed. by G. Bauer, F. Kuchar, and H. Heinrich (Springer, Berlin 1984), p. 183

[29]   W. Beinvogl, A. Kamgar, and J. F. Koch, Phys. Rev. **B14**, 4274 (1986)

[30]   U. Merkt, Phys. Rev. **B32**, 6699 (1985)

[31]   C. Kittel, Quantum theory of Solids (John Wiley & Sons, New York 1963)

[32]   L. D. Landau and E. M. Lifshitz, Course of theoretical Physics, Vol. 9, part 2, Statistical Physics, ed. by E. M. Lifshitz and L. P. Pitaevskii (Pergamon Press, Oxford 1980)

[33]   L. D. Landau, E. M. Lifshitz, Course of theoretical Physics, Vol. 3, Quantum Mechanics (Pergamon Press, Oxford 1977)

[34]   a) A. B. Pippard, Phil. Trans. Roy. Soc. **A256**, 317 (1964)
       b) A. B. Pippard, in: The theory of metals, ed. by J. Ziman (Cambrdige University Press, Cambridge 1969), Ch. 3

[35]   P. G. Harper, Proc. Phys. Soc. **68**, 879 (1955)

[36]   R. Tsui and J. Janak, Phys. Rev. **B9**, 404 (1974)

[37]   a) G. E. Zil'berman, Soviet Phys. JETP **5**, 208 (1957)
       b) G. E. Zil'berman, Soviet Phys. JETP **6**, 299 (1958)

[38]   G. Bastard, Phys. Rev. **B24**, 5693 (1981)

[39]   a) M. Altarelli, Phys. Rev. **B28**, 842 (1983)
       b) M. Altarelli, in: Applications of High Magnetic Field in Semiconductor Physics, ed. by G. Landwehr (Springer, Berlin 1983)

[40]   The Kronig-Penney-bandstructure is discussed thoroughly in: R. A. Smith, Wave Mechanics of Crystalline Solids (John Wiley & Sons, New York 1961)

[41]   a) R. E. Doezema and J. F. Koch, Phys. Rev. **B5**, 3866 (1972)
       b) R. E. Doezema and J. F. Koch, Phys. Rev. **B6**, 2071 (1972)

[42]   M. Wanner, R. E. Doezema, and U. Strom, Phys. Rev. **B12**, 2883 (1975) and references cited therein

[43]   T. Ando, J. Phys. Soc. Japan **39**, 411 (1975)

[44]   K. F. Bergren and D. J. Newson, Semicond. Sci. Technol. **1**, 327 (1986)

[45]   J. C. Maan, G. Belle, A. Fasolino, M. Altarelli, and K. Ploog, Phys. Rev. **B30**, 2253 (1984)

[46]   R. C. Miller, D. A. Kleinman, W. T. Tsang, and A. C. Gossard, Phys. Rev. **B24**, 1134 (1981)

[47]   G. Bastard, E. E. Mendez, L. L. Chang, and L. Esaki, Phys. Rev. **B26**, 1974 (1982)

[48]   P. Dawson, K. J. Moore, G. Duggan, H. I. Ralph, and C. T. B. Foxon, Phys. Rev. **B34**, 6007 (1986)

[49]   A. Fasolino and M. Altarelli, in: Springer Series in Solid State Sciences, Vol. 53, ed. by G. Bauer, F. Kuchar, and H. Heinrich (Springer, Berlin 1984), p. 176

[50]   J. N. Schulman and Y. C. Chang, Phys. Rev. **B31**, 2056 (1985)

[51]   R. Sooryakumar, D. S. Chemla, A. Pinczuk, A. C. Gossard, and W. Wiegmann, Proc. of the 17th Int. Conf. on the Physics of Semiconductors, San Francisco 1984 (Springer, New York 1985), p. 523

[52]   Y. C. Chang and G. D. Sanders, Phys. Rev. **B32**, 5521 (1985)

[53]   F. Ancilotto, A. Fasolino, and J. C. Maan, Proc. of the 2nd Int. Conf. on Superlattices, Gotenborg, 1986, J. of Superlattices and Microstructures, to be published

[54]   M. Kriechbaum, Proc. of the 2nd Int. Conf. on Superlattices, Gotenborg, 1986, J. of Superlattices and Microstructures, to be published

[55]   D. C. Rogers, J. Singleton, R. J. Nicholas, C. T. Foxon, and K. Woodbridge, Phys. Rev. **B34**, 4002 (1986)

[56]   W. Andreoni and R. Car, Phys. Rev. **B21**, 3334 (1980)

[57]   E. Carruthers and P. J. Lin-Chung, Phys. Rev. **B17**, 2705 (1978)

[58]   A. Ishibashi, Y. Mori, M. Itabashi, and N. Watanabe, J. Appl. Phys. **58**, 2691 (1985)

Festkörperprobleme 27 (1987)

Structure and Reactivity of Solid Surfaces

Gerhard Ertl

Fritz-Haber-Institut der Max-Planck-Gesellschaft, D-1000 Berlin (West) 33, Germany

Summary: The geometric configuration of the atoms in the surface of a solid is correlated with their valence electronic properties and thereby also with their chemical reactivity towards molecules interacting from the gas phase. Several aspects of these interactions are briefly reviewed: The atomic structure of clean single crystal surfaces and their changes by bond formation (chemisorption), the formation of chemisorbed phases with long-range order and associated phase transitions, the dynamics of the gas-surface interaction processes, as well as temporal oscillations in a catalytic reaction coupled to periodic structural transformations of a surface.

1 Introduction

The ideal crystal with infinite three-dimensional periodicity will never be realized. Apart from bulk defects a crystal will always exhibit a strong distortion of this periodicity at its termination: the surface. The ratio of atoms in the surface to those in the bulk increases with decreasing particle size or film thickness. The overall properties of such systems may then be essentially determined by those of the surface.

The surface atoms are missing part of their nearest neighbors which causes variations of the valence electronic properties as well as of the equilibrium positions of the nuclei. The former effect is reflected in the chemical reactivity of the surface by which 'dangling bonds' may become saturated, while the latter manifests itself in structural parameters deviating from those of the bulk. The present contribution intents to illustrate by means of a few selected examples our present knowledge of these phenomena and their mutual interplay.

2 The Structure of Clean Surfaces

The geometric location of the atoms in the outermost layer may differ from those of a corresponding bulk plane in two respects, namely, alterations of the interlayer spacings (relaxation) and lateral displacements connected with changes of the unit cell within the surface layer (reconstruction) [1].

As a first example Fig. 1 shows the structure of the clean Ni(110) surface which exhibits the same lateral periodicity as the bulk, but where the spacing between the first and second layer is contracted by 8.5 %, while that between the second and third is expanded by 3.5 % [2]. These findings are quite general and have now

169

Ni (110) - clean

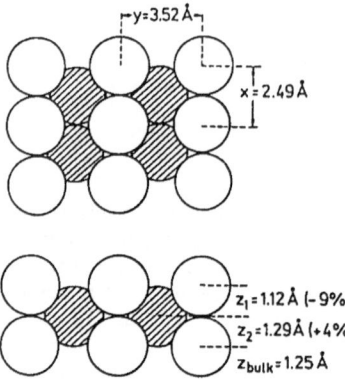

Fig. 1 Structure model of the (110) surface of fcc metals without reconstruction, such as Ni(110).

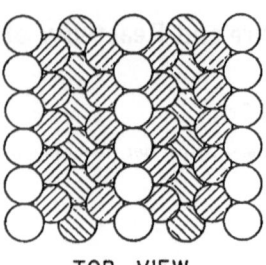

TOP VIEW

SIDE VIEW

Fig. 2 'Missing row' structure of the Pt(110) surface.

been established for a large series of metal surfaces [1]. The effects are more pronounced with the more open planes for which Δd_{12} of up 15 % were reported, while the atoms in the most densely packed planes exhibit usually only very minor deviations from their regular bulk positions.

The surface atoms will generally have the tendency to minimize their free energy by surrounding themselves by as many nearest neighbors as possible, i.e. by forming a most densely packed plane. This tendency is counterbalanced by the resulting mismatch between surface and bulk planes. This qualitative argument indicates why surfaces may undergo reconstruction: While the Ni(110) surface is unreconstructed, the Pt(110) surface is reconstructed (Fig. 2) [3]: Every second row in [1$\bar{1}$0]-direction is missing (leading to a 1 × 2 superstructure), and the surface may now be considered as existing of microfacets of the most densely packed (111) plane.

Similarly, the (100) planes of Ir, Pt, and Au are reconstructed in a way as illustrated by Fig. 3 [3]. The atoms of the topmost layer exhibit a quasi-hexagonal ('hex') configuration similar to that of the (111) plane placed on the square lattice of the (100) plane forming the second and deeper layer. The mismatch between first and second layer is in this case reflected by the fact that the topmost atoms do not form a perfectly flat plane, but are slightly 'buckled'. This buckling may be nicely made visible by the scanning tunneling microscope (STM). The STM image in Fig. 4 from a clean Pt(100) surface shows two domain orientations of the 'hex' surface with its corrugation periodicity of 13.5 Å, whereby the step in the center serves as domain boundary [4].

While most of the clean metal surfaces are not reconstructed, the situation is opposite with semiconductor surfaces where reconstruction is the rule. The 7 × 7-

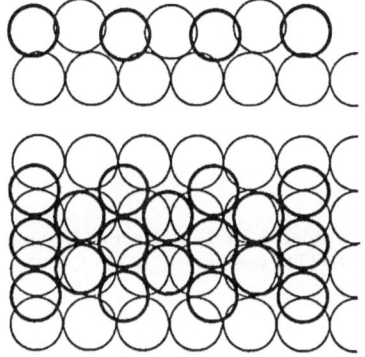

Fig. 3
Reconstructed Ir (100) surface where the atoms of the topmost layer form a hexagonal configuration yielding a 5 × 1-superstructure. Similar structures are found with clean Pt (100) and Au (100) surfaces.

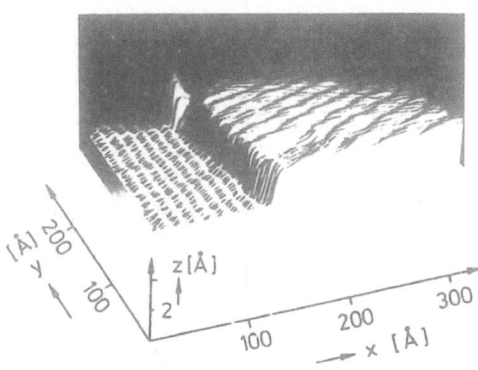

Fig. 4
Scanning tunneling microscope (STM) image of a clean Pt (100) surface with a step [4].

structure of the annealed Si (111) surface is probably the most famous example which is now — after many years of intense research — considered to be solved [5].

3 Structure of Adsorbate Covered Surfaces

The energy of surfaces may be lowered by the formation of chemical bonds with suitable particles arriving from the gas phase (chemisorption). We will restrict our discussion to cases in which this reactivity does not extend into deeper layers, eventually leading to the formation of new bulk compounds such as oxides etc. However, also those processes are initiated by chemisorption steps at the outermost atomic layer.

Fig. 5 shows the configuration of H atoms formed on a Ni (111) surface at $T \leqslant 200$ K and at a coverage $\theta = 0.5$ *) [6]. The H atoms prefer three-fold coordinated sites

---

*) The coverage θ is defined as the ratio of the density of adsorbed particles over that of the substrate atoms in the topmost layer.

171

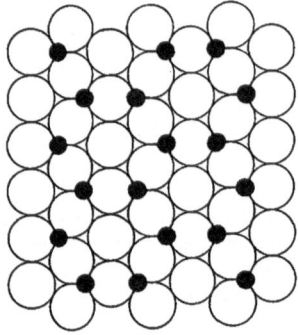

Fig. 5

Structure of the ordered 2 × 2-overlayer of H atoms adsorbed on a Ni(111) surface at T ≤ 200 K with a coverage $\theta = 0.5$.

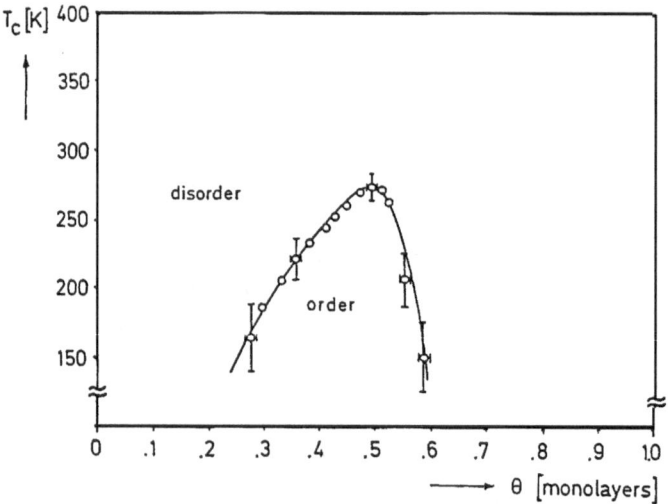

Fig. 6 Phase diagram of the 2 × 2-structure of the H/Ni(111) system.

(i.e. as many nearest neighbors as possible) to which they are attached by an energy of about 2.5 eV. The mutual configuration is under these conditions characterized by pronounced long-range order, giving rise to 'extra' beams in low energy electron diffraction (LEED). This long-range order is obviously due to the operation of interactions between the adsorbed particles, which in the present case are of the order ≲ 0.1 eV and are of the 'indirect' type, i.e. mediated through the valence electrons of the substrate metal. Increasing the temperature leads to continuous order-disorder transitions as can be followed through the variation of the respective LEED intensities. Determination of the transition temperatures at varying coverages enables to establish a phase diagram as reproduced in Fig. 6. Such two-dimensional phase diagrams have now been determined for different systems for which in turn theoretical simulation yielded good agreement if the interaction parameters were properly adjusted [7].

172

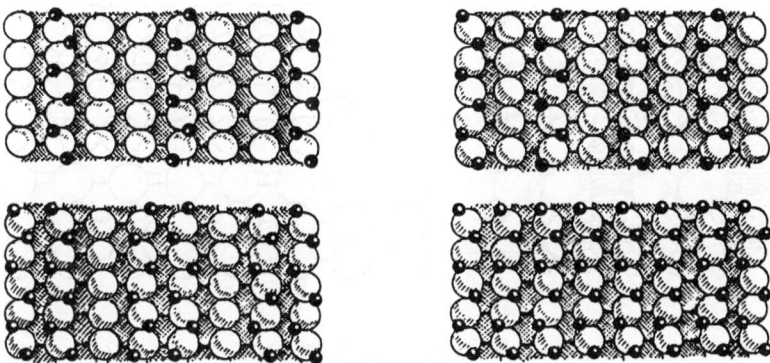

Fig. 7 Lattice gas structures of the system H/Ni (110) formed at T ≤ 180 K with increasing coverage up to $\theta = 1.0$.

The wealth of lattice-gas structures of the just described type is reflected by the sequence of phases formed at T ≤ 200 K on a Ni(110) surface with increasing coverage (Fig. 7). The H atoms tend to form 'zig-zag' rows along the $[1\bar{1}0]$-direction (due to attractive interactions), while repulsive interactions act in $[001]$-direction and keep parallel rows apart from each other until at $\theta = 1$ all rows are equally occupied. In this 2×1-structure the H-atoms occupy again three-fold coordinated sites (formed by two Ni atoms from the topmost and one atom from the second layer) with three equal Ni-H distances of 1.7 Å [2]. Interestingly, the expansion of the spacing between the first and second Ni layer is reduced from 8.5 to 3.5 % in the presence of the chemisorbed overlayer. Saturation of the 'dangling' bonds of the surface atoms by chemisorption obviously tends to restore the bulk situation, but also weakens the bond strength between surface atoms. (This has e.g. been verified recently by the observation of a 'softening' of the surface phonons due to chemisorption [9].)

Inspection of Fig. 7 reveals that at $\theta = 1$ by no means all of the highly-coordinated sites are occupied. However, further filling is prevented by the repulsions between the then resulting short H–H distances. Instead, the system finds another way to lower its total energy by further uptake of additional H atoms: In a first-order phase transition parallel rows of Ni atoms are moved together ('row pairing') thus enabling the uptake of another 0.5 monolayer of H atoms as shown in Fig. 8a. The surface reconstructs under the influence of an adsorbate, whereby not only the topmost atomic layer is affected by this process which takes place without any noticeable activation energy. The atoms in the second layer are also displaced from their previous positions ('buckling') whereby the Ni–Ni distances are restored to bulk-like values and internal strain is removed [10] (Fig. 8b).

Displacements of surface atoms under the influence of adsorbed particles may equally take place in the opposite direction, i.e. the reconstruction of a clean surface is removed. If, for example, on a Pt (100) surface (exhibiting reconstruction

173

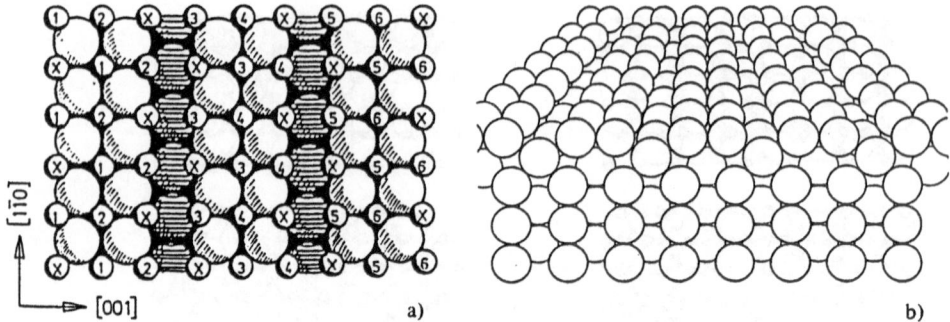

Fig. 8 Reconstructed 'row pairing' structure of the system H/Ni (110) with $\theta = 1.5$.

Pt (100)

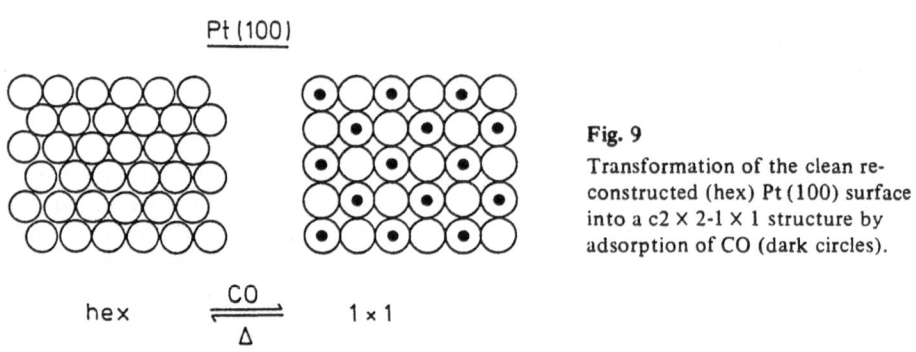

hex $\underset{\Delta}{\overset{CO}{\rightleftharpoons}}$ 1 x 1

Fig. 9

Transformation of the clean reconstructed (hex) Pt (100) surface into a c2 X 2-1 X 1 structure by adsorption of CO (dark circles).

into a 'hex' structure as outlined above) adsorbed CO reaches a critical coverage of about $\theta = 0.08$, patches of the non-reconstructed 1 X 1-surface with a local CO-coverage of 0.5 (associated with a c2 X 2-superstructure) are formed as illustrated by Fig. 9 [11]. The driving force for this first-order phase transition is the CO adsorption energy which is higher on the 1 X 1 than on the hex phase. If, on the other hand, the coverage on the 1 X 1-surface drops below another critical value of about 0.3, the surface transforms back into the hex phase. Pronounced hysteresis effects are observed if the surface is heated up or cooled down in a CO atmosphere, and these effects are of crucial importance for the oscillatory phenomena to be described later.

4 Dynamics of Molecule / Surface Interaction

Fig. 10 shows the schematic (one-dimensional) interaction potential for a molecule approaching a surface. In order to be bonded to the surface ('trapped'), i.e. to reach the bottom of the potential well, the incoming molecule has to transfer sufficient energy to the heat bath of the phonons of the solid. The sticking coefficient s

174

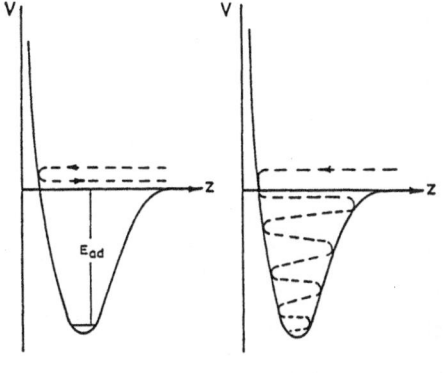

Direct - inelastic   Trapping / Desorption

Fig. 10 One-dimensional interaction potential of a particle approaching a surface and schematics of the two channels of interaction dynamics.

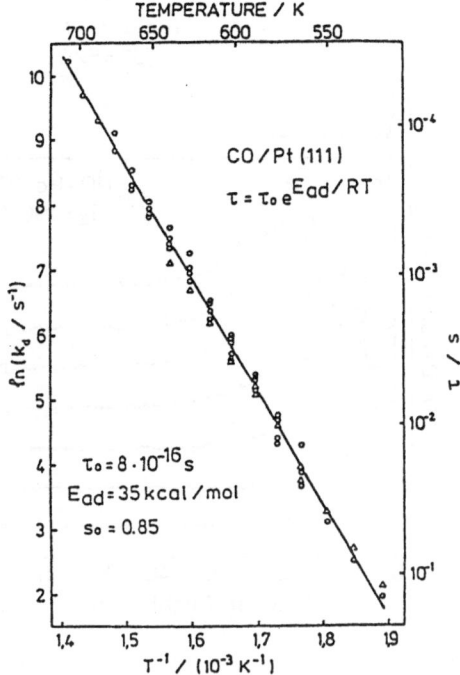

Fig. 11 Mean surface lifetime τ of CO adsorbed on a Pt(111) surface as a function of temperature as determined by modulated molecular beam experiments.

denotes the fraction of impinging particles which undergoes this transition, while the rest is inelastically scattered back into the gas phase (see Fig. 10).

Once adsorbed the particles spend a mean lifetime $\tau = \tau_0 \exp(E_{ad}/RT_s)$ on the surface (T_s = surface temperature) after which they are released back into the gas phase. As an example Fig. 11 shows τ as a function of T_s for CO adsorbed on a Pt(111) surface as determined by modulated molecular beam experiments [12]. Under steady-state conditions thus an equilibrium coverage (depending on gas pressure and temperature) will be established determined by the balance between adsorption and desorption rates.

The dynamics of the energy exchange processes associated with the molecule-surface interaction can be studied in quite detail by combining molecular beam and laser spectroscopic techniques. As an example Fig. 12 represents results characteristic for the interaction of rotationally cold NO molecules exhibiting a well-defined initial translation energy with an oxidized Ge surface [13]. The data are time-of-flight (TOF) spectra (determining the translational energy distri-

175

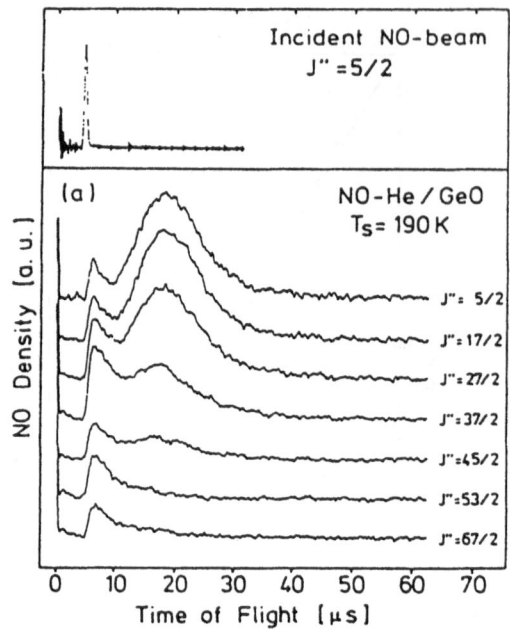

Fig. 12
Time-of-flight spectra of NO molecules in various rotational states (J'') after scattering at an oxidized Ge surface.

butions) for NO molecules coming off the surface in various rotational states from which a wealth of information can be extracted:

i) The TOF data consist of two kinds of particles: Slow ones (with long flight times) arising from desorbing particles and fast molecules which had undergone direct-inelastic scattering.

ii) The distribution of the desorbing particles over the various rotational states is Boltzmann-like, allowing the assignment of a rotational temperature T_{rot}. However, T_{rot} increases with the surface temperature T_s only to a certain limiting value (≈ 400 K). At higher T_s the molecules come off the surface with $T_{rot} < T_s$, even if they had been there in thermal equilibrium ('rotational cooling in desorption') [14].

iii) The rotational distribution of the inelastic scattering-fraction is non-Boltzmann, but exhibits characteristic overpopulation of higher rotational levels. This effect arises from the transformation of translational into rotational energy during the collision event ('rotational rainbow' [15]).

iv) The translational energy distribution of desorbing molecules is Maxwell-Boltzmann, but again T_{trans} becomes smaller than T_s at higher surface temperatures, independent of the rotational energy of the desorbing molecule.

v) In direct-inelastic scattering those molecules which are reflected back with higher rotational energy transfer less of their primary translational energy to the solid.

This simple listing of experimentally observed phenomena is just intended to illustrate how complex the molecule-surface interaction dynamics are in fact. At least qualitative theoretical understanding of these effects has, however, already been obtained, and rapid further progress is to be expected for the near future [16].

Frequently, the interaction of a molecule with a surface is associated with internal bond-breaking (dissociative chemisorption) such as in the case of the H_2/Ni systems discussed above. The energetics of this process is illustrated schematically by Fig. 13: A diatomic molecule approaching the surface experiences a relatively shallow potential energy minimum of the molecular 'precursor' $A_{2,ad}$. (This state can often be isolated, e.g. as $O_{2,ad}$ on Pt [17].) If instead the A_2 molecule would be dissociated in the gas phase, the interaction of the A atoms with the surface would be associated with a much deeper potential minimum. The crossing between the molecular and atomic interaction potentials obviously allows a molecule to become dissociatively chemisorbed without the need for surmounting a large activation barrier. The overall energy balance: $E_{ad} = 2 E_{M-A} - E_{diss}$, demonstrates that the dissociation energy of A_2 is overcompensated by the energy gain due to the formation of *two* bonds between the surface and the atoms A.

The sticking probability for dissociative chemisorption is of course strongly affected by the remaining activation barrier, which may in turn be sensitively influenced by the structure of the surface. It is generally higher at surface defects such as mono-atomic steps, or on more open single crystal surfaces than on the close-packed low-index planes. The room temperature sticking coefficient for H_2 is 0.95 on Ni(110), but only ≈ 0.05 on Ni(111) [18]. Dissociative O_2 chemisorption occurs on the non-reconstructed (metastable) Pt(100) surface with a probability of about 0.1, but with less than 10^{-3} on the reconstructed hex phase [19]. These pronounced differences demonstrate how sensitively the reactivity of a surface may be influenced by its structure.

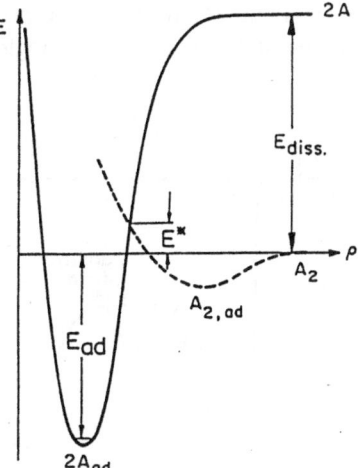

Fig. 13
One-dimensional potential diagram illustrating dissociative chemisorption.

177

5 Catalysis and Non-Linear Dynamics

Reversing dissociative chemisorption leads to recombination and desorption, viz. $A_{ad} + A_{ad} \rightarrow A_2$. In a similar way new kinds of particles may be formed by reaction in the adsorbed state and subsequent desorption. This phenomenon is called heterogeneous catalysis. As an example we consider a reaction which is also of practical relevance (automotive exhaust control), namely, the oxidation of carbon monoxide on platinum surfaces: $2\,CO + O_2 \rightarrow 2\,CO_2$.

This reaction proceeds through the following individual steps [20]:

$$CO + * \rightleftharpoons CO_{ad}$$
$$O_2 + 2* \rightarrow 2\,O_{ad}$$
$$O_{ad} + CO_{ad} \rightarrow CO_2 + 2*$$

The * denotes schematically a free adsorption site on the surface which has, however, a different meaning for CO and O_2: Dissociative O_2 chemisorption requires several empty neighboring surface atoms which is not the case for CO adsorption. As a consequence a CO coverage beyond a critical value blocks O_2 adsorption, while the rather open structures formed by O_{ad} are only of minor influence for the adsorption of CO.

The overall energetics of the catalytic reaction under discussion is depicted schematically in Fig. 14. Under stationary conditions the catalyst is typically held at a fixed temperature and is exposed to constant partial pressures of the reacting molecules in a flow system as illustrated by Fig. 15. As a consequence a constant flux of product molecules will leave the reactor due to the steady-state catalytic reaction whose rate is determined by the interplay between the various elementary processes at the surface. However, under certain conditions this is no longer the case: The reaction rate as well as certain surface properties exhibit temporal oscillations, despite the fact that the external parameters are kept constant.

Fig. 16 shows how such oscillations develop after fixing one of the external parameters (the O_2 partial pressure) to a new constant value. In this case the variation of the state of a Pt(110) single crystal surface as reflected by the work function

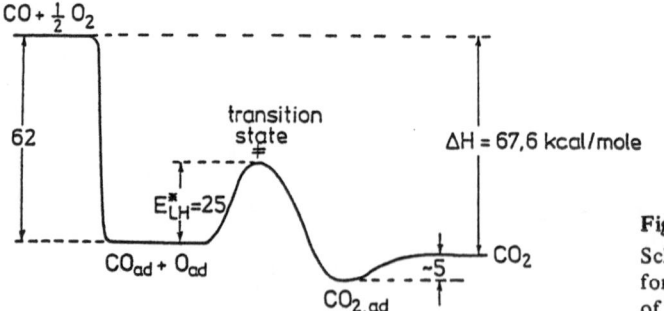

Fig. 14

Schematic potential diagram for the catalytic oxidation of CO on Pt.

change $(\Delta\varphi)$ is recorded [21]. Obviously, the whole surface area $(\approx 30 \text{ mm}^2)$ reaches its state of synchronized self-organization only over a larger number of periodic cycles.

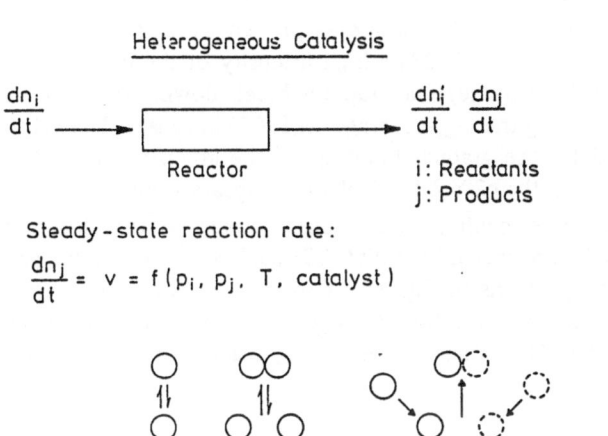

Fig. 15

Principle of steady-state catalysis in a flow system.

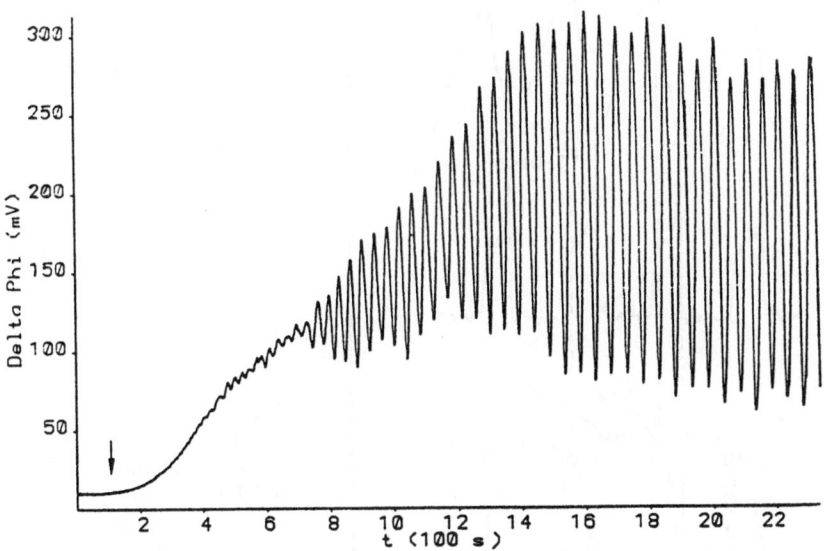

Fig. 16 Development of sustained temporal oscillations of the work function (= reaction rate) of a Pt (110) surface during steady-state catalysis of CO oxidation. T = 470 K, $p_{CO} = 2.3 \cdot 10^{-5}$ Torr; at the point marked by an arrow the O_2 pressure is rapidly adjusted from $1.5 \cdot 10^{-4}$ Torr to a new constant value of $2.0 \cdot 10^{-4}$ Torr.

The origin of these oscillations has been explored in great detail with the Pt(100) surface and can be traced back to a close coupling between structure and reactivity of a surface [22]. It turned out that temporal oscillations of the reaction rate or of the work function (which integrate over the whole surface area) are paralleled by periodic changes of the surface structure between the reconstructed hex and the non-reconstructed, CO-covered c2 X 2-1 X 1 phase as probed by LEED on a 0.1 mm² area. As can be seen from Fig. 17 [23] a high intensity of the c2 X 2-beam (≙ high CO coverage on the 1 X 1-phase) may rapidly break down, while the intensity of the (1̄1)-beam (reflecting the O_{ad}-coverage on 1 X 1) increases. The decay of the latter is accompanied by the gradual build-up of the intensity of the hex phase which finally collapses, and the c2 X 2-1 X 1 phase reappears again.

The mechanism underlying these ocillations is closely linked to the different reactivity of the 1 X 1- and hex-phases of the Pt(100) surface, respectively, and will be qualitatively outlined by means of Fig. 18: If we start with a surface close to the completion of the c2 X 2-1 X 1 structure, O_2 chemisorption will be restricted to few sites (X). A chemisorbed O atom will then rapidly react with a neighboring

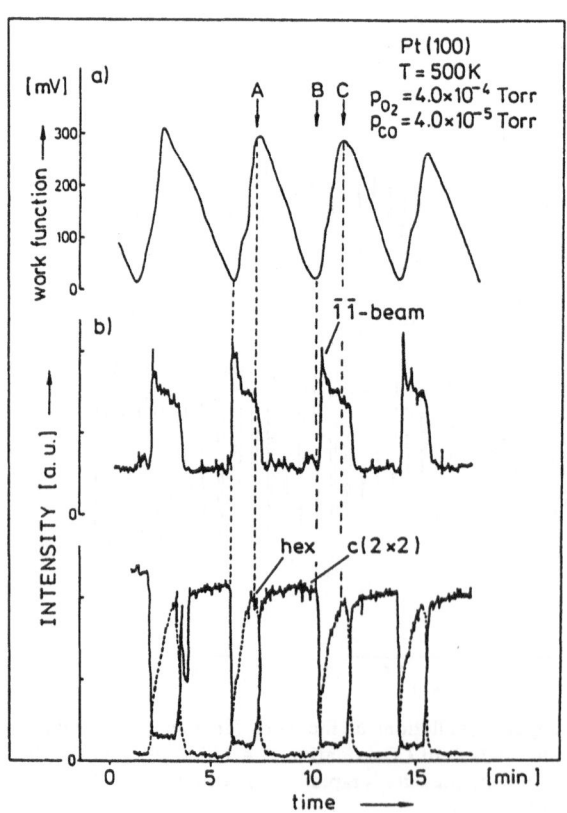

Fig. 17

Kinetic oscillations during catalytic oxidation of CO on a Pt(100) surface.

a) Variation of the work function (= reaction rate),
b) variations of the intensities of different LEED spots.

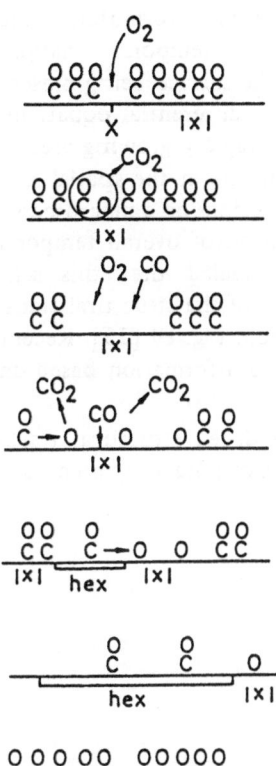

Fig. 18

Outline of the various stages involved in the periodic structural transformations on Pt (100).

CO to CO_2 which molecule is then immediately released into the gas phase. As a consequence *two* adsorption sites will be left, and in an autocatalytic process the $c2 \times 2$-phase will be dissolved. The bare 1×1-sites have a high sticking coefficient for oxygen and will therefore be rapidly occupied by O_{ad} (note that the O_2 partial pressure exceeds by far that of CO). That is why the $(\overline{1}1)$ spot intensity in Fig. 17 rises steeply while that of the $c2 \times 2$-structure drops. Due to the continuous reaction between O_{ad} and CO molecules arriving either from the gas phase or by diffusion from neighboring $c2 \times 2$-patches, the steady-state O-coverage will, however, not be high enough to prevent slow transformation of the underlying 1×1-structure into the hex phase. The $(\overline{1}1)$ intensity decreases while simultaneously that of the hex structure grows. The hex phase will only be covered by CO (the oxygen sticking coefficient is too small), which particles may diffuse to neighboring O-1×1 sites where they react off and cause further growth of the hex patches. With increasing domain size this process will, however, become less probable. As a consequence the CO coverage on the hex surface rises and finally exceeds the critical value for the hex $\rightarrow 1 \times 1$ transformation, and the surface switches rapidly back into the $c2 \times 2$-1×1 structure.

Quantitative description of the temporal oscillations is achieved by formulation and numerical solution of the differential equations describing the temporal variations of the CO and O coverages, as well as of the fraction of the surface being present as hex or 1×1-phase [24]. These are coupled, non-linear differential equations, and that is why the observed phenomena belong into the rapidly growing area of 'non-linear dynamics' or 'synergetics'. Another aspect consists in the spatial self-organization which synchronizes the behavior of different regions on a macroscopic surface which is a necessary prerequisite for the observation of overall temporal oscillations. Spatially resolved LEED measurements revealed that this self-organization is achieved through the wave-like propagation of the structural transformation over the whole surface area as can be seen from Fig. 19 [25]. Recent computer simulations reproduced this kind of spatial pattern formation based on the underlying microscopic reaction steps [26].

The Pt(110) surface may also undergo an adsorbate induced structural transformation (from 1×2 to 1×1). In this case the oscillations are restricted to a

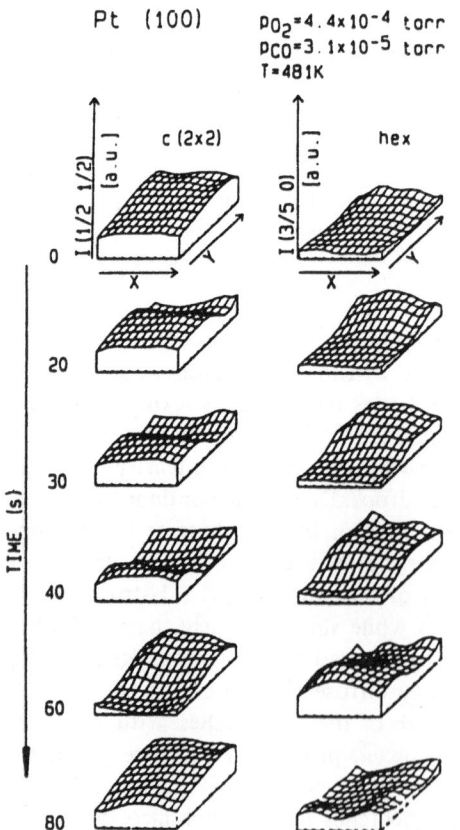

Fig. 19

Scanning LEED data of Pt (100) during the kinetic oscillations showing the wave-like propagation of the structural transformation across the whole surface area of 4×7 mm^2.

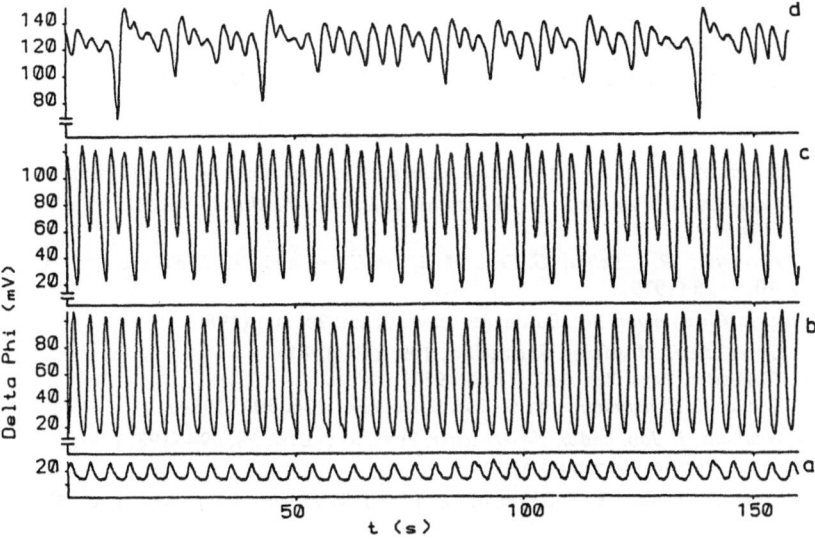

Fig. 20 Period doubling and transition to chaos in the oscillations of the work function of Pt (110) during catalytic CO oxidation by varying the CO pressure.

much narrower range of external parameters and are, on the other hand, much more regular in time. In addition another interesting aspect of non-linear dynamics may be studied (Fig. 20) [21]: Slight variation of an external parameter causes the full development of regular oscillations, followed by a sequence of period-doubling into a state of irregular ('chaotic') oscillations.

6 Conclusions

This brief review did not cover a series of other important aspects of the reactivity of solid surfaces, such as the role of defects or the penetration of particles below the surface initiating dissolution and the formation of bulk compounds. Systems of the type described here offer possibilities for fundamental studies in different areas which are also of general interest to solid state physics as a whole, such as phase transitions or synergetics. Apart from this they represent models for a series of phenomena in applied research, such as thin film technology, corrosion, catalysis, etc. The continuing development of experimental and theoretical tools provides now information on a level of sophistication which was unforeseen only a few years ago.

References

[1] *K. Müller*, Ber. Bunsenges. Phys. Chem. 90, 184 (1986)

[2] *W. Reimer, V. Penka, M. Skottke, R. J. Behm, G. Ertl*, and *W. Moritz*, Surface Sci. (in press)

[3] See e.g. *P. J. Estrup*, Springer Series Chem. Phys. 35, 205 (1984)

[4] *R. J. Behm, W. Hösler, E. Ritter*, and *G. Binnig*, Phys. Rev. Lett. 56, 228 (1986)

[5] *K. Takayanagi, Y. Tanishiro, M. Takahashi*, and *S. Takahashi*, J. Vac. Sci. Techn. A3, 1502 (1985)

[6] *K. Christmann, R. J. Behm, G. Ertl, M. A. van Hove*, and *W. H. Weinberg*, J. Chem. Phys. 70, 4168 (1979)

[7] See e.g. *K. Christmann*, Ber. Bunsenges. Phys. Chem. 90, 307 (1986)

[8] a) *T. Engel* and *K. H. Rieder*, Surface Sci. 109, 140 (1981)
    b) *K. H. Rieder*, Phys. Rev. B27, 7799 (1983)
    c) *V. Penka, K. Christmann*, and *G. Ertl*, Surface Sci. 136, 307 (1984)

[9] a) *D. Neuhaus, F. Joo*, and *B. Feuerbacher*, Phys. Rev. Lett. 58, 694 (1987)
    b) *H. Ibach*, personal communication

[10] *G. Kleinle, V. Penka, R. J. Behm, G. Ertl*, and *W. Moritz*, Phys. Rev. Lett. 58, 148 (1987)

[11] a) *R. J. Behm, P. A. Thiel, P. R. Norton*, and *G. Ertl*, J. Chem. Phys. 78, 7437; 7448 (1983)
     b) *T. E. Jackmann, K. Griffiths, J. A. Davies*, and *P. R. Norton*, J. Chem. Phys. 79, 3529 (1983)

[12] *C. T. Campbell, G. Ertl, H. Kuipers*, and *J. Segner*, Surface Sci. 107, 207 (1981)

[13] *A. Mödl, T. Gritsch, F. Budde, T. J. Chuang*, and *G. Ertl*, Phys. Rev. Lett. 57, 384 (1986)

[14] *A. Mödl, H. Robota, J. Segner, W. Vielhaber, M. C. Lin*, and *G. Ertl*, J. Chem. Phys. 83, 4800 (1985)

[15] *A. W. Kleyn, A. C. Luntz*, and *D. J. Auerbach*, Phys. Rev. Lett. 47, 1169 (1981)

[16] a) *J. A. Barker* and *D. J. Auerbach*, Surface Sci. Rep. 4, 1 (1985)
     b) *J. C. Polanyi* and *R. J. Wolf*, J. Chem. Phys. 82, 1555 (1985)
     c) *W. Brenig, H. Kasai*, and *H. Müller*, Surface Sic. 161, 608 (1985)
     d) *C. W. Muhlhausen, L. R. Williams*, and *J. C. Tully*, J. Chem. Phys. 83, 2594 (1985)

[17] *J. L. Gland, B. A. Sexton*, and *G. B. Fisher*, Surface Sci. 95, 587 (1980)

[18] a) *A. Winkler* and *K. D. Rendulic*, Surface Sci. 118, 19 (1982)
     b) *H. J. Robota, W. Vielhaber, M. C. Lin, J. Segner*, and *G. Ertl*, Surface Sci. 155, 101 (1985)

[19] *P. R. Norton, K. Griffiths*, and *P. E. Bindner*, Surface Sci. 138, 125 (1984), and references therein to earlier work

[20] *G. Ertl*, in: Catalysis. Science and Technology, ed. by *J. R. Anderson* and *M. Boudart* (Springer, Berlin 1983), Vol. 4, p. 209

[21] *M. Eiswirth* and *G. Ertl*, Surface Sci. 177, 90 (1986)

[22] *R. Imbihl, M. P. Cox*, and *G. Ertl*, J. Chem. Phys. 84, 3519 (1986)

[23] *M. P. Cox, G. Ertl, R. Imbihl*, and *J. Rüstig*, Surface Sci. 134, L517 (1983)

[24] *R. Imbihl, M. P. Cox, G. Ertl, H. Müller*, and *W. Brenig*, J. Chem. Phys. 83, 1578 (1985)

[25] *M. P. Cox, G. Ertl*, and *R. Imbihl*, Phys. Rev. Lett. 54, 1725 (1985)

[26] *P. Möller, K. Wetzl, M. Eiswirth*, and *G. Ertl*, J. Chem. Phys. 85, 5328 (1986)

Festkörperprobleme 27 (1987)

The Microstructure of Technologically Important Silicon Surfaces

Martin Henzler

Institut für Festkörperphysik, Universität Hannover, D-3000 Hannover, Federal Republic of Germany

Summary: The increasing requirements of modern silicon technology ask for structural analysis of non-periodic features down to atomic resolution. New instrumental developments in surface physics provide such informations both with imaging (TEM) and diffraction (LEED) techniques. The atomic steps at the Si/SiO_2 interface, which are produced during oxidation, are studied with respect to vertical and lateral distribution and to correlation with device performance (like interface state density and mobility). Also for epitaxial growth atomically smooth interfaces are required for best electrical and optical performance. LEED results show the two-dimensional nucleation and the layer-by-layer growth in detail.

1 Introduction

The discovery of the periodic arrangement of atoms in a crystal was the starting point for solid state physics. As soon as surface atom arrangement could be studied by electron diffraction [1], surface physics was possible. Here the relevant distances (atom-atom distance) are around 0.3 nm. From quite large single crystals semiconductor devices have been developed with dimensions of millimeters down to submicrons (Fig. 1). Therefore the typical distances as many other parameters

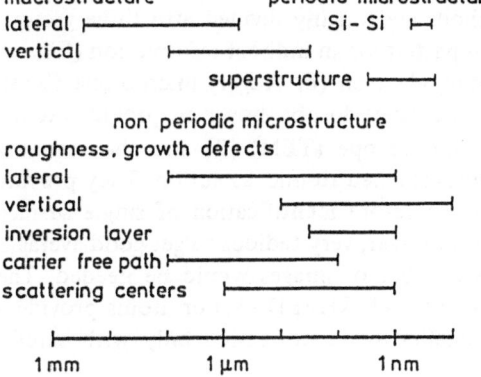

Fig. 1

Important distances in usual technology and usual surface physics and the overlapping range of microstructure.

185

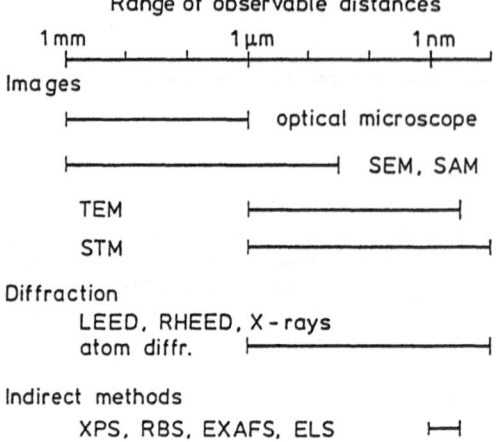

Fig. 2

Methods for study of non-periodic features at surface. The abbreviations are explained in the text.

(like atmosphere or cleanliness) were not really overlapping. With increasing requirements with respect to size and reliability of devices more microscopic informations were needed for technology. On the other hand surface physics developed more and more possibilities to investigate also larger, especially non-periodic features in surface and thin film physics. It is therefore now possible to study directly those parameters which are important for performance of present day devices and may become essential for future developments. Since both silicon devices and silicon surface physics have reached a high standard, it is the purpose of this paper to describe one overlapping aspect: the microstructure of technologically important silicon surfaces.

2 Methods of Investigation

Many methods have been developed to study any structural detail from macroscopic down to atomic dimensions. The methods are roughly divided into three groups: those providing an image or a diffraction pattern or an indirect information (Fig. 2). The optical microscope and the scanning electron (or Auger) microscope (SEM, SAM) are indispensible for macrostructure, they do not, however, provide atomic resolution. The transmission electron microscope (TEM) [2] and the scanning tunnel microscope (STM) [3] have recently reached atomic resolution. They provide direct images after special preparation. Therefore identification of single surface and interface features is easily done. It is, however, very tedious to get good averages or statistics, since then a tremendous number of images would be needed. The diffraction techniques with electrons (LEED [4], RHEED [5]) or atoms provide a diffraction pattern, which provides a qualitative identification only with careful analysis. It is, however, already a perfect average of a fairly large portion of the surface (a fraction of a square millimeter) so that diffraction is superior in quantitative analysis. It is found in literature that LEED should be insensitive to defects. This is true as long as the integral over a spot is the only information to be used, since just the details of spot profile can provide all informations on non-periodic, large-distance features. The indirect methods (x-ray photoelectron spectroscopy XPS, Rutherford backscattering spectroscopy RBS, extended x-ray absorption fine structure EXAFS, electron energy loss spectroscopy ELS) provide informations on the immediate neighborhood of surface atoms (distance, direction, or chemical bonding), which may be used for additional evaluation.

3 Description of a Real Surface

In surface physics mostly ideal surfaces are investigated: all surface atoms are close to regular sites of one lattice plane in strictly periodic, two-dimensional arrangement. Periodic deviations from the lattice sites may form a superstructure. The atom positions within a unit mesh provide a full description. Since such surfaces are only produced and kept in ultra high vacuum, real device surfaces have to be described with other parameters. The top layer may be incomplete, the atoms may deviate from regular even periodic lattice sites, contaminations may be present. For a complete description all atom positions would be needed individually. More useful are statistic numbers like the autocorrelation (or the probability to find two surface atoms in a given distance) or probability distributions for island sizes, terrace widths or step heights (Fig. 3). Some assumptions simplify description considerably: all surface atoms are on regular sites given by bulk arrangement (any deviation is neglected), then only distances with combinations of integer multiples of the lattice vectors are possible. From autocorrelation the average deviation of surface atoms from the average surface (asperity height) and the average lateral correlation (correlation length) are derived. For a more detailed description the autocorrelation may be specified. If the occurrence of a step is random (Markov chain), then the autocorrelation and the terrace width distribution are exponential, the spot profile of the diffraction spot is Lorentzian [4 ... 6]. For computational reasons frequently a Gaussian autocorrelation is used [7], which cannot be transformed into a terrace width distribution with strictly positive probabilities.

The parameters for description may be derived both from images and diffraction patterns. From images with atomic resolution the autocorrelation and its Fourier

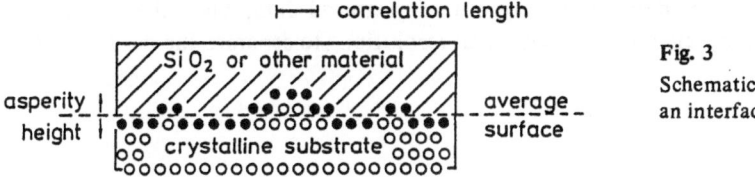

Fig. 3
Schematic cross-section of an interface.

187

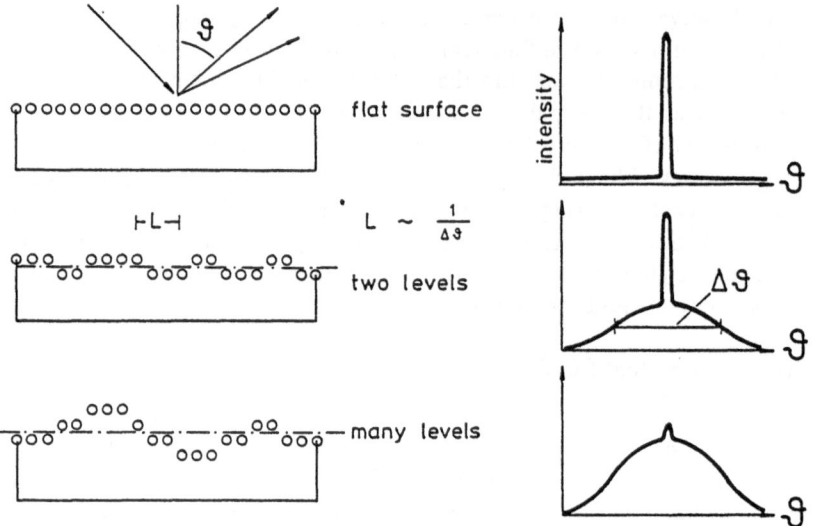

Fig. 4 Observed spot profile of any diffraction technique for a flat surface, a stepped surface within two levels and with many levels. The half width of the shoulder is related to an average terrace width L (approximately $L \sim 1/\Delta \vartheta$). The wave length is chosen to provide a phase shift between neighboring terraces close to 2π.

transform are derived from the actual surface atom positions. From diffraction patterns the spot profile analysis provides all informations. The profile is separated into a central spike and a shoulder. The spike represents the fraction of in-phase scattering of all surface atoms (depending on electron energy and instrumental resolution). The shoulder is due to out-of-phase scattering of surface atoms on different levels. As indicated in Fig. 4, the shoulder (which is usually strongest for out-of-phase condition for neighboring terraces) yields directly the lateral distribution, including edge atom density and correlation length. Additional information is derived out of the fraction of the central spike (compared with total spot intensity), since for ratios of intensities dynamical effects are eliminated. The ratio is described by kinematic theory just by the probabilities p_i of finding a surface atom in level i. For more layers contributing to the surface the central spike decreases in intensity. Especially the expansion close to in-phase condition ($K \approx 0$) provides the asperity height directly (Fig. 5). Therefore lateral and vertical distributions may be derived directly out of measured data, when the profiles are recorded for many energies and with high resolution (to separate the central spike).

188

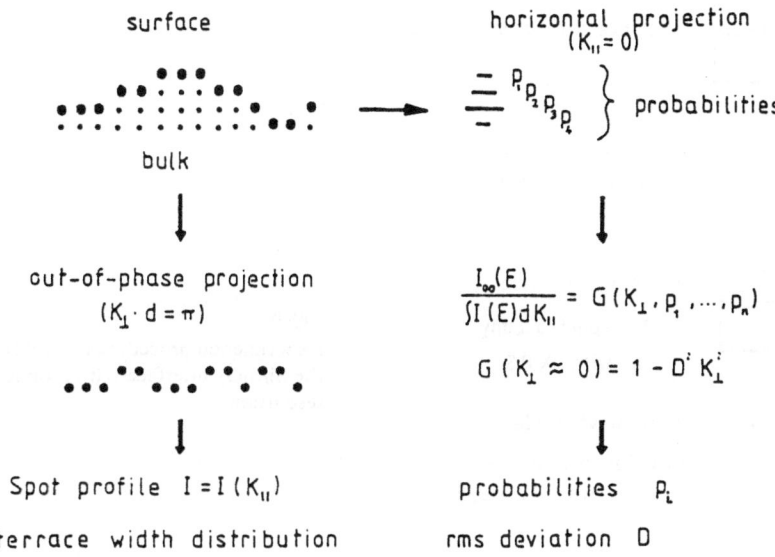

Fig. 5 Schematic procedure to derive terrace width distribution and vertical distribution out of the energy dependence of a spot profile.

4 The Structure of the Si/SiO$_2$-Interface

Among the numerous investigations here only those are of interest which provide informations down to atomic resolutions. The main problem is that the interface is not accessible with the usual techniques (Fig. 6). For TEM thin cross-sections (20 ... 40 nm) are produced so that a projection of the interface is visible with the oxide in place [2]. For quantitative analysis such a projection is digitized with respect to occurrence of steps, so that an autocorrelation and its Fourier transform (= scattering function) is obtained (Fig. 7, from [6]). Due to the limited resolution (≈ 0.2 nm) only one sublattice of the silicon lattice is visible (layer distance 0.27 nm) so that single step heights (0.135 nm) are not recognized. For other techniques the oxide is thinned down to less than 2 nm (for XPS, [8]) or completely removed (for LEED and STM, [9]). Reoxidation is avoided by a quick transfer from HF etch to uhv under a protecting droplet of methanol [9]. It has been shown that the surface has less than a monolayer of contamination (oxygen and carbon), which should not effect the evaluation. Since the measured quantities are not changed by repeated etching and correlate strongly with treatment before, during and after oxidation, the reported data are at least relevant, if not numercially completely applicable.

Only for x-ray analysis the oxide may stay in place completely. So far only preliminary results are available with x-rays [10] and STM [11].

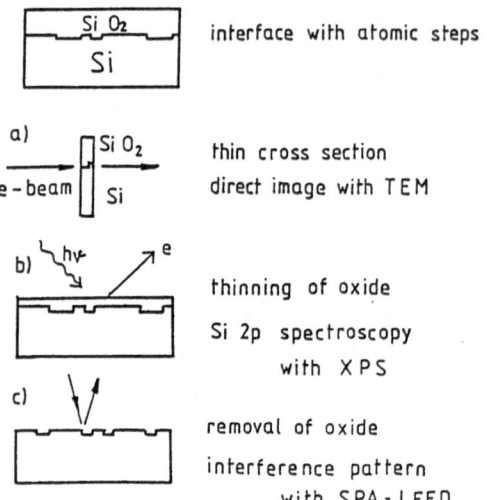

interface with atomic steps

a) thin cross section
direct image with TEM

b) thinning of oxide
Si 2p spectroscopy
with XPS

c) removal of oxide
interference pattern
with SPA-LEED

Fig. 6
Experimental procedures to study
the Si/SiO₂ interface with atomic
resolution

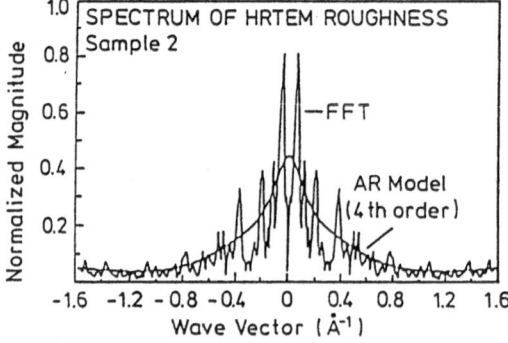

Fig. 7
Fourier transform (FFT) of an
autocorrelation function (spectrum
of roughness) as derived from a
TEM micrograph by visual digiti-
zation and additional smoothing
(AR model), (from |6|).

Several groups have reported high resolution TEM images of Si/SiO₂ interfaces
[6, 12]. Always a sharp interface within one layer is observed. Goodnick et al. [6]
evaluated quantitatively out of micrographs the autocorrelation and its Fourier
transform with respect to an inclined average plane as a reference (Fig. 7) which
is noisy due to the limited size of the micrograph. Nevertheless he could derive
an asperity height (≈ 0.2 nm) and a correlation length (1.0 to 2.2 nm depending
on oxidation parameters). They also estimated that the measured asperity height
is half of the real one due to projection along the interface. Hall effect measurements on the same wafer have been described quantitatively with those structural
data.

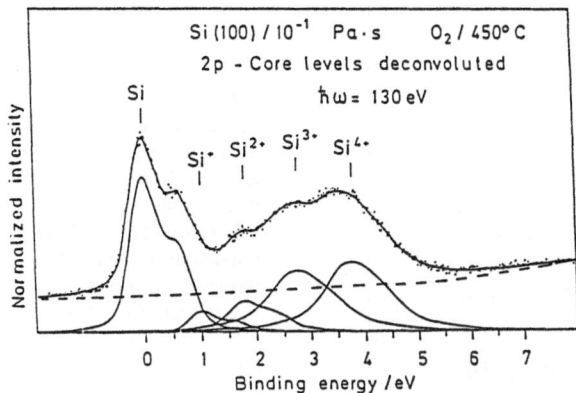

Fig. 8

Photoemission out of the Si 2p-level for Si (100) with 0.68 nm oxide. The five deconvoluted (double) peaks correspond to bulk silicon (Si) and silicon with 1 to 4 oxygen bonds (Si^+ to Si^{4+}), (from [14]).

The chemical shift of the Si 2p-level in photoemission yields the number of oxygen atoms close to the emitting silicon atom [8, 13, 14]. An example for a very thin oxide is shown in Fig. 8. The measured signal is deconvoluted into a signal of substrate silicon with one to four neighboring oxygen atoms (with nearly 1 eV shift per oxygen atom). Since silicon has four oxygen neighbors in quartz, the interface is characterized by the number of silicon atoms with one to three oxygen atoms. The position close to the interface is determined either by variation of photon energy [14] or by measurements after progressive thinning of the oxide [8]. It could be shown that the fraction of Si^+, Si^{2+}, and Si^{3+} is related to orientation of the substrate and to oxidation and treatment parameters, although a direct interpretation in structural terms is difficult.

The most extensive studies have been made with LEED [9, 15]. With low resolution measurements at Si(111) interfaces it has been demonstrated that the edge atom density is drastically dependent on pretreatment, oxidation, and especially post-oxidation annealing [9, 16]. Measurements of CV and Hall effect (at T = 4 K) on the same wafers show that carrier mobility, state density, and Coulomb scattering centers depend directly on edge atom density [9, 17]. The development of high resolution instruments [4, 18, 19] not only extended the range of detectable distances up to distances of 1 μm. Due to the separation of a central spike the vertical distribution is now available. The first measurements just included wide terraces [15]. Now the first systematic study shows both the lateral and vertical distribution for Si/SiO$_2$ interfaces of 10 nm oxide thickness after different annealing in nitrogen [20]. A typical spot profile is shown in Fig. 9 for an out-of-phase condition. The in-phase condition is shown as a dashed curve, which reproduces the instrumental broadening (= shape of central spike). By fitting with a sum of this spike and two Lorentzian shoulders the full profile is reproduced. The evaluation yields a strong variation of the shoulder with energy. The data are fitted with only three layers contributing to the surface. The evaluation for all samples

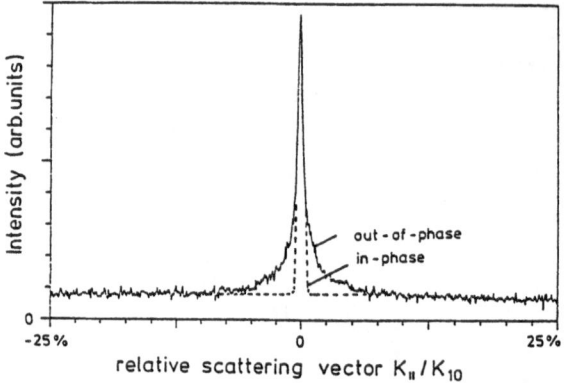

Fig. 9
LEED spot profile of a Si(111) with 10 nm oxide after oxide removed at an out-of-phase condition. The profile for in-phase condition is included to indicate the shoulder due to terraces (from [20]).

Table 1 Summary of results on roughness at the Si(111)/SiO$_2$ interface

Annealing in N$_2$	Si(111) with 10 nm oxide (dry oxidation at 800 °C)			
	none	800 °C 10 min	1000 °C 10 min	1000 °C 1 h
asperity height (nm)	0.20	0.16	0.15	0.15
contributing layers	3	2 ... 3	2	2
correlation length (long and short)	6.3 1.1	6.3 1.3	6.3 1.8	6.3 2.5
step atom density %	15	16	11	6

is shown in Table 1. It is seen that only 2 to 3 layers contribute depending on the annealing temperature and time. The edge atom density decreases simultaneously. It is therefore quantitatively shown that the interface has a root mean square deviation of less than a layer distance, which is reduced by annealing. For the Si(100) layer a special problem arises from the two sublattices, which may form the surface. Due to the small distance they are not resolved by TEM. For diffraction they have the problem of not being equivalent for most directions of incidence [21]. Only for normal incidence and inclination towards a [010]-direction the scattered intensity from the two levels is equivalent so that single and double steps are distinguished. A study with 10 nm oxides on Si(100) and identical annealing as before with Si(111) shows that always the minimum step height is observed (d = 0.125 nm) and always four layers contribute to the surface [20]. Due to the small layer distance the asperity height is only a bit larger. With annealing only the edge atom density is reduced. So far no electrical measurements are available to

192

show how electrical parameters are correlated with the high resolution LEED data. Calculations of mobility [22] would enable a direct comparison of measured structural and electrical data, as done preliminary in [6].

5 Epitaxial Surfaces

Modern technology requires thin layers of low conductivity or sequences of different doping, which are best produced by epitaxy. The thinner the layers the more important is the flatness of a growing layer. It is therefore desirable to study the microstructure during growth. With reflection electron microscopy growth and dissolution by migration to and evaporation from step edges at high temperatures have been observed in real time [23]. Also low energy electron microscopy enables such studies [24]. Here again a small area is studied in detail. Average data are available with diffraction. Many experiments have been performed with RHEED in molecular beam epitaxy (MBE), since the gracing incidence of primary and diffracted beam is compatible with the molecular beam at normal incidence [5] and available in all commercial MBE-systems. The layer-by-layer growth has been derived from the oscillation of the diffracted intensity with growth in real time for silicon [25, 26] and more frequently for III-V compounds [5, 27]. One period corresponds to growth of one monolayer. The minimum is due to destructive interference of half a monolayer (LEED) or by an additional change of the dynamic scattering factor due to islands (RHEED). Therefore quantitative analysis of LEED profiles is possible without dynamic calculations by use of intensity ratios as described in section 3. As an example the epitaxy of silicon on Si(111) is shown in Fig. 10. To understand the growth mode, the growth has been interrupted at different stages and studied carefully by measurement of full profiles at different

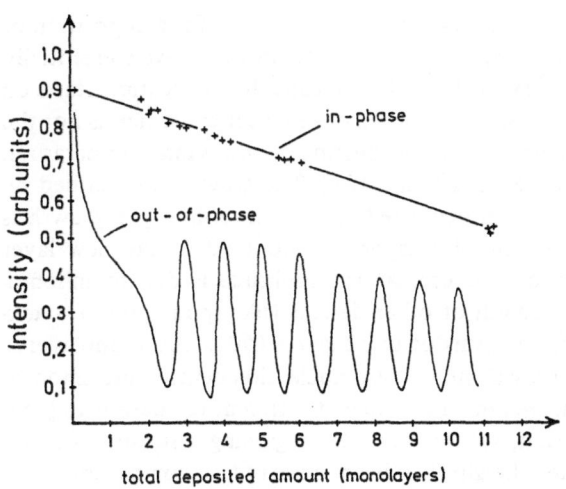

Fig. 10

Intensity of the center of the 00-beam from a Si(111) surface (out-of-phase condition) during epitaxial growth of Si (from [28]).

193

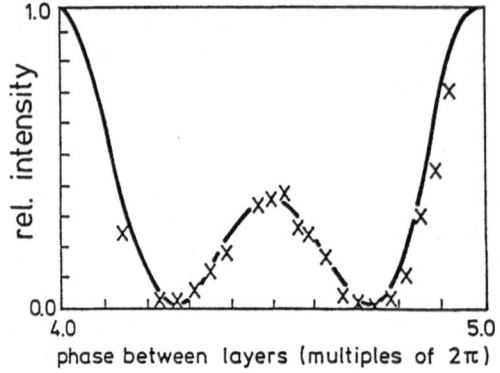

Fig. 11
Ratio of central spike to total spot intensity for a Si(111) surface after deposition of one double layer at 520 °C (from [28]).

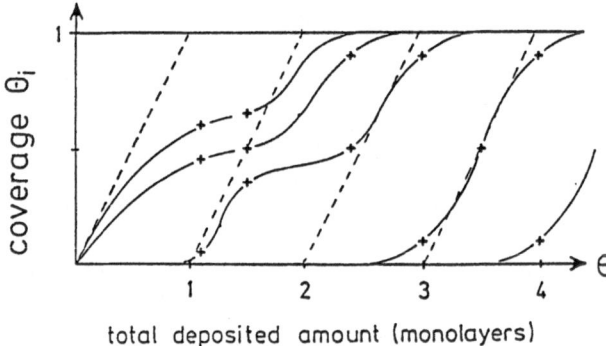

Fig. 12
Schematic presentation of the growth of the first layers of Si on Si(111) at T = 520 °C (from [28]).

energies. An example of the vertical analysis is shown in Fig. 11 for a deposition of one double layer. The evaluation reveals an arrangement in three levels essentially with islands, which are two double layers high. That means that nucleation starts on a well annealed 7×7 superstructure with an anomalous nucleation, which is not the case for deposition at room temperature and heating to the same temperature, where a complete, nearly perfect layer is found [28]. The growth is described by Fig. 12, which shows that after several deposited layers a layer-by-layer growth is established, where before completion of a layer (at about 90 %) the new layer already nucleates. The decrease of the oscillation amplitude is due to inhomogeneous evaporation and due to growth of other defects (like spiral growth, stacking faults) as derived from careful analysis of thick layers (50 ... 100 monolayers). The variation in nucleation and growth mode, also nucleation density and smoothing after deposition has been measured quantitatively at different temperatures [28]. For Si(100) no delayed nucleation, rather from the beginning a steady layer-by-layer growth with minimum step height has been observed at comparable (not

194

lower) temperatures [29]. LEED is therefore especially suited for quantitative analysis of homoepitaxial growth in situ without deterioration of the sample.

For heteroepitaxial growth a lot of new features have to be considered. Depending on the surface energy of the new layer the growth may be layer-by-layer (Frank-v.d.Merve), in three-dimensional islands (Vollmer-Weber) or after one or a few layers in islands (Stranski-Krastanov). Any misfit in lattice constants may produce layers of fixed or variable orientations, with strained layers (pseudomorphic, no misfit dislocations) or with an independent lattice constant (due to misfit dislocations). A lot of studies have been made for Ge on Si [30, 31] and $NiSi_2$ on Si [32]. Here also a lot of related electrical and optical properties have been found, which cannot be covered here.

6 Conclusion

The expansion of surface physics to the range of the recent and coming device technology provides a lot of insight into the processes, which form the micro-structure at technologically important interfaces. It is a challenge in scientific respect, since it describes processes effective over many atomic distances. It may help to understand and improve the electrical device parameters connected with the microstructure of the silicon interfaces.

Acknowledgements

The cooperation with Wacker-Chemitronic (which supplied also all silicon crystals) is gratefully acknowledged. The investigations have been supported by the Deutsche Forschungsgemein-schaft and the US Army through its European Research organization.

References

[1]   C. Davisson and L. H. Germer, Phys. Rev. **30**, 705 (1927)

[2]   O. L. Krivanek and J. H. Mazur, Appl. Phys. Lett. **37**, 392 (1980)

[3]   G. Binnig and H. Rohrer, Surf. Sci. **152/153**, 17 (1985)

[4]   a) M. Henzler, Appl. Surf. Sci. **11/12**, 450 (1982)
      b) M. G. Lagally, Appl. Surf. Sci. **13**, 260 (1982)

[5]   a) P. K. Larsen, B. A. Joyce, and P. J. Dobson in: Springer Series in Surf. Sci. **3**, ed. by F. Nizzoli, K. H. Rieder, R. F. Willis, Springer, Berlin 1985
      b) J. M. van Howe, C. S. Lent, P. R. Pukite, and P. I. Cohen, I. Vac. Sci. Techn. **B1**, 741 (1983)
      c) J. M. van Howe, C. S. Lent, P. R. Pukite, and P. I. Cohen, I. Vac. Sci. Techn. **A1**, 546 (1983)
      d) J. M. van Howe, C. S. Lent, P. R. Pukite, and P. I. Cohen, I. Vac. Sci. Techn. **B2**, 243 (1984)
      e) J. M. van Howe, C. S. Lent, P. R. Pukite, and P. I. Cohen, I. Vac. Sci. Techn. **B3**, 563 (1985)

[6]   a) S. M. Goodnick, R. G. Gann, J. R. Sites, D. K. Ferry, C. W. Wilmsen, D. Fathy, and O. L. Krivanek, J. Vac. Sci. Techn. **B1**, 803 (1983)
      b) S. M. Goodnick, D. K. Ferry, C. W. Wilmsen, Z. Lilienthal, D. Fathy, and O. L. Krivanek, Phys. Rev. **B32**, 8171 (1985)

[7] *T. Ando, A. B. Fowler,* and *F. Stern,* Rev. Mod. Phys. **54**, 437 (1982)

[8] *F. J. Grunthauer, P. J. Grunthauer, M. H. Hecht,* and *D. Lawson* in: Insulating Films on Semiconductors, ed. by *J. J. Simonne* and *J. Buxo* (North Holland, Amsterdam 1986)

[9] a) *P. O. Hahn* and *M. Henzler,* J. Appl. Phys. **54**, 6492 (1983)
   b) *P. O. Hahn* and *M. Henzler,* J. Vac. Sci. Techn. **A2**, 574 (1984)

[10] *I. K. Robinson,* priv. comm.

[11] *R. J. Behm* and *P. O. Hahn,* priv. comm.

[12] *C. d'Anterroches,* J. Micr. Specr. El. **9**, 147 (1984)

[13] a) *G. Hollinger* and *F. J. Himpsel,* Phys. Rev. **B28**, 3651 (1983)
   b) *G. Hollinger* and *F. J. Himpsel,* J. Vac. Sci. Techn. **A1**, 640 (1983)

[14] *W. Braun* and *H. Kuhlenbeck,* Surf. Sci. **180**, 279 (1987)

[15] *M. Henzler* and *P. Marienhoff,* J. Vac. Sci. Techn. **B2**, 346 (1984)

[16] a) *M. Henzler* in: INFOS, ed. by *J. J.Simonne* and *J. Buxo* (North Holland, Amsterdam 1986)
   b) *M. Henzler* in Solid State Devices 1986 (Inst. Phys. Conf. Series **82**, London 1986)

[17] *P. O. Hahn, S. Yokohama,* and *M. Henzler,* Surf. Sci. **142**, 545 (1984)

[18] *M. G. Lagally* and *J. A. Martin,* Rev. Sci. Instr. **54**, 1273 (1983)

[19] *M. Henzler,* Surf. Sci. **132**, 82 (1983)

[20] *J. Wollschläger,* Diplomarbeit, Hannover 1986

[21] *J. A. Martin, C. A. Aumann,* and *M. G. Lagally,* J. Vac. Sci. Techn. in press

[22] a) *A. Gold,* Phys. Rev. Lett. **54**, 1079 (1985)
   b) *A. Gold* and *W. Götze,* Phys. Rev. **B33**, 2495 (1986)

[23] a) *N. Osakabe, Y. Tanishiro, K. Yagi, G. Honjo,* and *K. Takayanagi,* Surf. Sci. **97**, 393 (1980)
   b) *N. Osakabe, Y. Tanishiro, K. Yagi, G. Honjo,* and *K. Takayanagi,* Surf. Sci. **102**, 424 (1980)
   c) *N. Osakabe, Y. Tanishiro, K. Yagi, G. Honjo,* and *K. Takayanagi,* Surf. Sci. **104**, 527 (1981)

[24] *W. Telieps* and *E. G. Bauer,* priv. comm.

[25] *K. D. Gronwald* and *M. Henzler,* Surf. Sci. **117**, 180 (1982)

[26] *T. Sakamoto, N. J. Kawai, T. Nakagawa, K. Ohta,* and *T. Kojima,* Surf. Sci. **174**, 651 (1986)

[27] *P. J. Dobson, J. H. Neave,* and *B. A. Joyce,* Surf. Sci. **119**, L339 (1982)

[28] a) *M. Horn* and *M. Henzler,* J. Cryst. Gr. **81**, 428 (1987)
   b) *R. Altsinger,* Diplomarbeit, Hannover 1986

[29] *U. Gotter,* Diplomarbeit, Hannover 1987

[30] *H. J. Gossmann* and *L. E. Feldmann,* Appl. Phys. **A38**, 171 (1985)

[31] *E. Kasper,* this volume

[32] a) *R. T. Tung, J. M. Gibson,* and *J. M. Poate,* Phys. Rev. Lett. **50**, 429 (1983)
   b) *J. C. Hensel, R. T. Tung, J. M. Poate,* and *F. C. Unterwald,* Surf. Sci. **142**, 37 (1984)

Dedicated to Prof. Karlheinz Seeger on the occasion of his 60th birthday

Hot Phonons

Peter Kocevar

Institut für Theoretische Physik, Universität Graz, A-8010 Graz, Austria

Summary: The conventional theories of electrical transport in solids neglect any carrier-induced perturbation of the phonon system from its thermal equilibrium. While this "Bloch-assumption" is well justified within the linear response regime, it is expected to fail in cases of strong carrier excitation and highly effective carrier-phonon interactions.

The present survey summarizes the experimental and theoretical indications of noticeable such nonequilibrium-phonon effects in the energy dissipation of laser-pulse excited hot electron-hole plasmas in semiconductors and discusses some further theoretical evidence for similar effects in nonohmic d.c. semiconductor transport, with main emphasis on the strongly coupled carrier-LO phonon system in polar materials such as bulk GaAs or GaAs-based heterostructures.

1 Introduction

The ultrashort time response of electrically or optically excited charge carriers in solids, and in particular in semiconductors, has recently become an intensively investigated and disputed topic. One source for this growing interest is the rapidly evolving time-resolved laser spectroscopy which for the first time has opened the pathway to direct studies of even the fastest carrier relaxation processes.

New insight into these fundamental carrier processes is not only of scientific relevance but might soon have direct technological implications. As the switching times of high-speed electronic devices become comparable to the carrier relaxation times, the question arises, whether some hitherto neglected or irrelevant details of the carrier dynamics might under such highly transient conditions noticeably modify the device performance.

Since carriers loose their energy through optical phonon emission, one such expectation would be that effects of nonequilibrium optical phonons might influence the electrical and optical high-field response of semiconductors.

As only small regions around band extrema are occupied by the carriers, the conservation of energy and momentum for each carrier-phonon (c-ph) collision restricts the electronically active lattice modes to small regions in $\vec{q}$-space. Because of the marked $\vec{q}$-dependence of the couplings a still much smaller fraction of modes will dominate the c-ph dynamics. Even then, efficient energy and momentum relaxation of the carriers is only provided by those phonons which rapidly decay into the "heat bath" of the electronically non-active phonons or decay at the crystal boundaries. Otherwise the most active modes will soon form a bottleneck for the lattice relaxation of the carriers and start to feed, through increased reabsorption processes, the initially received momentum and energy back to the carriers.

To quantify this "heat bath" condition, which finally decides, whether such nonequilibrium or "hot" phonon effects can be expected under given experimental conditions, we note that the lifetimes of optical phonons are generally of the order of several picoseconds, with corresponding phonon thermalization rates of several 10^{11} s^{-1}. Under hot-electron conditions, however, the amplification rates of long-wavelength LO phonons through the Coulombic polar-optical coupling to electrons and holes can in compound semiconductors reach 10^{12} s^{-1}, and only slightly slower maximal amplification rates of both LO and TO phonons are found for the short-range optical deformation potential coupling to holes and L-valley electrons in the compound as well as elemental tetrahedral semiconductors.

So experimentally measurable consequences of LO-phonon disturbances on the high-field response of semiconductors had for some time been expected and were indeed theoretically predicted for the transient current-field characteristics as well as for the collective dielectric breakdown of n-InSb [5].

It should be noted that even earlier theoretical work on phonon disturbances was concerned with the electric-field induced amplification of acoustic phonons, mainly in connection with acousto-electric [6] and phonon-focussing phenomena [7]. But as the acoustic electron-phonon couplings are weak, the generation times for acoustic phonon disturbances are typically of the order of microseconds or longer and therefore of no relevance for the ultrafast effects of our present concern.

The purpose of this contribution is to give a critical survey of current hot (optical) phonon research. In chapter 2 we shall summarize the direct and indirect experimental evidence for effects of optical phonon disturbances in the dynamics of laser-pulse excited hot carrier plasmas in bulk semiconductor materials. Chapter 3 will state the basic transport model and the three most effective methods for its solution. In chapter 4 we shall discuss direct applications of these approaches to experimental data or device-related simulations. Chapter 5 is devoted to the question of possible hot-phonon effects in micron or submicron semiconductor structures, with some key references to experimental and theoretical work related to the hot-phonon problematics. The last chapter 6 will summarize the present state of hot-phonon research.

2 Experiment

Strongly nonequilibrium populations of optical phonons were first experimentally detected in time-resolved pulse-and-probe Raman spectroscopy of polar semiconductors [8], whereas the systematic experimental research on the detailed hot carrier-hot phonon dynamics started 1979, when J. Shah and collaborators interpreted their experimental finding of a strongly retarded energy relaxation of highly laser-excited electron-hole (e-h) plasmas in GaAs as an effect of a nonequilibrium LO-phonon distribution [9]. These relaxation rates were obtained from time-resolved transmission spectra, whose shape could be reasonably well fitted with Fermi-distributions for electrons and holes at a common carrier temperature T_c. The decrease of T_c with time after a 0.5 ps excitation pulse of energy flux densities of several ten to hundred MW/cm^2 was compared with standard hot-electron calculations for equilibrium phonons. The measured plasma-cooling rate was found to be several times slower than the theoretical prediction, even after properly accounting for the screening of the long-range polar coupling by the photo-excited free carriers.

Similar findings of a strongly reduced cooling rate of laser generated carrier plasmas in GaAs and other polar materials involved the same model of a common e-h-temperature for the analysis of time-resolved transmission or hot luminescence spectroscopy with ultrashort excitation and probe laser pulses. Depending on material and experimental conditions, the effect was either attributed to the dominance of free-carrier screening [10 ... 14] or of LO-phonon heating [15, 16] or to a comparable combined action of the two mechanisms [17, 18].

At the same time new direct experimental evidence for strong laser-pulse induced LO-phonon amplification accumulated from further refined picosecond and subpicosecond pulse-and-probe Raman techniques [19 ... 21].

These investigations were additionally stimulated by the intensification of the "laser-annealing debate" about the possibility of a nonthermal melting of ion-implanted semiconductor surfaces, especially of silicon, during illumination with an extremely intense laser pulse. This controversy had originally started in 1980 from a misinterpretation of Raman data indicating a cold instead of hot optical phonon population at the illuminated surface, which led to the strongly disputed proposal of a softening of the still cold lattice through the breaking of interatomic bonds in the presence of an extremely hot, dense, and long-lived carrier plasma [22].

3 Theory

3.1 Transport Formulation

Most of our following discussion will be formulated in the framework of semiclassical transport theory, which is concerned with the dynamics of individual particles under the influence of external forces and of instantaneous scattering events with quantum mechanical transition probabilities. We shall consequently consider carriers in or near band extrema with effective mass m, crystal momentum $\vec{p}$, and energy $E(\vec{p})$, interacting with phonons treated as ballistic particles of energy $\hbar\omega_b(\vec{q})$ and crystal momentum $\vec{q}$, where ω denotes the frequency and b the branch index of the vibrational mode.

Strictly speaking, such a description of phonons as ballistic particles requires that the frequency of a phonon is much higher than its collision or decay rate, which is well satisfied for our following transport description of optical phonons. Much more restrictive is the corresponding criterion for the use of the carrier-transport equation, that the mean free time between carrier-phonon processes should be much longer than the vibrational period of the participating phonons. Since the frequency of optical lattice vibrations ($\approx 10^{13}$ s^{-1}) is comparable to the rates for the dominant optical phonon scattering of hot carriers in polar materials, all semi-classical transport descriptions of highly excited carriers in polar semiconductors are at the limits of their applicability. For a deeper discussion of these fundamental questions we refer to the introductory literature on nonlinear quantum transport [23 ... 25].

As optical phonons are very short-lived and of very small frequency dispersion, their group velocity and resulting spatial diffusion can be neglected. For optical excitation they are therefore confined to the light-absorption layer (with a depth of the order of one micron), within which the e-h plasma is being created. In this way one can treat the carrier-phonon system as spatially homogeneous and could even include effects of rapid carrier diffusion out of the excitation layer [26] by using a properly increased effective pair-recombination coefficient [27].

For d.c. transport the assumption of spatial homogeneity in theoretical hot-phonon calculations has been mainly motivated by the strongly reduced computational complexity. But in spite of the decisive role of contact effects and of the field distribution on the space-time dependence of the current density within the sample [28, 29], the strong localization of the amplified optical phonons should allow their eventual straightforward inclusion in one of the spatially coarse-grained transport codes for small devices.

Due to our restriction to spatial homogeneity, the evolution of the coupled carrier-phonon system is governed by the time dependence of the momentum distribution functions $f(\vec{p})$ for all types of carriers and $N(\vec{q})$ for the electronically active optical phonons. These distributions are the solutions of their respective Boltzmann equations

$$\frac{df}{dt} = \frac{df}{dt}\bigg|_{field} + \frac{df}{dt}\bigg|_{c\text{-}ph} + \frac{df}{dt}\bigg|_{c\text{-}c} + \frac{df}{dt}\bigg|_{c\text{-impurity}} \tag{1}$$

and

$$\frac{dN}{dt} = \frac{dN}{dt}\bigg|_{ph\text{-}c} + \frac{dN}{dt}\bigg|_{ph\text{-}ph} . \tag{2}$$

These two nonlinear integro-differential equations are coupled through the c-ph collision and ph-c collison integrals, which for each scattering event contain the mean occupation probability f_i of the initial and the mean non-occupation probability $(1 - f_f)$ of the final carrier state, as well as the statistical weights $(N + 1)$ or N for a phonon emission or absorption. It is the appearance of these products of the unknown distribution functions which makes the solution of the coupled transport equations a formidable computational problem.

As the Boltzmann equation for charge carriers in semiconductors is well documented in the textbook literature, we shall concentrate on the phonon equation, with special emphasis on our later applications to polar materials.

200

3.2 Carrier-Phonon Couplings

In the tetrahedral semiconductor compounds the electronic optical-phonon processes are dominated by the polar-optical (po) coupling of electrons and holes to long-
wavelength LO phonons, the optical deformation potential (odp) scatterings of
(zone-center) holes or L-valley electrons with both LO and TO phonons, and the
deformation potential couplings of zone-boundary phonons and of electrons in intervalley (iv) transfer processes. In lack of experimental indications of perturbations of
these intervalley phonons [20], they are generally treated as in thermal equilibrium.

The ph-c collision term in the transport equation for each type of electronically
active phonons contains all its possible couplings to the various types of carriers:

$$\frac{dN(\vec{q})}{dt}\bigg|_{ph\text{-}c} = \sum_{carriers} \sum_{couplings} \frac{2\pi}{\hbar} B(q) \cdot$$

$$\cdot \sum_{\vec{p}} \{f(\vec{p}) \cdot (1 - f(\vec{p} - \vec{q})) \cdot (N(\vec{q}) + 1) \cdot \delta(E(\vec{p} - \vec{q}) - E(\vec{p}) + \hbar\omega) -$$

$$- f(\vec{p}) \cdot (1 - f(\vec{p} + \vec{q})) \cdot N(\vec{q}) \cdot \delta(E(\vec{p}) - E(\vec{p} + \vec{q}) + \hbar\omega)\}, \quad (3)$$

where the coupling constants $B(q)$ are obtained from the quantum-mechanical
transition amplitudes for the phonon emission or absorption processes

$$|\langle \vec{p} \mp \vec{q}; N(\vec{q}) \pm 1 | H_{c\text{-}ph} | \vec{p}; N(\vec{q}) \rangle|^2 = B(q) \cdot (N(\vec{q}) + \frac{1}{2} \pm \frac{1}{2}). \quad (4)$$

The range of the allowed $\vec{q}$ vectors for a given carrier-phonon process is fixed by
the energy and momentum conservation but can be further restricted by the
momentum dependence of the various c-ph couplings.

The polar-optical coupling is given by [30]

$$B_{po}(q) = \frac{1}{V} 2\pi e^2 \hbar^3 \omega_{LO} \left(\frac{1}{\epsilon_\infty} - \frac{1}{\epsilon_0}\right) \frac{1}{q^2} \left(\frac{q^2}{q^2 + q_s^2}\right)^2, \quad (5)$$

where the screening of this long-range Coulombic interaction by the free carriers
has been introduced through the standard static Thomas-Fermi or Debye-Hückel
correction with the screening parameter q_s. V denotes the crystal volume, and ϵ_∞
and ϵ_0 are the optical and static dielectric constants.

More refined dynamical screening models have lately been developed for highly
photo-excited e-h plasmas [31,32], with the possibility of antiscreening as consequence of intracollisional field effects. This finding revives Doniach's old idea of
possible plasma-induced antiscreening effects of the polar interaction [33], which
also anticipates recent attempts to include effects of the formation of plasmons or
of mixed phonon-plasmon modes into the polar interaction [34 ... 36].

It is seen from Eq. (5) that the momenta of the most strongly coupling LO phonons will be of the order of the screening parameter q_s, which is a slowly increasing function of the carrier density n_c (for nondegeneracy $q_s \sim n^{1/2}$); LO phonons of smaller q will be screened out.

The nonpolar optical deformation potential (odp) couplings are given by [30]

$$ B(q) = \frac{D^2 \hbar}{2V\rho\omega} , \qquad (6) $$

with ρ the mass density and D the deformation potential for intravalley or intervalley scattering, typically of the order of 10^9 eV/cm. Due to the short range of these couplings, free-carrier screening is of secondary importance and can be even further reduced by symmetry. Nevertheless, some unexpected novel properties of odp screening have been recently proposed for photo-excited e-h plasmas in Si and Ge [37]. The momentum independence of the coupling and the available phase-space have the consequence that high q-values are preferred in odp scatterings.

3.3 Phonon-Phonon Interaction

The dominant dissipation channels for nonthermal zone-center optical phonons are decay processes, whose (weak) temperature dependence is governed by the phonon creation coefficients $N + 1$ of the decay products, which by energy and momentum conservation belong to electronically inactive zone-boundary modes. As long as this second phonon generation and its descendants decay into still lower lying phonons, this cascade mechanism into the thermal reservoir of the crystal lattice will act as an energy and momentum sink for the coupled carrier-phonon system, at least for the short times of our present interest. Time and space resolved spectroscopy of the high-frequency acoustic phonons involved in the early stages of this phonon cascade has become a very active field of phonon research and leads to interesting new insights into the decay and propagation dynamics of some zone-boundary modes, e.g. of the TA branches in GaAs [7].

There is a wide spread of experimentally determined LO-phonon lifetimes, mainly from Raman linewidths and excite-probe Raman spectra. Although the dependence of the lifetimes on the lattice temperature T_L is reasonably well described by the (material independent) Klemens-formula [38]

$$ \tau_{op}(T_L) = \tau_{op}^0 \, \frac{1}{1 + 2(\exp\frac{\hbar\omega}{2k_B T_L} - 1)^{-1}} , \qquad (7) $$

the zero-temperature lifetimes τ_{op}^0 extrapolated from the measurements vary between 8 ps [39, 19] to more than 20 ps [40, 8]. The possibility of a strong influence of the quality of the irradiated surface on the phonon-decay rates within the thin excitation layer, as found in II-VI compounds [41], has to our knowledge never been systematically investigated in GaAs. But the more recent measurements seem to converge towards the lower τ_{op}^0 values around 8 ps [42].

Although TO-phonon lifetimes in compound semiconductors are found to be slightly longer, this difference has been generally neglected in theoretical work on non-equilibrium TO phonons in polar materials.

The dominance of the decays of zone-center optical phonons into pairs of undisturbed zone-boundary phonons has the great advantage of allowing the use of a "single-mode" relaxation time for the nonelectronic thermalization rates of amplified LO and TO phonons, in analogy to the description of the decay of externally induced acoustic phonons in ultrasonic attenuation or in heat-pulse experiments:

$$\frac{dN(\vec{q})}{dt}\bigg|_{ph\text{-}ph} = -\frac{N(\vec{q}) - N_L}{\tau_{op}} \, , \tag{8}$$

where N_L is the thermal Planck distribution

$$N_L = \frac{1}{\exp \dfrac{\hbar\omega}{k_B T_L} - 1} \tag{9}$$

and τ_{op} is given by Eq. (7).

3.4 The Hot-Phonon Concept

Similar to the notion of hot carriers, one speaks of "hot phonons", whenever the mean phonon occupation number of a lattice mode deviates strongly from its thermal equilibrium Planck distribution N_L.

As for hot carriers, the property "hot" does not imply that the isotropic part of the disturbed distribution is just a thermal distribution at an elevated temperature. In general the perturbing action of the hot carriers will lead to an enhancement and also to a deformation of the phonon distribution in $\vec{q}$-space. The interplay between this carrier-induced perturbation and the thermalizing action of the nonelectronic phonon processes will ultimately determine the amplification and shape of the phonon distribution. So the decisive factors for the development of nonequilibrium phonon effects are: (i) the degree of field-induced carrier "heating" and drift, (ii) the strength of the c-ph couplings, and (iii) the efficiency of the nonelectronic phonon dissipation processes.

3.5 The Carrier Temperature Models

Among the three major techniques for solving the coupled transport equations for the various types of carriers and phonons the approximate but well established carrier-temperature models have for many years dominated the theoretical research on hot phonons. They assume that carrier-carrier scattering is by far the most efficient carrier process and should be able to always maintain a heated Maxwell (or Fermi) distribution for each type of photogenerated carrier or to establish a heated and drifting Maxwellian (HDM) or Fermi distribution for high d.c. field-excited carriers. The corresponding parameters of each of these distribution functions are assumed to vary in time: the carrier temperature T_c, the chemical potential μ_c, and for d.c.

field excitation also the mean drift velocity $\vec{v}_d$. In this way the functional form of the distributions remains fixed and allows the full analytical solution of the ph-c collision integrals in Eq. (2), which then assume the highly desired simple form of the usual relaxation-time approximation

$$\frac{dN(\vec{q})}{dt}\bigg|_{ph\text{-}c} = -\frac{N(\vec{q}) - N_c(\vec{q})}{\tau_c} .$$

(10)

For HDM carriers the ph-c scattering rate $g_c = 1/\tau_c$ is given by [27]

$$g_c(\vec{q}) = 2\,B(q) \cdot \sqrt{\frac{2m_c\,\pi}{\hbar^2 k_B}} \cdot n_c \cdot \frac{1}{\sqrt{T_c}} \cdot \frac{1}{q} \cdot \exp\left(-\frac{q^2}{8m_c\,k_B T_c}\right) .$$

$$\cdot \exp\left(-\frac{m_c\left(\dfrac{\hbar\omega}{q} - v_d\cos\beta\right)^2}{k_B T_c}\right) \cdot \sinh\frac{\hbar\omega - q v_d\cos\beta}{k_B T_c} ,$$

(11)

and N_c is a Planck distribution, heated to the carrier temperature T_c and Doppler-shifted about the frequency $\vec{v}_d \cdot \vec{q}/\hbar$,

$$N_c(\vec{q}) = \left(\exp\frac{\hbar\omega - q v_d\cos\beta}{k_B T_c} - 1\right)^{-1} .$$

(12)

For photo-excitation it is advantageous to calculate the ph-c rates for Fermi distributions to include both possibilities of degeneracy as well as nondegeneracy of the electron and of the hole plasma component. The result is again of a relatively simple form [43], where we prefer to write Eq. (10) in terms of the emission and absorption probabilities W_e and W_a of a phonon:

$$\frac{dN(\vec{q})}{dt}\bigg|_{ph\text{-}c} = W_e(\vec{q}) \cdot (N(\vec{q}) + 1) - W_a(\vec{q}) \cdot N(\vec{q}),$$

(13)

so that $g_c = W_a - W_e$ and $N_c = W_e/(W_a - W_e)$, where

$$W_e = \frac{m_c\,k_B\,T_c}{\pi\hbar^4} \cdot \frac{\exp(-\Omega)}{1 - \exp(-\Omega)} \cdot \ln\frac{\kappa \cdot \exp\dfrac{(\Omega - z^2)^2}{4z^2} + 1}{\kappa \cdot \exp\dfrac{(\Omega - z^2)^2}{4z^2} + \exp(-\Omega)} \cdot \frac{B(q)}{q}$$

(14)

and

$$W_a = W_e \exp\Omega$$

(15)

with $\Omega = \hbar\omega/k_B T_c$, $z = q/(2\,m_c k_B T_c)^{1/2}$, and $\kappa = \exp(-\mu_c/k_B T_c)$. In this case of no mean carrier drift, N_c is the heated Planck distribution at the temperature T_c of the plasma component under consideration.

204

Summing up the partial rates g_c and $g_{ph} = 1/\tau_{op}$ for carrier and phonon scattering and introducing the total scattering rate $g_c + g_{ph}$, the phonon-Boltzmann equation assumes the form of a first order linear inhomogeneous differential equation

$$\frac{dN}{dt} = -g_c(N - N_c) - g_{ph}(N - N_L), \tag{16}$$

with time-dependent coefficients through the explicit dependence of N_c and g_c on the time-varying carrier parameters T_c, μ_c, and $\vec{v}_d$. If these were known, the solution of Eq. (16) would give the time development of the phonon distribution.

Eq. (16) shows the tendency of the c-ph interaction to drive the phonons into a mutual equilibrium with the hot carriers, with the corresponding ph-distribution N_c, and the competing action of the ph-decay processes to thermalize the disturbed phonons. It is helpful for the quantitative analysis of phonon amplifications to observe from Eq. (10) that $N = N_c$ is just the detailed-balance condition for an isolated c-ph system in mutual equilibrium and therefore the borderline between predominant ph-emission (for $N < N_c$) and predominant ph-absorption (for $N > N_c$) by the carriers.

There have been two approaches to obtain the time variation of the carrier parameters in consistency with the simultaneous evolution of the phonon system. As the c-ph relaxation rates for carrier distributions are generally one order of magnitude faster (for LO phonon scattering of the order of $10^{13}/s$) than the corresponding rates g_c for the phonon distributions, the conventional method assumes an instantaneous adaption of the carrier distribution to the more slowly varying phonon populations. For photo-excitation this is achieved by fitting the carrier parameters to the momentary concentration and energy of the e-h plasma during and after the laser pulse, by calculating at each time step the gain from the laser, the recombination losses and Auger contributions, and the lattice power losses of the carriers

$$\frac{dE}{dt}\bigg|_{c\text{-ph}} = -\sum_{\vec{q}} \hbar\omega \cdot \frac{dN(\vec{q})}{dt}\bigg|_{ph\text{-}c} \tag{17}$$

from Eq. (13), with the phonon distributions obtained from the parallel integration of Eq. (16) [44]. For the d.c. field case the carrier parameters are calculated from the assumed instantaneous energy and momentum balance of the carrier system between the gains from the field and the losses to the lattice, again for the momentary phonon distributions from Eq. (16) [5]. The intervalley transfers, whose time scale is comparable to the time evolution of the phonon disturbances, are, through a parallel integration of their rate equations, treated on the same footing as the phonon evolution in Eq. (16) [45].

The second method couples the HDM formulation of the ph-c term in the phonon transport equation with a simultaneous Monte Carlo simulation of the carriers, which is used to provide the instantaneous mean energy ("temperature") and drift velocity of the carriers in the presence of the phonon disturbances from the parallel integration of Eq. (16) [46].

3.6 The Phonon-Cerenkov Effect for Displaced Maxwellian Carriers

A specific feature of the HDM-carrier model is the explicit demonstration of the possibility of resonant Cerenkov-type phonon emissions. It is seen from Eqs. (11) and (12), that whenever the component of the mean carrier drift $v_d \cdot \cos\beta$ along a phonon momentum $\vec{q}$ becomes larger than the phase velocity of the phonon,

$$v_d \cdot \cos\beta > \hbar\omega/q, \qquad (18)$$

g_c and N_c become negative and can no longer be interpreted as (nonnegative) relaxation rate and distribution function. Since a negative $g_c = 1/\tau_c$ in Eq. (16) acts as an amplification rate, all phonons within the Cerenkov cone defined by Eq. (18) would in the absence of ph-ph processes ($g_{ph} = 0$) be exponentially amplified. But due to the counteraction of the phonon decay processes a net phonon amplification requires

$$g_c + g_{ph} < 0, \qquad (19)$$

which is much more restrictive than the Cerenkov condition $g_c < 0$ for infinitely long-lived phonons. It is obvious that the onset of such an exponential phonon build-up would very soon either drive the carriers back to a subcritical HDM state (with nonnegative $g_c + g_{ph}$) or otherwise lead to a collective breakdown of the spatially homogeneous carrier-phonon system [27].

Equation (19) and the expression

$$N_{as}(\vec{q}) = \frac{g_c N_c + g_{ph} N_L}{g_c + g_{ph}} \qquad (20)$$

for the asymptotic steady-state solution of the phonon transport equation (16) provide a simple estimate of possible nonequilibrium phonon effects for a given experimental situation. By using a reasonable guess for the phonon and HDM carrier parameters in Eqs. (7) and (11) one can first examine the condition $g_c + g_{ph} < 0$ for the onset of a phonon instability and, for subcritical conditions, the degree of phonon build-up by comparing the asymptotic phonon distribution, Eq. (20), with the thermal equilibrium distribution N_L.

3.7 Monte Carlo Simulations

Monte Carlo (MC) techniques have for a long time provided the most powerful methods for solving the hot-carrier Boltzmann equation, because they contain no assumptions about the form of the carrier distribution and directly follow the microscopic particle dynamics. For a detailed description of the computational procedures for steady-state and time-dependent ensemble MC simulations we refer to the literature [47, 48].

To include nonequilibrium phonons into these conventional MC codes, one performs the usual carrier simulations, but keeps track of the varying phonon populations by setting up an appropriate grid in $\vec{q}$-space and counting the number of phonon emissions into and absorptions out of each of the so created $\vec{q}$-cells. The resulting coarse-grained phonon occupation histogram is updated after each carrier-phonon process and scaled by the ratio of the actual and simulated carrier numbers and by the number of $\vec{q}$ states per $\vec{q}$-cell to obtain the properly normalized phonon distribution function. To account for the nonelectronic phonon losses the simulation time is divided into short timesteps $\delta t \ll \tau_{op}$. After each of these subhistories the phonon distribution is corrected for the ph-ph processes during the preceding time interval δt by using Eq. (8) and adding $(dN/dt|_{ph\text{-}ph}) \delta t$. With this completely updated phonon distribution the next subhistory is started.

For optical excitation (and neglect of band anisotropies and carrier diffusion effects) the isotropy of the c-ph system allows the use of concentrical shells in $\vec{q}$-space as the grid for the phonon histograms. For d.c. excitation (and spherical bands) the cylindrical symmetry about the field direction requires a two-dimensional grid in q and in the angle between $\vec{q}$ and the field.

By following the time evolution of the phonon distributions and of the density, mean energy and mean momentum of the simulated carrier ensembles, including generation and recombination processes of carriers by adding or subtracting particles during the simulation, one can perform a detailed study of the transient response of the coupled c-ph systems [49 ... 51].

In these first applications of MC techniques to hot phonons carrier-carrier scattering and plasmon formation are being neglected for simplicity, although techniques for their MC treatment have been developed for conventional carrier simulations [35, 36].

3.8 Direct Numerical Solutions of the Coupled Transport Equations

The most complete analysis of the initial transients immediately after the onset of a subpicosecond laser pulse, before and during the internal thermalization of the carrier system, has been recently achieved by Collet and coworkers through a direct step-by-step solution of the coupled transport equations for electrons, holes, and LO phonons with the exact inclusion of the various (statically screened) c-c and c-ph collision integrals [52].

The solution of the coupled system of nonlinear integro-differential equations is achieved by sampling the carrier distributions within an energy grid of $\delta E \ll \hbar \omega_{LO}$ and the phonon distributions in momentum space and, if necessary, interpolating between the discretization points. The resulting system of typically hundred coupled equations is then stepwise solved in time, with most of the computer time consumed by the carrier-carrier scatterings.

4 Applications

4.1 Carrier Temperature Model: D.C. Transport and Hot Luminescence

Before applying the HDM carrier formulation of chapters 3.5 and 3.6 to the realistic many-valley conduction in n-GaAs we first demonstrate the phonon-Cerenkov effect for the artificial example of a nonequilibrium LO phonon-induced collective breakdown in a hypothetical one-valley model of n-GaAs.

For a first estimate of possible LO phonon instabilities we follow the simple approach of guessing a hopefully realistic range of n_c, v_d, and T_c values to search for possible regions of negative total phonon rates $g = g_c + g_{ph}$. Indeed, by choosing a high but still practically interesting carrier density of $3 \cdot 10^{17}$ cm^{-3} we find in Fig. 1 a quite pronounced range of forward modes ($\beta = 0$) with $g < 0$ for $T_c = 1000$ K and v_d-values $\geq 3 \cdot 10^7$ cm/s, well within the range of the velocity-overshoot predictions [29]. To further investigate the possibility of an LO-phonon avalanche and of a collective c-ph instability, we start from a conventional (ph-equilibrium) steady state HDM solution as initial condition and then apply the techniques of chapter 3.5 to allow

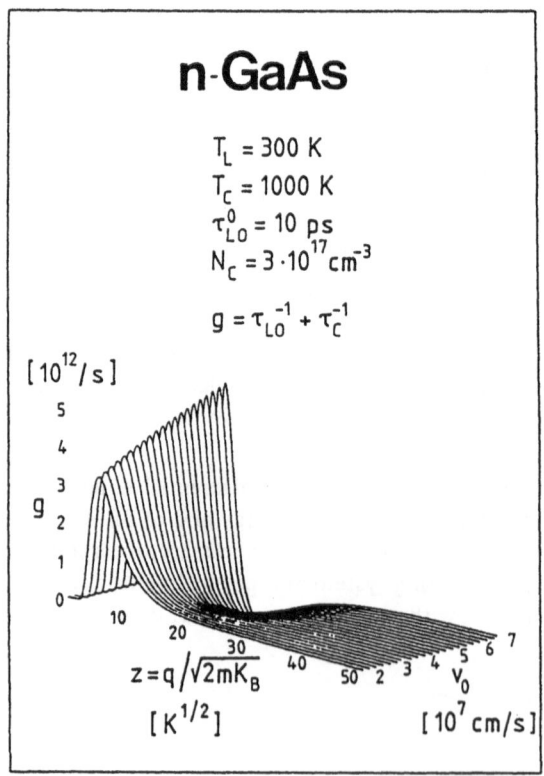

Fig. 1

Total phonon rate as function of $\vec{q}$ and v_d in the central valley of n-GaAs.

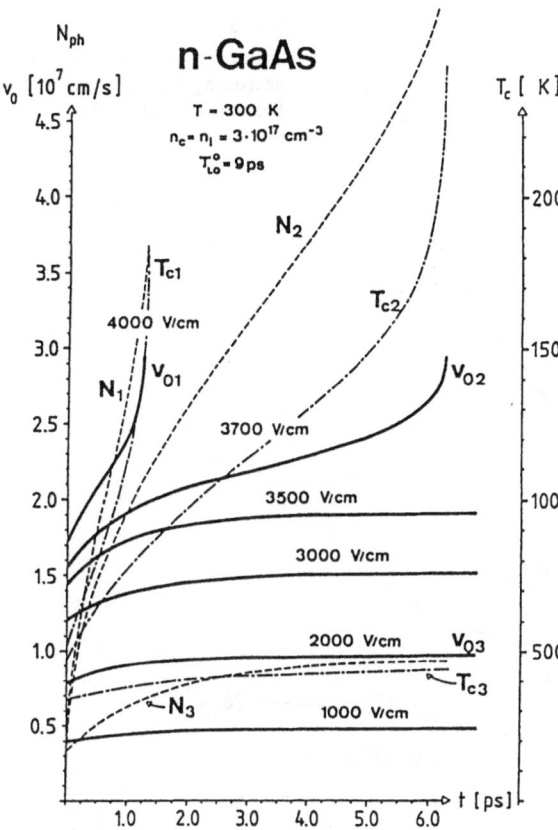

Fig. 2
LO-phonon distributions, v_d and T_c during LO-phonon build up after the establishment of a conventional HDM steady state for the one-valley model of Fig. 1 (from [27]).

the phonons to evolve in time. Fig. 2 shows the results, with the development of a collective breakdown for fields $F \geq 3.7$ kV/cm, initiated by a dramatic increase of the most strongly coupling LO distributions, of the electron temperature T_c (reduced cooling efficiency of the phonons), and of the mean electron drift velocity (carrier drag by the initially amplified forward phonons) [27]. The rather subtle interplay of the phonon-Cerenkov effect, requiring high v_d and high n_c, and the decelerating effect of ionized impurity scattering can be seen from Fig. 3 for the supercritical field $F = 3.7$ kV/cm and for a compensated material. Although the ph-disturbances and the resulting carrier instabilities should increase with a doubling of n_c in Eq. (11), the frictional effect of the also increased ionized-impurity scattering sufficiently reduces v_d to prevent the Cerenkov-induced onset of the breakdown.

This decisive role of the phonon-Cerenkov mechanism shows the insufficiency of all attempts to introduce nonequilibrium phonon effects into high d.c. field transport by the use of a phonon temperature T_{ph}, somewhere between T_L and T_c, which

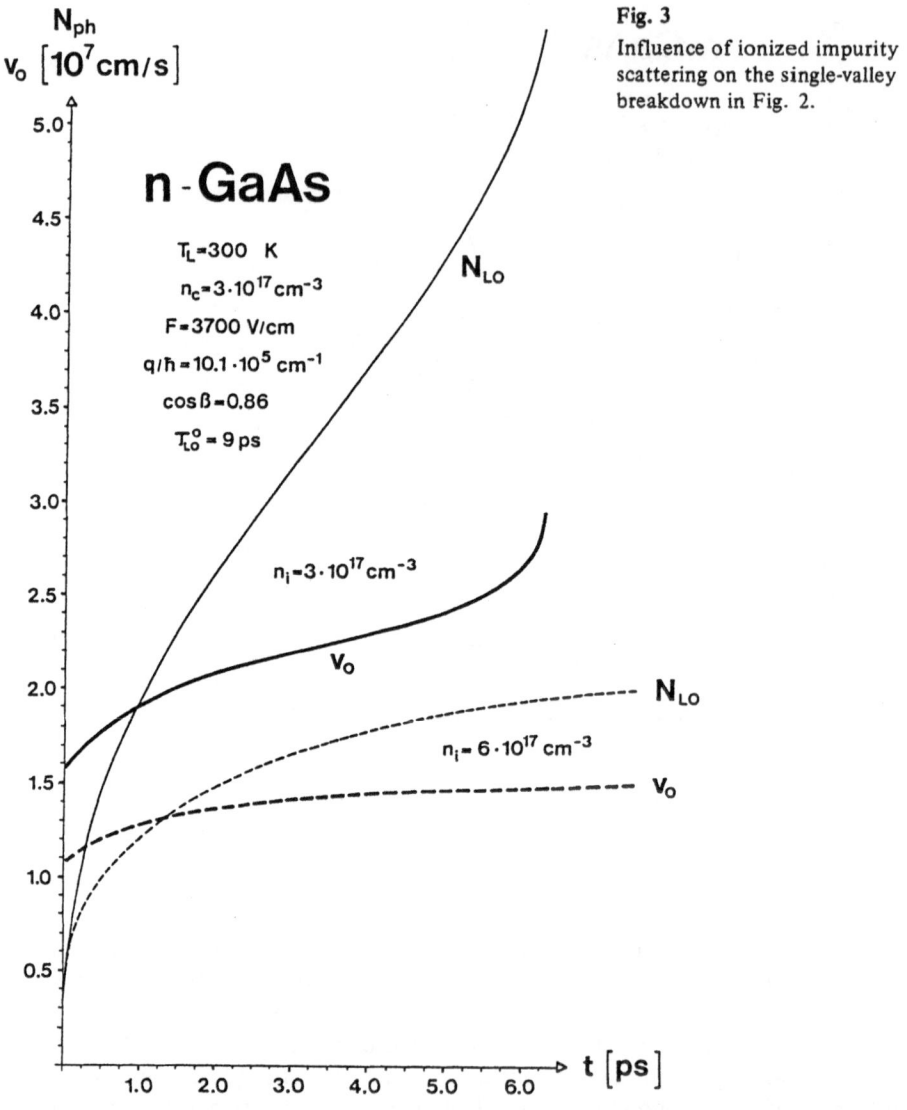

Fig. 3
Influence of ionized impurity
scattering on the single-valley
breakdown in Fig. 2.

procedure would completely miss the drag contributions from the high $\vec{q}$-space anisotropy of the disturbed phonon distributions.

The question of direct practical relevance is, whether such a nonequilibrium central-valley breakdown could also occur for the actual many-valley structure of GaAs, since for fields ≥ 4 kV/cm the Γ-L valley transfer will strongly reduce the number of the fast drifting Γ-electrons responsible for the LO-phonon avalanching.

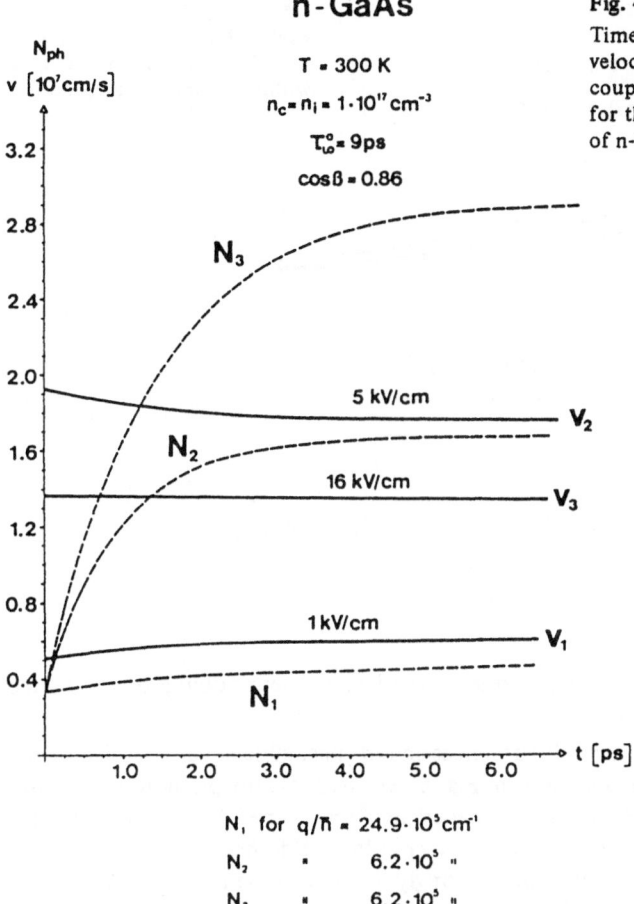

n-GaAs

N_{ph}
$v\ [10^7 cm/s]$

T = 300 K

$n_c = n_i = 1 \cdot 10^{17} cm^{-3}$

$\tau_{LO}^0 = 9\,ps$

$\cos\beta = 0.86$

N_3

N_2

5 kV/cm $\qquad V_2$

16 kV/cm $\qquad V_3$

1 kV/cm $\qquad V_1$

N_1

Fig. 4
Time evolution of the mean drift velocity v and of some strongly coupling LO-phonon distributions for the many-valley bandstructure of n-GaAs.

N_1 for $q/\hbar = 24.9 \cdot 10^5 cm^{-1}$

N_2 " $6.2 \cdot 10^5$ "

N_3 " $6.2 \cdot 10^5$ "

To see whether this transfer is able to stop the initial development of the breakdown we have recently extended the HDM model to the realistic band structure of GaAs [45]. The results are shown in Fig. 4. As was to be expected from the missing experimental evidence for a very fast central valley breakdown in GaAs, the Γ-L transfer is indeed sufficient to ensure the establishment of an asymptotic steady state even at fields much higher than the critical fields of Fig. 2. Only moderate modifications of the velocity-field characteristics are found in spite of strong LO-phonon amplification.

Carrier drag by the initially amplified forward phonons dominates at fields $\leq 2\,kV/cm$, below the onset of noticeable valley transfer, whereas a reduction of the mean carrier drift velocity is found at higher fields as consequence of the increased popu-

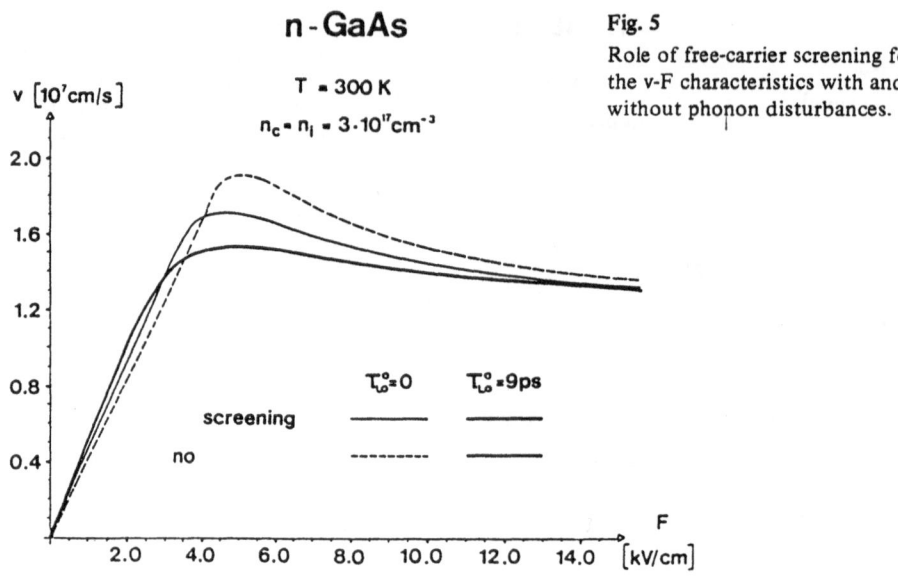

n - GaAs

T = 300 K

$n_c = n_i = 3 \cdot 10^{17} cm^{-3}$

v [10^7cm/s]

Fig. 5

Role of free-carrier screening for the v-F characteristics with and without phonon disturbances.

lation of the low mobility L-valleys due to the less efficient cooling of the Γ-electrons by the hot LO phonons.

The decisive role of the phonon disturbances and the negligible net effect of screening on the v-F characteristics are seen in Fig. 5. Let us first, for phonon equilibrium ($\tau_{op} = 0$), compare the results for screened and unscreened c-ph couplings. The results for no screening show a reduction of v_d at low fields, because of the increased impurity scattering of the now more strongly phonon-cooled Γ-electrons and an increase of v_d at higher fields due to the reduced transfer of the now cooler Γ-electrons into the L-valleys. However, when the perturbation of the LO phonons is taken into account, these screening effects completely disappear, as seen from the coinciding v-F curves for screened and unscreened couplings for a finite phonon thermalization time. The reason for this quite general property of phonon disturbances is easily understood: the (fictitious) case of no screening means stronger couplings and thereby stronger phonon disturbances and in turn a reduced relaxation efficiency of the phonons for the energy and momentum of the carriers, balancing the increase of the carrier-phonon scattering rates.

We have found very similar results in a carrier-temperature model calculation of hot luminescence after 25 ps laser pulse excitation of GaAs with energy flux densities between 2 and 44 MW/cm², as experimentally obtained by Göbel and coworkers [15]. Amplification of LO phonons is seen to cause the strong retardation of carrier cooling during and after the excitation pulse. Again the net effects of screening on the carrier relaxation were very small, as seen in the lower part of Fig. 6, by com-

212

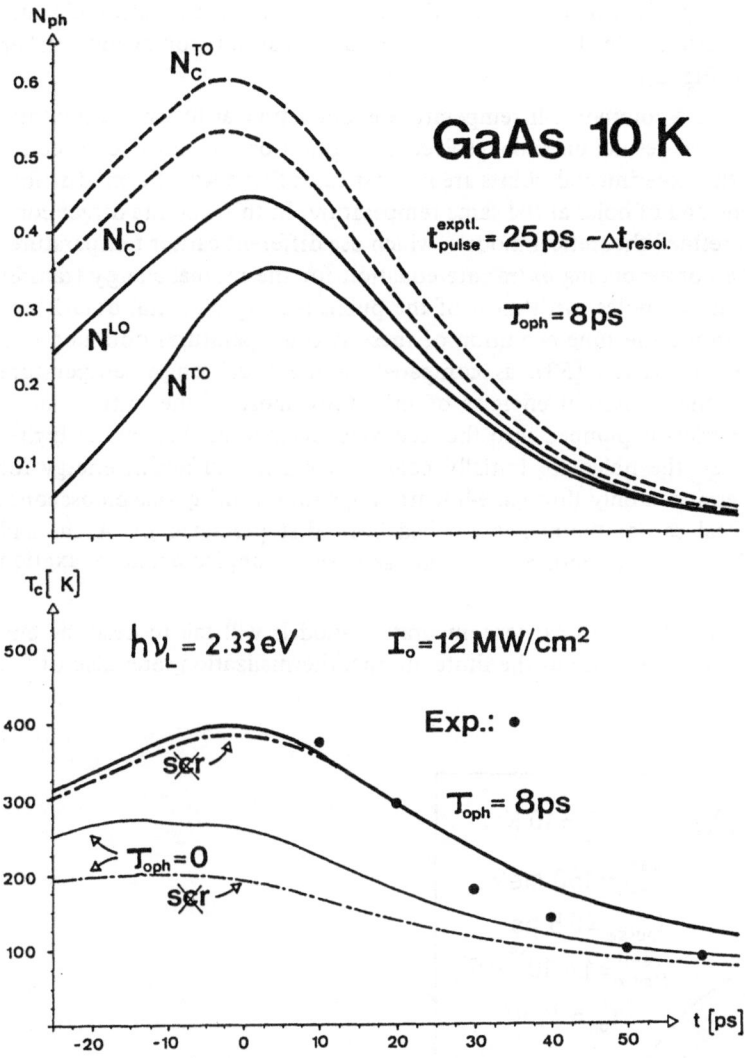

Fig. 6 Electron-hole plasma temperature, some strongly coupling LO and TO phonon distributions, and their corresponding heated Planck distributions during and after a 25 ps laser-pulse excitation of GaAs, and the role of free carrier screening on the plasma cooling (from [27]).

paring its profound influence on the plasma temperature for a calculation without phonon-disturbances and its much smaller influence on the results for finite phonon thermalization times.

The unexpected finding in these and similar calculations was that the strong amplification of the LO phonons so strongly reduces their cooling efficiency that the much weaker odp coupling of the holes becomes the dominant energy dissipation channel for the carriers, resulting in the development of additional, but somewhat smaller TO disturbances (Fig. 6).

The assumption of a common e-h temperature is quite typical for the representation of experimental results on time-resolved absorption or luminescence spectroscopy, because the experimental points are in most cases fitted with Fermi distributions of electrons and of holes at the same temperature. In this way the direct comparison of more refined theoretical models, which use different carrier temperatures T_e and T_h and a corresponding extra rate equation for the mutual energy transfer between electrons and holes, with most of the published experimental data is not possible. Fig. 7 shows the time evolution of these two temperatures obtained in a recent hot phonon analysis [53], as compared to the usual single temperature model [44]. For the excitation energies of this study most of the initial kinetic energy of the carriers is pumped into the electrons, because of their much higher band curvature. So the holes are initially cold, and reach a sufficient energy for optical phonon emission only through e-h scattering within roughly one picosecond. This retarded total carrier thermalization had been first predicted by Asche and Sarbei [54] and is of great importance for any analysis of subpicosecond relaxation phenomena.

But quite generally the use of carrier-temperature models will fail to describe any early transient effects, because of the finite internal thermalization rates also of the

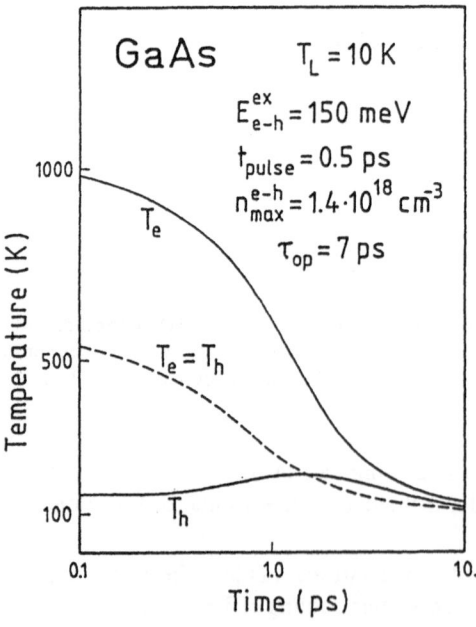

Fig. 7

Comparison of a two-temperature cooling with the conventional single plasma temperature cooling (from [53]).

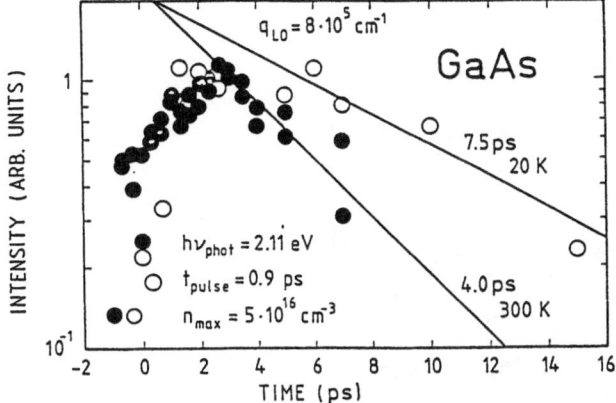

Fig. 8

Time-resolved antistokes Raman intensity of non-equilibrium LO Phonons in GaAs (from [21]).

separate electron and hole subsystems, which for lower carrier concentrations can become of the order of a few picoseconds. For the faster subpicosecond response one has to rely on the model independent techniques of chapters 3.6 and 3.7.

4.2 Monte Carlo Simulations: Excite-and-Probe Raman Spectroscopy

As an example of MC applications to hot phonons we discuss very recent results of a simulation of the time-resolved picosecond Raman experiments of J. Kash and coworkers [21].

Fig. 8 shows the measured antistokes probe intensities as functions of the pump-probe delay. The rise and fall of these intensities reflect the build-up and decay of nonequilibrium LO phonons during and after the 0.9 ps excitation pulse, and their long-time tails are used for the determination of the LO-phonon lifetimes, as indicated in the figure.

To investigate the possibility of contributions of phonon reabsorptions by the carriers to these apparent decay rates of the nonequilibrium phonon populations the following MC procedure has been set up [49, 50]. For unscreened polar couplings and neglecting the holes because of their small injection energies, the dynamics of the laser-generated electrons and of the LO phonons is followed under conditions resembling the experimental excitation. Fig. 9 shows some simulated LO-phonon distributions as function of time. The exponential decay with a thermalization time of 7 ps for the small-q modes, including the Raman-sampled modes around $8 \cdot 10^5$ cm^{-1} is in excellent agreement with the experimentally determined LO-phonon lifetimes. These decay rates can be well interpreted by considering the changing phase-space restrictions from energy and momentum conservation during the cooling of the electrons, as summarized in Figs. 10 and 11. In Fig. 10 the initially excited LO-phonon distribution is compared to the minimal possible phonon wavevectors for emission and absorption by electrons of energy E_e. During the cooling of the carriers

GaAs

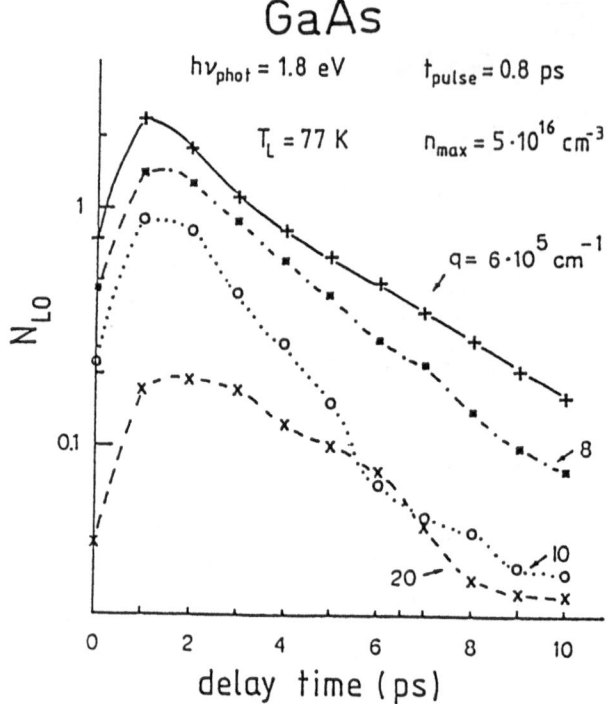

$h\nu_{phot} = 1.8$ eV $\qquad t_{pulse} = 0.8$ ps

$T_L = 77$ K $\qquad n_{max} = 5\cdot10^{16}$ cm^{-3}

N_{LO}

$q = 6\cdot10^5$ cm^{-1}

delay time (ps)

Fig. 9
Time evolution of LO-phonon distributions from a Monte Carlo simulation of Fig. 8 (from [50]).

GaAs

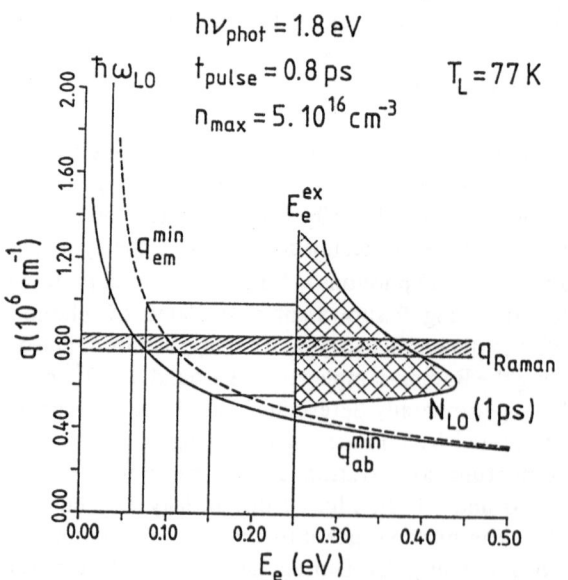

$h\nu_{phot} = 1.8$ eV

$t_{pulse} = 0.8$ ps $\qquad T_L = 77$ K

$n_{max} = 5.10^{16}$ cm^{-3}

$\hbar\omega_{LO}$

$q(10^6$ cm$^{-1})$

q_{em}^{min}

E_e^{ex}

q_{Raman}

$N_{LO}(1ps)$

q_{ab}^{min}

E_e (eV)

Fig. 10
Phase-space analysis of the electron-phonon dynamics of Fig. 9.

216

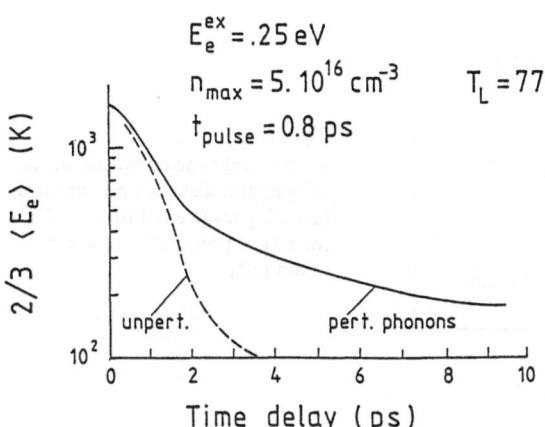

n–GaAs

$E_e^{ex} = .25\,eV$

$n_{max} = 5.\,10^{16}\,cm^{-3}$ $T_L = 77\,K$

$t_{pulse} = 0.8\,ps$

$2/3 \langle E_e \rangle$ (K)

unpert.   pert. phonons

Time delay (ps)

Fig. 11
Electron cooling with and without LO-phonon disturbances for the situation of Figs. 9 and 10.

from their injection energy E_e^{ex} at 0.25 eV (Fig. 11), the small-q phonons can soon no longer be emitted and reabsorbed by the carriers. So already very soon after the excitation pulse the decay of the small-q modes will be governed by nonelectronic processes and the corresponding decay time τ_{op}, Eq. (7). The mode of $q = 1 \cdot 10^6\,cm^{-1}$ will no longer be emitted, when most of the carriers have cooled to energies below 0.08 eV, but can still be reabsorbed during the next few picoseconds of carrier cooling, explaining the initially higher apparent decay rate of these phonons. The dynamics of the higher-q modes is less restricted by phase space and therefore contains less information about the separate c-ph and ph-ph dynamics.

A further useful information from this type of MC simulations is the information about the evolution of the initially spike-like carrier excitation spectrum towards a thermalized distribution, in the present example within the first few picoseconds.

4.3 Direct Numerical Solutions: Excitation by Ultrashort Laser Pulses

Our example for the application of the direct numerical solution technique of chapter 3.8 is devoted to the calculation of the response of the coupled electron-hole and LO phonon system in GaAs to extremely short (84 fs) and high power laser pulses [52]. The results are shown in Figs. 12 and 13 for the time evolution of the LO-phonon distributions and in Fig. 14 for the mean kinetic energy of the e-h plasma. Comparing these curves with the examples of the foregoing chapters, we again notice the important role of LO-phonon amplification on the energy relaxation of the carriers. As these calculations contain screened e-e, e-h, h-h, and carrier-LO phonon scattering, as well as plasma-induced gap renormalization and band filling effects, they can be considered as the hitherto most complete theoretical

217

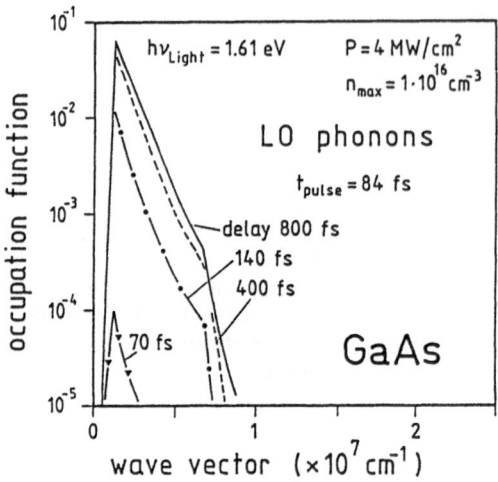

Fig. 12

Theoretical time evolution of the LO-phonon distribution for ultra-fast e-h plasma excitation in GaAs for a laser power of 4 MW/cm² (from [52]).

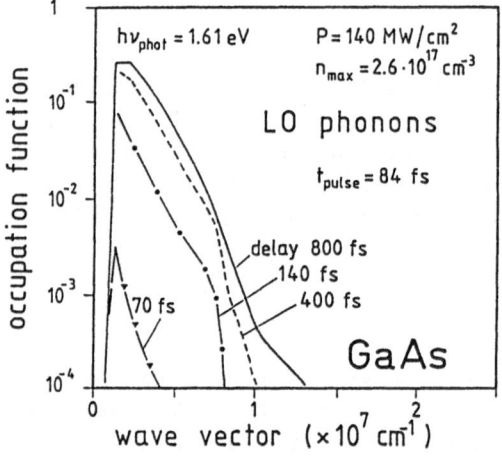

Fig. 13

As in Fig. 12, but for a laser power of 140 MW/cm².

analysis of highly nonequilibrium e-h plasmas and LO phonons. Moreover, they could, at least in principle, be generalized to include the optical deformation potential interaction of holes with LO and TO phonons, as found to be important for longer pulses [44, 27].

5 Ultrasmall Semiconductor Structures

Many of the concepts developed and applied in the preceding chapters are likely to become questionable, when straightforwardly applied to micron or submicron structures. For ultrasmall structures the most doubtful concept is the classical

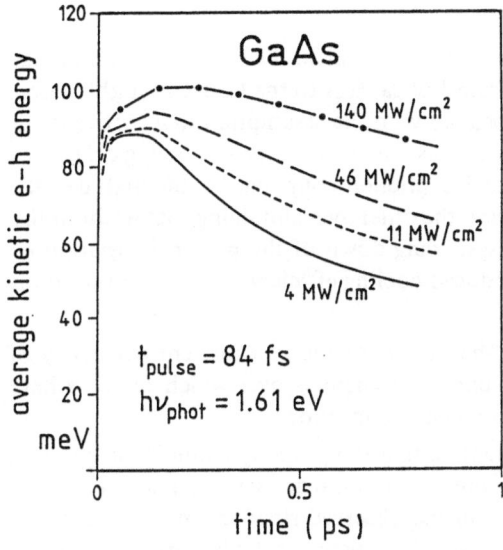

Fig. 14
Energy relaxation of the e-h plasma for the cases of Figs. 12 and 13.

transport assumption of instantaneous scattering events and otherwise free particle trajectories under the influence of external forces. There are some possibilities to extend the limits of this kinetic approach by stepwise modifications of the Boltzmann formalism, but most of the recent fundamental theoretical work on high-field effects in submicron structures has used a quantum transport description [23 ... 25, 55].

Most of the present-day experimental material on semiconductor quantum-well structures is generally believed to be still within the limits of the conventional transport description, with the essential new features of the size-quantization of the carrier and phonon states and of the corresponding changes of their energy spectra and mutual couplings.

A great number of experimental findings of reduced cooling rates of photo-excited carriers in quantum-well structures has been either interpreted by invoking non-equilibrium-phonon effects or at least their possibility [56 ... 61], whereas some authors have claimed no evidence of hot-phonon effects [62] or proposed different mechanisms to explain the slowed energy relaxation [63].

Theoretical work on hot phonons in quantum-well structures includes some general considerations about the novel aspects of the dynamics of nonequilibrium phonons [64, 65], and some specific applications within the carrier-temperature [66] or the Monte Carlo approach [67] to GaAs-AlGaAs quantum wells with strong indications of important effects of disturbed polar optical modes in the existing experimental data.

6 Summary

In essence there have been two false attempts to explain the experimental findings of a strongly retarded energy transfer from hot carriers to the lattice in highly laser-excited polar semiconductors: one associated with the assumption that, without the need to invoke nonequilibrium-phonon effects, the reduced carrier cooling originates from the screening of the polar carrier-LO phonon couplings by the high-density photo-generated electron-hole plasma, and the other by maintaining that without the need for including screening effects the slowing down of the power dissipation of the carriers is the consequence of the reduced cooling efficiency of a nonequilibrium LO phonon system.

It is this "either-or" terminology and the believe of the independent additivity of the two effects to the "bare" carrier-phonon dynamics over which scholars have spread and are still spreading much fundamental confusion.

The present analysis has tried to demonstrate that the properly formulated question is, whether the interplay of (the ever present) free-carrier screening and (the ever present) phonon disturbances can explain the observed slow energy dissipation of hot carriers. And it is found in this type of consistent calculations that for the moderately high carrier densities (of 10^{17} to several 10^{19} cm^{-3}) generated by present-day ultrashort laser pulses the decisive mechanism for the cooling retardation is, apart from an initial subpicosecond internal carrier thermalization, the amplification of optical phonons. The reason for this dominance of the nonequilibrium-phonon effects is the fact that they will always oppose and outweigh any change in the free-carrier screening, as e.g. caused by the changes of the plasma density during and after laser pulses of different pulse intensities. This irrelevance of the details of the carrier-phonon couplings in the presence of phonon disturbances is the more true for the many recent attempts to improve theoretical results through refinements of the free-carrier screening models.

These findings for the transient dynamics of bulk materials can be expected to be of similar relevance for the role of screening and of size-induced changes of c-ph couplings in smaller dimensional structures, mainly because of the strong localization and spatial stationarity of optical phonon disturbances. Such expectations have been confirmed in a few preliminary theoretical calculations which again find an important role of phonon-disturbances under the fundamentally changed conditions for the quasi-two dimensional carrier dynamics in layered micron and submicron semiconductor structures.

Acknowledgement

Support by the European Research Office is gratefully acknowledged.

References

[1] Physics of Nonlinear Transport in Semiconductors, NATO Advanced Study Institute Series Vol. 52, ed. by *D. K. Ferry, J. R. Barker,* and *C. Jacoboni* (Plenum Press 1980)

[2] Proc. of the 3rd Int. Conf. on Hot Carriers in Semiconductors (Montpellier 1981), J. de Physique 52, Colloque C-7 (1981)

[3] Proc. of the 4th Int. Conf. on Hot Electrons in Semiconductors (Innsbruck 1985), Physica 134 B (1985)

[4] Proc. of the ICTP-IUPAP Symp. on High Excitation and Short Pulse Phenomena (Trieste 1984); ed. by *M. H. Pilkuhn* (North Holland 1985)

[5] *P. Kocevar,* J. Phys. C 5, 3349 (1972), Acta Phys. Austriaca 37, 259 (1973)

[6] *P. Kocevar,* in [1], p. 401

[7] *R. G. Ulbrich,* in [2], p. 423

[8] *J. Shah, R. C. C. Leite,* and *J. F. Scott,* Solid State Comm. 8, 1089 (1970)

[9] *J. Shah,* in [2], p. 445

[10] *R. J. Seymour, M. R. Junnarkar,* and *R. R. Alfano,* Solid State Comm. 41, 657 (1982)

[11] *S. S. Yao, J. Buchert,* and *R. R. Alfano,* Phys. Rev. B 25, 6534 (1982)

[12] *W. Graudszus* and *E. O. Göbel,* Physica 117 B, 555 (1983)

[13] *H. J. Zarrabi* and *R. R. Alfano,* Phys. Rev. B 32, 3947 (1985)

[14] *M. R. Junnarkar* and *R. R. Alfano,* Phys. Rev. B 34, 7045 (1986)

[15] *W. Graudszus, Ph. D. Thesis,* Univ. Stuttgart 1985, unpubl.

[16] *K. Kash, J. Shah, D. Block, A. C. Gossard,* and *W. Wiegmann,* Physica 134 B, 189 (1985)

[17] *J. C. Tsang* and *J. A. Kash,* Phys. Rev. B 34, 6003 (1986)

[18] *K. Kash* and *J. Shah,* in [4], p. 333

[19] *D. von der Linde, J. Kuhl,* and *H. Klingenberg,* Phys. Rev. Lett. 44, 1505 (1980)

[20] *C. L. Collins* and *P. Y. Yu,* Phys. Rev. B 30, 4501 (1984).

[21] *J. A. Kash, J. C. Tsang,* and *J. M. Hvam,* Phys. Rev. Lett. 54, 2151 (1985)

[22] *A. Compaan,* in [4], p. 425

[23] *J. R. Barker,* in [1], p. 126

[24] *J. R. Barker,* in [2], p. 245

[25] *J. R. Barker,* in: Handbook on Semiconductors, ed. by *T. S. Moss,* Vol. 1, (North Holland 1982) p. 617

[26] *G. Mahler* and *A. Fourkis,* in [4], p. 18

[27] *P. Kocevar,* in [3] p. 155

[28] *H. L. Grubin, D. K. Ferry, G. J. Iafrate,* and *J. R. Barker,* in: VLSI Electronics: Microstructure Science, Vol. 3 (Academic Press 1982), p. 197

[29] *H. L. Grubin* and *J. P. Kreskovsky,* ibid., Vol. 10 (Academic Press 1985), p. 237

[30] *E. Conwell,* Solid State Phys. Suppl. 9 (1967)

[31] E. J. Yoffa, Phys. Rev. B 23, 1909 (1981)

[32] *D. Lowe* and *J. R. Barker,* J. Phys. C 18, 2507 (1985)

[33] *S. Doniach,* Proc. Phys. Soc. 73, 849 (1959)

[34] *J. Collet, A. Cornet, M. Pugnet,* and *A. Amand,* Solid State Comm. 42, 883 (1982)

[35] *P. Lugli* and *D. K. Ferry,* in [3] p. 364

[36] *M. A. Osman, U. Ravaioli, R. Joshi, W. Pötz,* and *D. K. Ferry,* to be published in the Proc. of the 18th Int. Conf. on the Phys. of Semic., Stockholm 1986

[37] *M. Combescot* and *J. Bok,* Phys. Rev. **B 35**, 1181 (1987)

[38] *P.G. Klemens,* Phys. Rev. 148, 845 (1966)

[39] *R.K. Chang, J.M. Ralston,* and *D.E. Keating,* Proc. of the Int. Conf. on Scattering Spectra in Solids, ed. by *G.B. Wright* (Springer, New York 1969), p. 369

[40] *A. Mooradian,* in: Laser Handbook, ed. by *F.T. Arecchi, E.D. Schulz-Du Bois* (North Holland 1972), p. 1410

[41] *T.C. Damen, R.C.C. Leite,* and *J. Shah,* Proc. of the 10th Int. Conf. on the Physics of Semic., Cambridge (Mass); ed. by *S.P. Keller, J.C. Hensel,* and *F. Stern* (U. S. Atomic Energy Comm. 1970) p. 735

[42] *E.O. Göbel,* priv. comm.

[43] *M. Pugnet, J. Collet,* and *A. Cornet,* Solid State Comm. 38, 531 (1981)

[44] *W. Pötz* and *P. Kocevar,* Phys. Rev. **B 28**, 7040 (1983)

[45] *P. Kocevar,* to be published

[46] *P. Bordone, C. Jacoboni, P. Lugli, L. Reggiani,* and *P. Kocevar,* J. Appl. Phys. 61, 1460 (1987)

[47] *W. Fawcett, A.D. Boardman,* and *S. Swain,* J. Phys. Chem. Solids 31, 1963 (1970)

[48] *C. Jacoboni and L. Reggiani,* Rev. Mod. Phys. 55, 645 (1983)

[49] *P. Lugli, C. Jacoboni, L. Reggiani,* and *P. Kocevar,* Appl. Phys. Lett., in print

[50] *P. Lugli, C. Jacoboni, L. Reggiani,* and *P. Kocevar,* to appear in SPIE Proc. Vol. 793 (1987)

[51] *M. Rieger, P. Kocevar, P. Bordone, P. Lugli,* and *L. Reggiani,* unpublished

[52] *J. Collet* and *T. Amand,* J. Phys. Chem. Solids 47, 153 (1986)

[53] *W. Pötz,* to be published

[54] *M. Asche* and *O.G. Sarbei,* phys. stat. solidi (b) 126, 607 (1984)

[55] *J. Barker, in* [1] p. 589

[56] *J.F. Ryan, R.A. Taylor, A.J. Turberfield, A. Manciel, J.M. Worlock, A.C. Gossard,* and *W. Wiegmann,* Phys. Rev. Lett. 53, 1841 (1984)

[57] *J. Shah, A. Pinczuk, A.C. Gossard,* and *W. Wiegmann,* Phys. Rev. Lett. 54, 2045 (1985)

[58] *K. Kash, D. Block,* and *J. Shah,* Phys. Rev. **B 33**, 8762 (1986)

[59] *K.T. Tsen* and *H. Morkoc,* Phys. Rev. **B 34**, 4412 (1986)

[60] *J.F. Ryan, R.A. Taylor, A.J. Turberfield,* and *J.M. Worlock,* Surf. Sci. 170, 511 (1986)

[61] *J.A.P. Da Costa, R.A. Taylor, A.J. Turberfield, J.F. Ryan,* and *W.I. Wang,* Proc. of the 18th Int. Conf. on the Physics of Semic., Stockholm 1986, to be published

[62] *C.H. Yang, J.M. Carlsson-Swindle,* and *S.A. Lyon,* Phys. Rev. Lett. 55, 2359 (1985)

[63] *H. Uchiki, Y. Arakawa, H. Sakaki,* and *T. Kobayashi,* Solid St. Comm. 55, 311 (1985)

[64] *P. Price,* Superl. and Microstr. 1, 255 (1985)

[65] *P. Price,* in [3], p. 164

[66] *W. Cai, M.C. Marchetti,* and *M. Lax,* Phys. Rev. **B 34**, 8573 (1986)

[67] *P. Lugli* and *S. Goodnick,* in: High Speed Electronics ed. by *B. Källbäck* and *H. Beneking* (Springer, Berlin 1986), p. 116

Festkörperprobleme 27 (1987)

The Physics of Electrical Breakdown and Prebreakdown in Solid Dielectrics

Hans-Rudolf Zeller, Thomas Baumann, Eduard Cartier, Helmut Dersch, Peter Pfluger, and Fredi Stucki

Brown Boveri Research Center, CH-5405 Baden, Switzerland

Summary: We discuss physical models of high field phenomena in solid dielectrics. After a brief treatment of classical (thermal and electrostrictive) instabilities we turn to the discussion of current instabilities. We show that the strong nonlinearities in the current vs. voltage relation greatly simplify the picture. This is also true for avalanche breakdown where the carrier generation is strongly affected by selffields. We present experimental results for the energy and momentum loss functions for model dielectrics which allow to compute the critical field for impact ionization.
The overall breakdown patterns are often tree-like and can be treated in terms of a fractal model.
The review concludes with a short discussion of field induced ageing effects.

1 Introduction

An insulator in any given electrode-insulator geometry can support only a limited voltage. If that voltage is exceeded then dielectric breakdown occurs. This is obviously a very loose and superficial definition of breakdown. Phenomenologically, the critical voltage depends on all sorts of electrical and nonelectrical parameters such as voltage pulse form, time constant, polarity, single or repetitive pulses, or ac load, temperature, superimposed mechanical stress, etc. [1].

On the physics side we have to distinguish between different breakdown mechanisms e.g. whether we are dealing with an instability in carrier density (impact ionization avalanche), current instability (field dependent mobility), thermal instability (Joule heating), electrostrictive instability, etc. Which one of the different instabilities limits the voltage in a given situation depends on electrical, materials, and environmental parameters.

In this paper we will give a qualitative review of the major breakdown and prebreakdown mechanisms. Since the microscopic models are in general intractable, we will use whenever possible phenomenological models which allow order of magnitude estimates of the quantities of interest.

Section 2 treats electromechanical instabilities. Electromechanical instabilities, i.e. instabilities due to field induced mechanical stress are in general not relevant in solids. They are important, however, in the form of hybrid instabilities in which the yield stress of the material is reduced by Joule heating, chemical deterioration or other phenomena or in situations with high mechanical prestress.

Section 3 treats phenomenologically an important class of instabilities which result from the strong temperature dependence of electrical conductivity in a dielectric. Above a critical voltage the solutions of the coupled heat flow and current flow equations become unstable and a thermal runaway situation occurs.

The instabilities discussed in Sect. 2 and 3 are classical instabilities in the sense that only low field parameters are needed. Section 4 treats true high field mobilities. The discussion is based on a very simple model, i.e. space charge controlled currents. The model is generalized in several steps in order to include high field effects. The first step (Sect. 4.2) consists in the introduction of a highly nonlinear high field mobility. Surprisingly this leads to a dramatic simplification of the otherwise virtually untractable trap dominated space charge controlled transport in wide band gap disordered materials.

For several reasons (e.g. thermal instability) a negative differential resistance regime will be reached above a certain critical current density (Sect. 4.3). This leads to the decay of the current distribution into filamentary structures. As in Sect. 4.2 where the threshold field for high mobility completely defines the space charge distribution the nonlinearities greatly simplify the problem. In a stable situation (stable prebreakdown structure) current flow will stop if the field inside the filaments has dropped to the "hold field", the minimum field required to keep the carriers in the negative differential resistance regime. This together with geometrical factors defines the charge distribution in the filaments and also provides a breakdown criterion.

Similar effects occur in the case of impact ionization discussed in Sect. 4.4. The carrier multiplication process quickly results in carrier densities which produce strong selffields. Since the carrier multiplication rate is a very steep function of local field the selffields will strongly affect the carrier multiplication. The net result is qualitatively speaking a peak levelling. Whenever the local field exceeds the threshold for rapid carrier multiplication then carriers are produced until the field drops due to selffield effects. Since we are not dealing with a stationary but a highly dynamical process, the above statement is an oversimplification.

Not only impact ionization but also chemical damage by bond scission requires the existence of hot electrons, i.e. electrons with kinetic energies much higher than thermal energies. A calculation of the kinetic energy distribution as a function of field requires as input parameters two scattering functions: The momentum loss rate $\gamma_p (E_{kin})$ and the energy loss rate $\gamma_u (E_{kin})$. In Sect. 5 we discuss a new experimental technique to determine scattering functions and review experimental results.

Section 6 treats a fractal model of breakdown which makes it possible to connect the overall pattern to three parameters with simple physical interpretation.

Economically the most important aspect of high electrical fields is dielectric ageing. Ageing can occur at nominal fields two orders of magnitude below the pulse breakdown field. This is due to a complex interplay between various factors. In Sect. 7 we are only able to give a short glimpse of the field and have to refer the reader to the literature.

2 Field Induced Mechanical Instabilities in Dielectrics

We consider a homogeneous and isotropic solid. σ_{ik} denotes the stress tensor, ρ the space charge and $\vec{F}$ the electrical field. The equation for the field induced stress then becomes [2]

$$\frac{\partial}{\partial x_k} (\sigma_{ik} - \sigma_{ik}^0) = -\frac{\epsilon_0}{2} \left(\frac{1}{2} + a_2 \right) \frac{\partial F^2}{\partial x_i} + \left(1 - \frac{a_1}{2\epsilon} \right) \rho \, F_i \tag{1}$$

where σ_{ik}^0 is the stress at zero field and a_1, a_2 are the two electrostriction coefficients of an isotropic solid.

In the absence of space charges Eq. (1) can be integrated and results in

$$\sigma_{ik} - \sigma_{ik}^0 = \epsilon_0 \left[\left(\epsilon - \frac{a_1}{2} \right) F_i F_k - \frac{1}{2} (\epsilon + a_2) F^2 \delta_{ik} \right]. \tag{2}$$

If the local stress exceeds the mechanical stability then plastic flow will occur. Depending on conditions this instability is locally confined or may spread and lead to macroscopic destruction. Electromechanical instabilities of this type have attracted little attention because in the absence of space charges and for reasonably strong dielectrics even at several MV/cm the electrical stress is small compared to the yield stress. For instance with $a_1 = a_2 = 0$, $\epsilon = 2.3$, and $F = 1$MV/cm the field induced stress is of the order of 1 bar.

However, even at modest overall fields, strong field enhancements occur at defects and electrode irregularities. This leads to local space charge injection [3] and to a strong enhancement of the mechanical stress [4]. The yield stress on the other hand is locally reduced by local heating and possibly by cumulative chemical damage due to hot electron impact [5]. As a result electromechanical instabilities play an important role in ageing and partial discharge phenomena [4]. They are unimportant, however, as a bulk instability in fairly homogeneous fields.

3 Thermal Instabilities

Phenomenologically, the conductivity of a solid dielectric can be written as

$$\sigma(T) = \sigma_0 \exp \left(-\frac{\Delta}{kT} \right) \tag{3}$$

where Δ is the activation energy and k the Boltzmann constant. In the stationary state the heat balance equation then becomes:

$$\text{div} \, (\chi \, \text{grad} \, T) = - \sigma(T) \cdot F^2 \tag{4}$$

where χ is the thermal conductivity. The insertion of Eq. (3) into Eq. (4) results in a very nonlinear equation for the temperature distribution in the dielectric. With increasing field the temperature inside the dielectric first increases smoothly. If ΔT exceeds the value of

$$\Delta T = c \left(\frac{kT}{\Delta} \right) T \tag{5}$$

225

then a thermal instability [1, 6] occurs. c is of order 1 and its exact value depends on geometry and boundary conditions. Of particular interest is the plane capacitor geometry with the electrodes kept at ambient temperature T_0. The thermal instability then occurs at a critical voltage V_c independent of the thickness d of the dielectric. V_c is given by

$$V_c = m \sqrt{\frac{kT_0^2 \cdot \chi}{\sigma(T_0) \cdot \Delta}} \qquad (6)$$

where m is of order one and depends on Δ. Eq. 6 then gives the maximum voltage a given material can support. For instance with $\chi = 10^{-4}$ VAs^{-1} m^{-1}, $\Delta = 1$ eV and $\sigma(T_0) = 10^{-14}$ $(\Omega\text{cm})^{-1}$ V_c becomes of the order of a few 100 kV independent of the thickness of the dielectric. We also conclude that thermal instabilities are important only for large dielectrics. For small dielectrics the fields may be high even at small voltages and thus true field instabilities become important. As a rule of thumb thermal instabilities are dominant if d $\gg$ 1 cm and unimportant for d $\ll$ 1 cm where d is a characteristic electrode spacing.

4 Current Instabilities

4.1 Low Field Charge Flow in a Dielectric

By nature dielectrics are wide band gap and usually also low mobility materials. As an example we consider polyethylene. Polyethylene has a band gap of about 9 eV. The conduction band edge is above the vacuum level [7]. In a chemists language this means polyethylene has a negative electron affinity. Since almost any conceivable impurity or defect has a positive electron affinity they will invariably lead to deep trap states inside the gap. As a result the low field mobility of polyethylene is extremely small and completely trap controlled [8]. This is qualitatively true also for most other polymers and has led to the general folklore that polymers do not have energy bands. It is in fact true that in general transport properties reflect trap and not band properties and also that optical results reflect mostly excitonic properties [7, 8] but nevertheless there is ample proof of wide conduction bands in polyethylene [7, 9].

The introduction of a π system in a side chain for instance in polystyrene does not change the picture. The π states are in the gap of the σ states. The width of the π bands is negligibly small and transport occurs by hopping to adjacent π states. The π states have thus to be treated as molecular states and not as energy bands [7].

With reasonable values for the electron-phonon coupling parameter in polyethylene a conduction-band mobility of the order of 100 cm^2/Vs would be expected [7]. Pulsed drift mobility experiments carried out with ultrapure liquid hydrocarbons do in fact show band type electron mobilities of that order for the tetrahedrally coordinated neopentane [10]. The linear n-pentane, however, has a much smaller and thermally activated mobility [10]. This is probably due to the strong intramolecular disorder in the flexible linear hydrocarbons and the lack of intramolecular

226

disorder in the rigid neopentane. Since also in the solid state polymers are either completely or partially amorphous, we have to deal with another complication besides traps. Disorder will in general lead to localization at the band edges (mobility edge).

To sum up, we can state that low field transport in dielectrics is an extremely complex phenomenon. Not only is the transport trap controlled, but also the charge carrier distribution is never in thermodynamical equilibrium. The residence time in deep traps is always larger than a reasonable experimental time scale. As a result most semiconductor type models fail and specific models which explicitly take into account the dynamics of the energy distribution have to be introduced [11].

4.2 Space Charge Limited Currents at High Fields

Experiments on various dielectrics have shown that above a field of the order of several 100 kV/cm a dramatic increase of mobility with field is observed [11]. We will discuss the effect of this on dielectric instabilities in terms of a very simple model which leads to analytical results.

First we simplify the situation by modelling the onset of high mobility by a sharp threshold [4]. The mobility μ is assumed to be $\mu = 0$ für $F < F_c$ and high for $F \geqq F_c$. F_c is the critical field for the onset of the high field mobility. Obviously, the homogeneous field situation is of no interest because by construction breakdown occurs at $F = F_c$. We thus consider inhomogeneous field situations. The simplest is the spherical geometry. We assume a metallic sphere on potential V_0 and with radius r_0 in an infinite dielectric. We then consider unipolar injection from the sphere and space charge limited transport in the absence of thermally generated charge carriers. As long as we are interested in the stationary situation only, there is no need to specify $\mu(F)$ above F_c. Evidently, for $V_0 > F_c \cdot r_0$ charge will flow until $F = F_c$ inside the space charge region and $F < F_c$ outside. $\mu(F)$ determines the dynamics of the space charge build-up but not the final distribution. The space charge cloud will extend from r_0 to $r = (V_0/F_c + r_0)/2$ and be of the form [4]

$$\rho(r) = \frac{2 \epsilon \epsilon_0 F_c}{r} . \tag{7}$$

Of course it is also easy to give analytical expressions for the field induced stress and for the Joule heating [4, 12]. Since the only parameter required to completely define the problem is the mobility threshold field, the model is called FLSC (field limited space charge) model.

For more complex electrode geometries numerical computations are needed [3]. Fig. 1 shows the boundaries of the space-charge cloud for a needle-plate geometry as a function of the parameter V_{max}/V_c where V_c is the voltage at which $F = F_c$ at the tip.

We have experimentally checked the model by measuring the charge injected from a metallic needle tip into a dielectric [3]. Since this is not a spherical geometry $\rho(r)$ has to be computed numerically. This is a rather difficult experiment since the

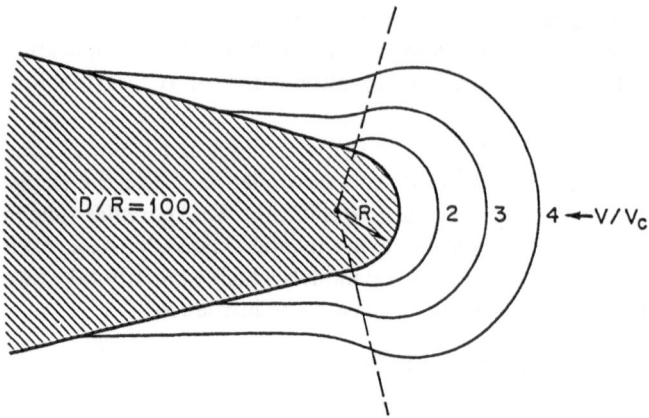

Fig. 1 Boundary of the space-charge cloud as a function of V/V_c in a needle-plate geometry. D/R is the ratio between tip-plate distance and tip radius.

capacitive charge on the needle is about six orders of magnitude above the required sensitivity. The needle was thus guarded exposing only the tip (r_0 typically 5 μm) to the dielectric. This reduces the capacitive charge by about three orders of magnitude. The remaining charge is compensated in a bridge-like circuit. The experimentally determined charge i.e. the detuning of the bridge is not directly the injected charge but the difference between injected charge and image charge on the needle.

Experimental results on epoxy resin with triangular pulses or sinusoidal voltages in the 0.1 to 10 ms range show excellent agreement with the model [3]. F_c is of the order of MV/cm and does not depend on dV/dt which shows that at each voltage ρ is quasi-stationary or that μ (F) has in fact a well defined threshold. This is remarkable because it implies that only one parameter is needed (F_c) to describe the charge flow at high fields.

An example is shown in Fig. 2. As predicted by the model, injection starts at V_c and ends at V_{max}. The measured charge varies as $(V - V_c)^2$ above V_c and remains trapped after the pulse maximum.

Two classes of complications may occur. The first is that the current flow is not limited by the critical field for mobility F_c but by a critical field for injection from the electrode F_{inj}. If $F_{inj} > F_c$ then ρ (r) detaches from the spherical electrode. It then extends from r_i to r_a [12] where

$$r_i = r_0 \sqrt{\frac{F_{inj}}{F_c}} \, , \tag{8}$$

$$r_a = r_i + \frac{V_0 - F_{inj} \cdot r_0}{2 F_c} \, . \tag{9}$$

228

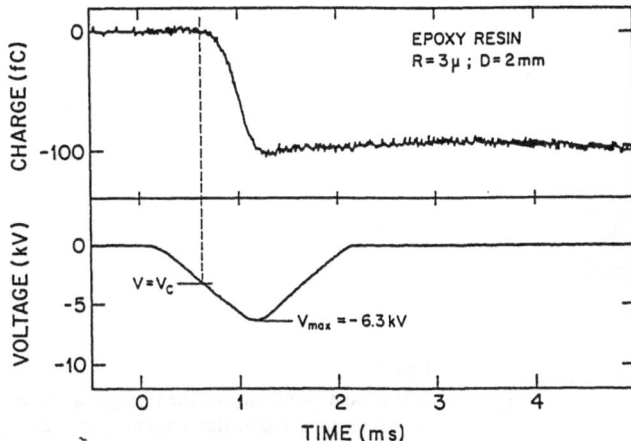

Fig. 2 Space-charge injection in epoxy resin. Above a critical voltage V_c a continuously increasing space-charge injection is observed. After the pulse maximum the charge stays constant even on a much longer time scale than shown in the figure (seconds).

The two cases can be experimentally easily distinguished. In the mobility limited case the apparent charge (injected minus image charge) varies initially as $(V - V_c)^2$ where $V_c = F_c \cdot r_0$ for $V > V_c$. In the injection limited case there is a term linear in $V - V_c$ where V_c is now $V_c = F_{inj} \cdot r_0$. In the experiments so far only the quadratic behaviour was observed [3].

It is also conceivable that $\mu(F)$ exhibits a current controlled negative resistance. In this case the measured charge would show a step at the critical voltage. We will discuss this in more detail in a later section.

We have recently extended our experimental technique to move the needle and indent it at any desired position on the surface of a dielectric. Since the needle radius can be of the order of 1 μm it is then possible to perform a sort of dielectric injection microscopy. This makes it possible to study composite dielectrics of technical importance [13]. In particular we have studied the mica-epoxy system. We find a pronounced charge flow in the plane of the mica flakes even at low fields (Fig. 3). For practical purposes the flakes can thus be considered to define equipotential surfaces. Since the flakes are perpendicular to the global field direction they are very efficient in reducing local field enhancements due to defects. Such field enhancements are the main cause of problems in technical dielectrics.

For reasonable values of the parameters in practical situations no instabilities will occur [12]. The fields are limited by space charges, the mechanical stress is smaller than a typical yield stress, and for typical ac frequencies also the Joule heating is small. Dielectric ageing thus requires the existence of weak spots (for instance at the electrode-dielectric interface) or cumulative chemical damage due to current flow.

229

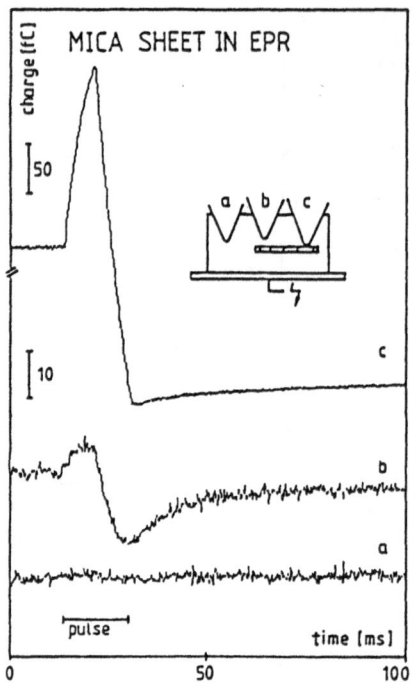

Fig. 3
The time dependence of the charge signal in response to a triangular voltage pulse of 250 V (3a, 3b) and 200 V (3c) amplitude in polyester resin. No signal was observed in the pure resin (3a). A transient due to polarisation shows up when testing some μm above the mica sheet (3b). With the needle directly positioned on the mica splitting this effect is greatly enhanced and an additional, long-lived signal shows up (3c).

4.3 Space Charge Limited Currents at Negative Differential Resistance

In this context it is more convenient to discuss the local field as a function of local current density than vice versa. For very small current densities this relation shows a threshold character, the field is about F_c and depends little on the current density. As the current density is increased, it is very likely that a current controlled negative resistance (NDR) regime is reached. There are numerous physical reasons for it. Examples are thermal instabilities, field induced detrapping, the mobility edge, etc.

Basically, any strong nonlinearity will eventually lead to negativ differential resistance because the increase in current density will lead to a thermal instability [1, 6].

A negative differential resistance has a qualitative influence on the current distribution. The smooth and continuous current distribution decays into filaments. In the simplest model two parameters are needed. The first is the critical field F_c required to reach the NDR regime. The second is the hold field F_s, the minimum field required to sustain the high mobility state. Whenever the current density reaches the NDR regime the current distribution becomes unstable and decays into filaments. This has a series of fundamental consequences. Let us assume we have a spherical electrode on potential V_0. For homogeneous injection of space charge and a critical field F_c (FLSC model), the space charge will penetrate a distance $\ell_{hom} = (V_0/F_c - r_0)/2$ into the dielectric. For filamentary injection with the filament dia-

230

meter $d \ll r_0$ the length of a single filament becomes $\ell_{fil} \approx V_0/F_s - r_0$ [12]. Since $F_s < F_c$ the prebreakdown currents will penetrate at least twice as far in the dielectric compared to the homogeneous (FLSC) case. Even more important, the local current densities are orders of magnitude higher which promotes thermal instabilities and other related phenomena.

If filamentary injection from an electrode does not lead to breakdown but to a stable prebreakdown structure then the axial field in the filaments will selfadjust to the hold field F_s. Together with radius and overall geometry this defines the charge distribution in the filament.

For a voltage pulse with constant amplitude the injection currents scale inversely proportional to the pulse time. The NDR regime is thus easily reached for fast pulses. In fact, homogeneous injection is invariably observed for ms pulses on epoxy [3], whereas μs pulses on liquid hydrocarbons can lead to filamentary injection [14]. Also we stress that once a dielectic is damaged then it is no longer the timescale of the external voltage pulse which is relevant. In a damaged dielectric we observe so-called partial discharges [15] which are gas discharges inside damage structures. The relevant time scale is now the time scale of the discharges which is in the submicrosecond regime.

4.4 Space Charge Limited Impact Ionization

If the field is sufficiently high then a high energy tail in the kinetic energy distribution will develop and impact ionization becomes possible. We postpone a discussion of the ionization rate $\beta(F)$ but focus on the phenomenological consequences of impact ionization.

Thermal destruction of a dielectric requires a pulse energy deposition of the order of 10^9 VAs/m^3. With a breakdown field of 10^8 V/m this results in a charge per area of 10 As/m^2. A capacitor with this surface charge and a typical dielectric constant would exhibit a field $F \gtrsim 10^{11}$ V/m which is impossible. Such arguments have traditionally [1] been used to show that the growth of an avalanche is limited by selffields before it has reached the critical size for thermal destruction. This is a strictly one-dimensional argument, however, and does not apply to filamentary breakdown.

But also for filamentary breakdown the impact ionization is selffield controlled whenever the thermal destruction limit is approached. Qualitatively speaking the situation is similar to the FLSC case [3]. The impact ionization rate has a threshold character. Whenever F becomes larger than F_{th} the threshold field, then charge carriers are generated. When they separate they create a dipole field which counteracts the external field. As a net result the field inside the avalanche is limited to the threshold field. Numerical simulations show that this is in fact the case but rather as a qualitative rule than a strict law. In particular there may be regimes where $F < F_{th}$.

An example of a numerical simulation is shown in Fig. 4. This applies to a filamentary avalanche starting from a spherical electrode. The details depend on injection currents, ionization rates, etc., but the field limiting effect is always evident.

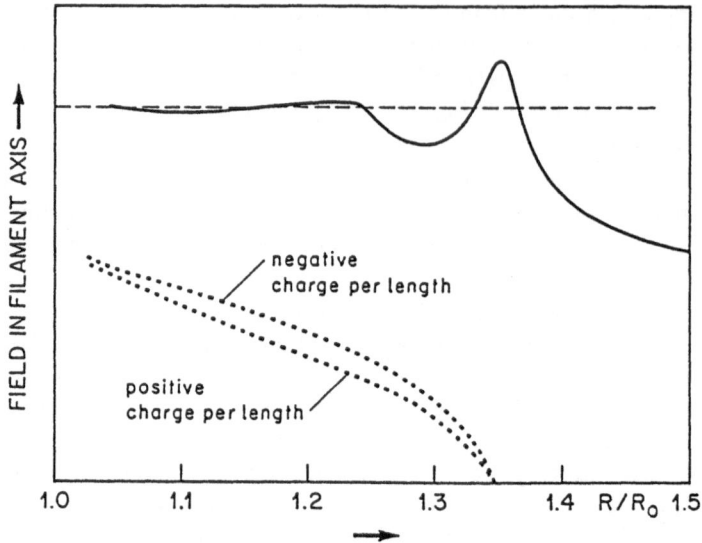

Fig. 4 The effect of impact ionization on the field and charge distribution. Electron injection occurs from a small spot on a sphere. The dashed line is the threshold field for impact ionization. Note the self-regulating effect on the field with the consequence that the maximum Joule heating occurs near the injection site.

We stress that the avalanche model does not predict an intrinsic breakdown field. The materials parameter which does not depend on geometry is the ionization rate as a function of local field $\beta\,(F_{loc})$. Breakdown occurs whenever the local power or energy deposition exceeds a critical value. The relation between $\beta\,(F_{loc})$ and the breakdown criterion always contains geometrical factors, and thus the avalanche breakdown field depends for instance on the thickness of a dielectric film. However, since $\beta\,(F_{loc})$ has a very pronounced threshold character this dependence is rather weak, and it is still possible to talk about an avalanche breakdown field in a qualitative sense. This field will be close to the threshold field for $\beta\,(F_{loc})$.

5 Scattering and Kinetic Energy Distribution of Hot Charge Carriers

The scattering of hot electrons is controlled by two scattering functions both of which depend on energy. The first is the momentum loss rate $\gamma_p\,(E_{kin})$. Within the average electron model the drift velocity v_d of a charge carrier is given by

$$v_d = \frac{e \cdot F}{m \cdot \gamma_p\,(E_{kin})}.$$ (10)

The kinetic energy on the other hand depends on the energy loss rate $\gamma_u\,(E_{kin})$ as

$$E_{kin} = \frac{e \cdot F \cdot v_d}{\gamma_u\,(E_{kin})} = \frac{e^2\,F^2}{m\,\gamma_p\,(E_{kin}) \cdot \gamma_u\,(E_{kin})}.$$ (11)

232

Of course the average electron model is inadequate to discuss impact ionization. A high energy tail in the kinetic energy distribution can cause avalanche breakdown even if the average electron energy is still small. Better approximations of the Boltzmann transport equation show that the energy distribution $n (E_{kin})$ depends on the integral [16]

$$\int_0^{E_{kin}} \gamma_u (E') \cdot \gamma_p (E') \, dE'. \tag{12}$$

We have designed a method to experimentally determine the scattering functions [17]. A metal surface is covered with a thin film of the dielectric. The film thickness d is of the order of a mean free path, i.e. of the order of 100 Å. Photoelectrons are then excited from the metal into the dielectric by UV irradiation. The electrons escaping from the dielectric into the vacuum are collected and their kinetic energy distribution is determined. For various reasons we have performed our experiments on long chain linear alcanes $C_n H_{2n+2}$ ($n \gg 1$). First, the alcanes are electronically very simple model compounds for polyethylene ($n = \infty$). Second, they have a negative electron affinity which means that electrons do not encounter a barrier when escaping into the vacuum. Also the films can be prepared sufficiently clean to avoid trapping of electrons and charging of the films. It turns out that films grow very well on Pt substrates, probably because Pt as a well known hydrocarbon catalyst strongly interacts with the C-H bond.

Injected photoelectrons may either ballistically cross the dielectric and escape into the vacuum or undergo scattering processes. Purely elastic processes lead to momentum randomization and will cause a fraction of the electrons to travel back into the metal instead of being emitted into the vacuum. The decrease in the number of emitted electrons as a function of d is thus a measure of γ_p. The downshift in kinetic energy of the emitted electrons on the other hand is a measure of the energy loss rate γ_u. Both processes can be easily seen from Fig. 5. With increasing film thickness the escaping electrons rapidly loose kinetic energy and decrease in number.

In his classical paper [18] Fröhlich has studied electron scattering by polar modes. He found a maximum in the scattering rate at a kinetic energy equal to the LO phonon energy and a decrease in the energy loss function $\gamma_u \sim E_{kin}^{-3/2}$ for $E_{kin} \gg \hbar\omega_{LO}$. Since in polymers the relevant polar modes are intermolecular vibrations, the momentum randomization at each scattering event is nearly complete. In his paper Fröhlich considered only two scattering channels. The first is polar mode scattering and the second impact ionization itself. Since at $E_{kin} \gg \hbar\omega_{LO}$ $\gamma_u \sim E_{kin}^{-3/2}$ and $\gamma_p \sim E_{kin}^{-1/2}$ [16] the solution for $E_{kin}(F)$ becomes unstable in this regime. In other words, once the field is sufficient to reach $E_{kin} > \hbar\omega_{LO}$ then the electrons will accelerate up to the impact ionization energy.

This instability is partly or completely suppressed by deformation potential scattering. This is a scattering mechanism based on nonpolar phonon emission and is mostly elastic. Model calculations on alcali halides [16] show it to be irrelevant at small energies but to increase rapidly above ≈ 1 eV.

233

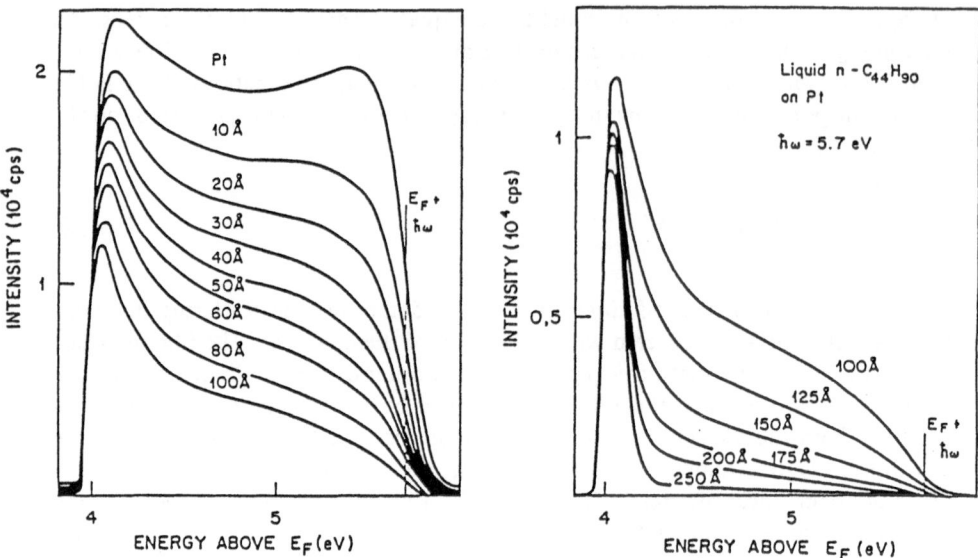

Fig. 5 Kinetic energy distribution of photoexcited electrons emitted across liquid film of n-C$_{44}$H$_{90}$ of various thickness into vacuum. Zero kinetic energy corresponds to about 4 eV.

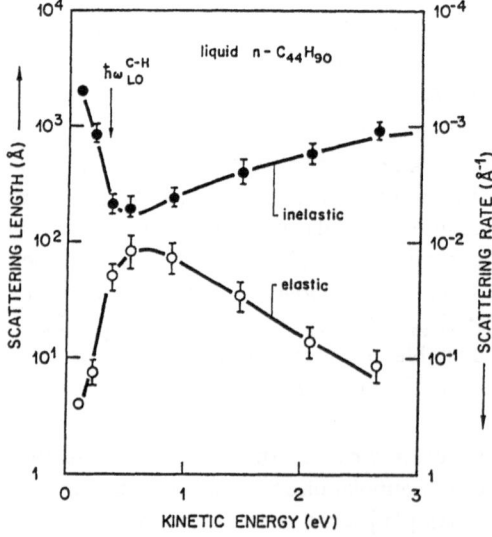

Fig. 6

Elastic and inelastic (LO phonon excitation) scattering processes in liquid n-C$_{44}$H$_{90}$. LO-phonon emission has a maximum near the phonon energy compatible with the Fröhlich model. At small energies the elastic scattering diverges (mobility edge) at high energies it is dominated by acoustic phonon scattering.

We have performed the scattering experiments on long chain linear alcanes also because we can study the behaviour of solid polycrystalline films and of liquid layers in order to get an idea on disorder scattering and the mobility edge. It turns out that there is little difference in the scattering functions between the ordered solid and the disordered liquid state above $E_{kin} \approx \hbar\omega_{LO} \approx 0.36$ eV. For very small kinetic energies γ_p diverges in the liquid indicative for the disorder localization already found for thermal electrons in other hydrocarbons.

In Fig. 6 the analysis is based on the assumption that we have a completely elastic process with scattering length ℓ_{el} and a LO-phonon emission process with energy loss $\hbar\omega_{LO} = 0.36$ eV, scattering length ℓ_{inel} and complete momentum randomization in both cases. We find ℓ_{inel} compatible with Fröhlich's model for polar mode scattering. ℓ_{el} is dominated by disorder scattering for small E_{kin} and by deformation potential scattering for large E_{kin}. A similar analysis for solid polycrystalline alcane films results in virtually identical scattering functions outside the regime of disorder scattering at small E_{kin}.

The observed scattering functions are compatible [17] with the observed breakdown fields of the order of ≈ 1 MV/cm for crystalline alcanes. We had speculated that the much higher breakdown field (≈ 6 MV/cm) in polyethylene is due to the amorphous fraction of the partially crystalline polymer. However, the difference in the scattering functions between crystalline and liquid $C_{44}H_{90}$ is too small to account for such a difference. In particular the width of the mobility edge is only a few kT. This means that there is always a reasonably high probability that a seed electron for an avalanche will be thermally excited into a high mobility state. The mobility edge alone thus cannot explain the difference in breakdown field between crystalline C_nH_{2n+2} and polyethylene. We believe that the difference in trap densities and possibly even a synergistic effect between disorder localization and capture in deep traps is responsible.

Qualitatively speaking two scattering regimes can be distinguished. For energies of the order of LO-phonon energies (typically a few tenth of an eV) polar mode scattering is dominant. Each scattering event thus basically leads to a nearly complete loss of energy and momentum. At energies large compared to $\hbar\omega_{LO}$ but smaller than the threshold for electronic processes, the scattering is quasi-elastic. The energy loss at each scattering event is small compared to the kinetic energy. In the limit of quasi-elastic scattering the Boltzmann transport equation can be mapped onto a diffusion equation in energy space. This greatly facilitates the calculation of impact ionization rates [16].

6 Fractal Models for Dielectric Breakdown

Phenomenologically breakdown patterns often have a tree-like structure. This is true for discharges in gases, for discharges at gas-dielectric interfaces (Lichtenberg figures), and for prebreakdown processes in liquids and solids. In solids tree-like structures which consist of hollow branches with diameters in the μm range, the so-called electrical trees, may grow over time periods from hours to years [19].

235

In the presence of humidity and electrical ac fields tree-like structures which are filled with water (water trees) develop [20]. The morphological features of water trees are beyond the resolving power of optical microscopes and electron microscopy has been plagued with artifacts. Hence little is known about the true microscopic structure of water trees.

All the different electrical breakdown, prebreakdown, and ageing phenomena have in common that they form tree-like, branched filamentary structures. The model which we will discuss in the following has no ambitions to explain the microscopic physics in each case. It should be understood rather as a "minimum number of ingredients" model which is able to phenomenologically reproduce trees.

A fractal model for tree growth has been introduced first by Niemeyer, Pietronero, and Wiesmann [21]. In their model the breakdown structure grows in steps on a lattice of lattice parameter a. The structure has zero internal resistance. The growth is controlled by the following rules:

1. The probability that a new bond to an adjacent lattice point is added to the breakdown structure is a function of the local field, for instance a power law $P(F_{loc}) \sim F_{loc}^\eta$.

2. After each growth step the Laplace equation has to be solved for the new boundary conditions.

The growth algorithm produces branched fractal structures but needs to be generalized in order to reproduce physical aspects of breakdown [22]. The first generalization consists in the introduction of a threshold field for growth. $P(F_{loc})$ is modified such that $P(F_{loc}) = 0$ for $F_{loc} < F_c$ and $P(F_{loc}) \sim F_{loc}^\eta$ for $F_{loc} > F_c$. This defines a breakdown condition in the sense that for a given electrode configuration no growth occurs below a critical voltage. In general, once growth starts it will proceed to the counterelectrode and no stable prebreakdown structures exist.

The second generalization consists in the introduction of an internal field F_s in the structure [22]. The voltage at a given lattice point of the breakdown structure is now $F_s \cdot s$, where s is the shortest path within the structure connecting the point to the electrode. This makes it possible to obtain stable prepreakdown structures in which growth proceeds to a certain size of the structure and then stops.

For the special case $F_s = F_c$ the prebreakdown structure is completely space filling, and the charge distribution exactly corresponds to the FLSC model of Sect. 4.2. With decreasing ratio F_s/F_c the structure breaks up into a tree-like branched pattern. The structure is no longer strictly speaking fractal because by introducing the parameter F_s and F_c we have introduced length scales. $F_s/F_c < 1$ corresponds to a negative differential resistance. F_c is the critical field to reach the NDR regime and F_s the "hold" field. Figs. 7 and 8 show examples of two-dimensional simulations. Fig. 7 corresponds to $F_s = F_c$ and results in a filled and stable prebreakdown structure. In Fig. 8 $F_s = 0$, and a branched pattern extends to the counterelectrode.

The physical significance of the probability law is at least qualitatively evident. For $\eta = 1$, i.e. $P(F_{loc}) \sim F_{loc}$ (and $F_s = F_c = 0$) the model is formally equivalent to the

236

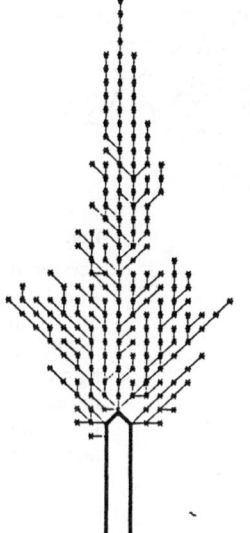

Fig. 7
For $F_S = F_C$ the structure becomes space filling with a charge distribution equal to the FLSC model.

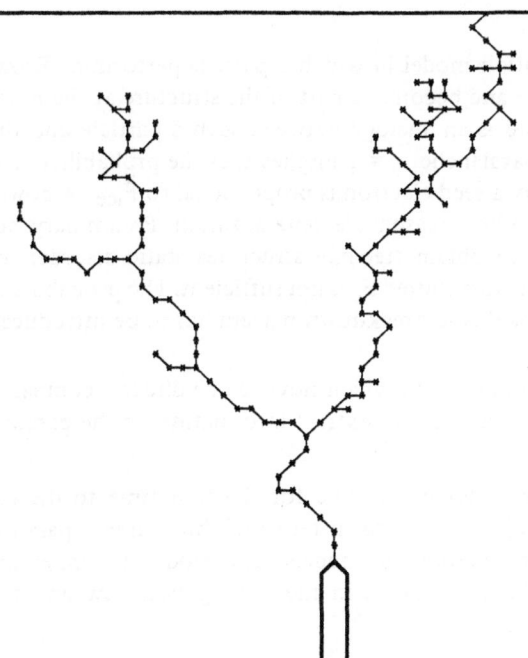

Fig. 8
The critical field for growth F_C is about equal to the original field at the tip. Initially no branching is possible. As the counterelectrode is approached, the local field increases leading to an increasing branching probability.

237

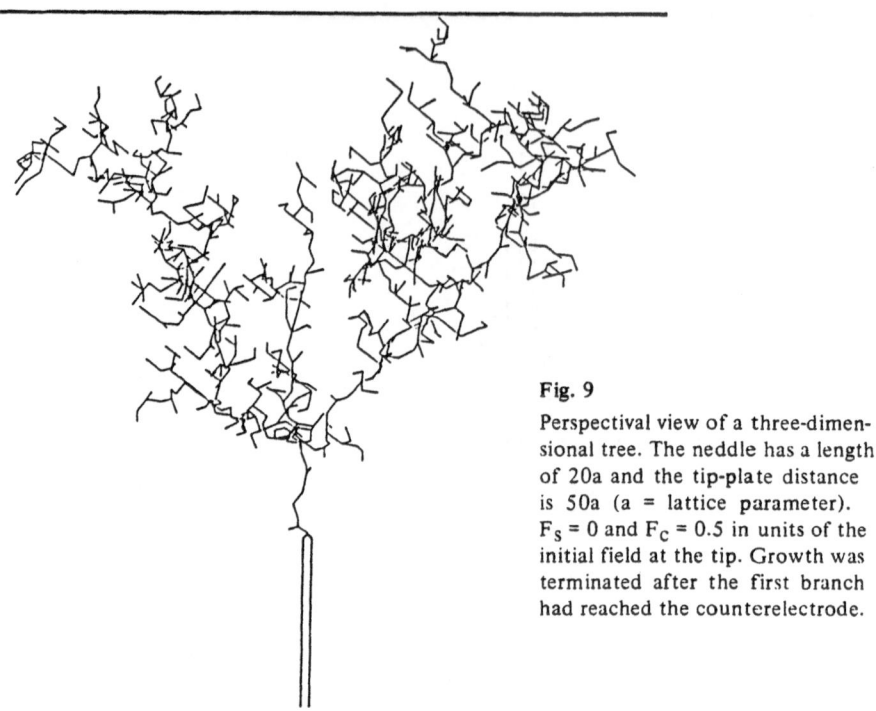

Fig. 9

Perspectival view of a three-dimensional tree. The neddle has a length of 20a and the tip-plate distance is 50a (a = lattice parameter). $F_s = 0$ and $F_c = 0.5$ in units of the initial field at the tip. Growth was terminated after the first branch had reached the counterelectrode.

so-called diffusion limited aggregation model in which a particle performing Brownian motion starts at large distance and becomes a part of the structure at the point where the particle hits first. There is an analogy between such a particle and the seed electron for forming a microavalanche. $\eta = 1$ implies that the probability that a given high field site is reached by a seed electron is proportional to F_{loc}. A power law type probability law seems to be a reasonable generalization. Preliminary numerical simulations indicate that to obtain tree-like structures static disorder for instance in the form of a statistically distributed F_c is not sufficient. The probabilistic element required to obtain a probabilistic breakdown pattern has to be introduced explicitly in the growth law.

Fig. 9 shows a simulation in three dimension which now can be directly compared with experimental discharge pattern in order to extract information on the parameters F_s, F_c, and η.

With the model discussed here, it becomes possible for the first time to discuss complex breakdown and prebreakdown patterns in terms of three simple parameters (F_c, F_s, η) which connect to microscopic models. The model illustrates the subtle interplay between probabilistic elements in the local growth law and the deterministic Poisson equation.

238

7 Dielectric Ageing

This is not a central subject in this review and will be treated only shortly and superficially. Although the breakdown field for a good dielectric is > 1 MV/cm, it is found that for nominal fields as small as 10 kV/cm slow ageing processes may occur which after years may lead to failure of an insulator. One large class of such phenomena is called electrical treeing [19]. An electrical tree invariably starts at a point of local field enhancement produced by material defects, electrode irregularities, etc. After the so-called initiation time, a microvoid forms which later develops into a tree-like hollow structure with branch diameter of order μm. Within the tree intermittent gas discharges take place.

A model for treeing has been introduced which states that a tree will grow if the gain in electrostatic energy is larger than the formation energy of the tree [4]. In its derivative form the criterion says that the tip of a branch will grow if the combined mechanical and dielectric stress at the tip is sufficient to cause plastic deformation. Depending on geometry this corresponds to about 3 to 4 times the yield stress σ_{yield}.

For a further discussion the reader is referred to the literature except for two comments. First, the growth of electrical trees is strongly affected by external mechanical stress field exactly in the way predicted by the growth criterion [23]. Second, the yield stress in the criterion is not the bulk yield stress but may be reduced by heating and cumulative chemical damage for instance bond scission by hot electrons [5]. Also the dielectric stress is mostly determined by the Coulomb forces excerted on the space charge distribution [4].

Another class of ageing phenomena are water trees [20]. Water trees are of economic importance mostly in polyethylene insulated cables. The damage structures are water filled, they can be made visible by dyeing. Water treeing requires relative humidities of at least about 60 % and ac fields. The morphology of water trees is very complex. Probably the branches usually consist of strings of microcavities (≤ 1 μm) filled with water. The cavities seem to be connected by pathways of enhanced diffusion but not by channels in a geometrical sense.

It has been thought for a long time that water trees form because of the large difference in the dielectric constants of water and polyethylene [24]. A closer inspection shows, however, that the dielectric stress is too small to fulfill the growth criterion [25]. Later it was independently found by two groups [25, 26] that electrochemical effects are the chief culprit. Space-charge injection from liquid water inclusions into the polymer leads to electrochemical processes such as oxidation at the interface. In polyethylene oxidation is an autocatalytic process. Free radicals and partial oxidation have been detected by several techniques in water treed regions [20].

After partial oxidation we are left with a macromolecular system in which the polar and hydrophilic oxidized groups are distributed in a nonpolar hydrophobic matrix [12]. Such a situation is thermodynamically unstable and a phase separation will occur. In this respect the system is very similar to the standard ionomeric polymers, and in fact the local morphology of water trees strongly resembles the morphology of ionomers [12].

Attempts to develop inhibiting additives based on either buffering the redox reaction or scavenging the free radicals have led to promising results. This is a beautiful example of the multidisciplinary nature of work in the area of dielectric ageing.

References

[1] J.J. O'Dwyer, The Theory of Electrical Conduction and Breakdown in Solid Dielectrics (Oxford University Press, London 1973)

[2] L.D. Landau and E.M. Lifshitz, Electrodynamics of Continuous Media (Pergamon Press, London, New York, Paris 1960)

[3] T. Hibma and H.R. Zeller, J. Appl. Phys. 59, 1614 (1986)

[4] H.R. Zeller and W.R. Schneider, J. Appl. Phys. 56, 455 (1984)

[5] E. Cartier and P. Pfluger, Proc. of the 2nd Int. Conf. on Conduction and Breakdown in Solid Dielectrics, Erlangen (1986), p. 308 (IEEE Service Center, Piscataway N.J, USA)

[6] H.J. Wintle, J. Appl. Phys. 52, 4181 (1981)

[7] J.J. Ritsko, in: Electronic Properties of Polymers, ed. by J. Mort and G. Pfister (Wiley, New York 1982)

[8] K.J. Less and E.G. Wilson, J. Phys. C: Solid State Phys. 6, 3110 (1973)

[9] J.J. Ritsko, J. Chem. Phys. 70, 5343 (1979)

[10] W.F. Schmidt, IEEE Transactions on Electr. Insul. EI-19, 389 (1984)

[11] G. Pfister and H. Scher, Advances in Physics 27, 747 (1978)

[12] H.R. Zeller, IEEE Transactions on Electr. Insul., to appear April 1987

[13] Th. Baumann, P. Pfluger, F. Stucki, and H.R. Zeller, Proceedings 1986 Conference on Electrical Insulation and Dielectric Materials (IEEE Service Center, Piscataway, N.J., USA 1986)

[14] R.E. Hebner, E.F. Kelley, E.O. Forster, and G.J. Fitzpatrick, IEEE Transactions on Electr. Insul. EI-20, 281 (1985)

[15] J.C. Devins, IEEE Trans. on Electr. Insul. EI-19, 371 (1984)

[16] M. Sparks, D.L. Mills, R. Warren, T. Holstein, A.A. Maradudin, L.J. Sham, and D.F. King, Phys. Rev. 24, 3519 (1981)

[17] P. Pfluger, H.R. Zeller, and J. Bernasconi, Phys. Rev. Lett. 53, 94 (1984) and IEEE Trans. on Electr. Insul. EI-19, 200 (1984)

[18] H. Fröhlich, Proc. Roy. Soc., Vol. A 160, 230, (1937)

[19] E.J. McMahon, IEEE Trans. on Electr. Insul., EI-13, 2177 (1979).

[20] M.T. Shaw and S.H. Shaw, IEEE Trans. on Electr. Insul., EI-19, 419, (1984)

[21] L. Niemeyer, L. Pietronero, and H.J. Wiesmann, Phys. Rev. Lett. 52, 1033 (1984).

[22] H.J. Wiesmann and H.R. Zeller, J. Appl. Phys. 60, 1770, (1986)

[23] J. Schirr, PhD Thesis, University of Braunschweig (Germany), (1974)

[24] S.L. Nunes, and M.T. Shaw, IEEE Trans. on Electr. Insul. EI-15, 437 (1980)

[25] H.R. Zeller, Proceedings of the 1985 International Conference on Properties and Applications of Dielectric Materials, Xian (China), Xian Jiaotong University Press, Xian (China), and IEEE Trans. on Electr. Insul., accepted for publication

[26] H.J. Henkel, N. Müller, J. Nordmann, W. Rogler, and W. Rose, Proc. 1986 Conference on Conduction and Breakdown in Solid Dielectrics, (IEEE Service Center, Piscataway, N.J., USA 1986)

Festkörperprobleme 27 (1987)

The Physics of Czochralski Crystal Growth

Werner Uelhoff

Institut für Festkörperforschung, Kernforschungsanlage Jülich, D-5170 Jülich, Federal Republic of Germany

Summary: Czochralski's crystal growth method is applied especially for the growth of nearly perfect single crystals. The resulting crystal shape is determined by the melt meniscus. Its actual shape allows the derivation of signals for controlling the growth process. Crystal perfection and shape of the growing crystal is determined by the heat flow in the crystal and by the matter and heat flow in the melt. The generation of the defects during crystal growth and possible improvements of the crystal perfection either by the application of growth in a magnetic field or in micro-gravity are discussed.

1 Johan Czochralski and his Crystal Pulling Method

In 1986 in Warsaw a symposium was arranged on the occasion of the 101th birthday of Johan Czochralski. At this scientific meeting a scientist was honored whose name is well known to crystal growers. But the personal data of his life are rather incomplete.

What do we know? According to [1], he came to AEG in Berlin in 1907, 22 years old. Though he was a self-made man he became the first head of the "Metall-Laboratorium". Main research topics of the laboratory were special alloys based on aluminium, lead, and cadmium as basic materials for plaines. He was well known in the field of deformation and recrystallization and was involved in the foundation of the "Deutsche Gesellschaft für Metallkunde". In 1926 he was the president of the society. The era of Czochralski ended very abruptly in Germany on Juli 1, 1928. Notes regarding the events do not exist. Shortly after having been promoted to Doctor honoris causae by the University of Warsaw he became director of the Institute of Metallurgy and professor there. A publication in a Swiss journal still shows his interest in the problems of crystallization and melting. According to a letter of his daughter he died on 22.4.1953 in Kcynia/Poland.

From all scientific successes of Czochralski, his crystal pulling method has influenced the posterity until our times in technological and social regards. In 1918 he published a method, as he assumed, for measuring the crystallization velocity of metals. In Fig. 1 the original drawing, showing the setup in his paper [2], is reproduced. An axle of the clock at the top pulls up a thread. A small hook of glas, fixed at its lower end, is immersed in the metal melt contained in the heated bowl. Then the thread slowly is pulled upward. So the melt solidifies unidirectionally at the line e-e above the melt meniscus.

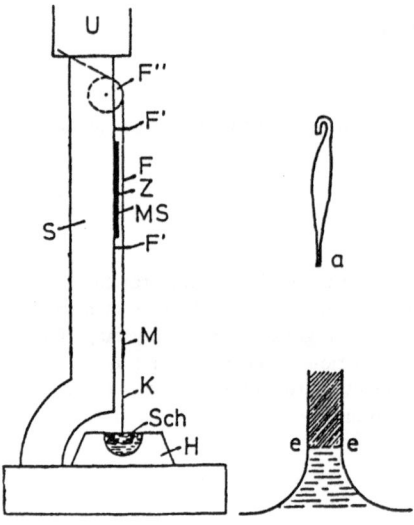

Fig. 1
Czochralski's setup for pulling metal single crystals. From [1].

Czochralski himself believed to be able to measure the maximum crystallization velocity, which was reached by separating the crystal from the melt, and to obtain informations concerning the atomistic mechanism of crystallization. This idea was not correct because the crystallization rate was limited in his setup by the amount of heat which was dissipated by the growing crystal. Of course he noticed that he was able to grow single crystalline metals by his method. So he performed deformation experiments with metal single crystals. He detected that deformation occurred on certain slip planes. However, a further understanding of the deformation process was not possible in those days: the concept of dislocations was developed some years later (in the early thirties) by Polanyi, Orowan, and others.

Czochralski could not have imagined in his dreams that nowadays nearly 10 000 tons of artificial crystals are grown per year. Some of them are listed in Tab. 1.

Table 1

Material	Application
Si (to 20 cm Ø, 2 m long)	Electronic devices
GaAs	Fast electronic devices, optoelectronics
Garnets	Laser, magnetic bubble memories
Halogenides	Optics (infrared and UV region)
Ni$_3$Al	High temperature material
fcc metals (to 1.5 cm Ø)	Nearly perfect: research purposes

In competition to the pulling technique some other crystal-growth methods have been developed. The most important techniques are (for details see Brice [3]):

the *Bridgman method* (inexpensive, directional solidification in a vertical mould, damage of crystal surface); the *zone melting in a horizontal boat* (a melt zone is moved horizontally through the bar of material, contact with the wall unavoidable, thus generation of dislocations); the *vertical zone melting without crucible* (no contamination, difficult to control the free meniscus, expensive, high thermal stresses), the *Verneuil technique* (inexpensive high temperature method, an oxy-hydrogen flame melts together, e. g., oxide particles).

The Czochralski method depends on a suitable crucible material. But lattice defects generated by contact of the hot crystal with the crucible wall are avoided. The second point finally led to the big success of Czochralski's method.

The physical problems of Czochralski growth will be discussed in the following sequence. Starting with defining more accurate the aimes to be reached, subsequently some essential topics will be stressed: the main features of the phase transition liquid/solid, the importance of the melt composition, and the atomic structure of the interface. In the next chapter the mechanism of steady crystal growth with so-called rough interfaces will be described. The measurement of growth quantities, the modelling and the automation of the growth process will be addressed. Own results of metal growth are presented. Unsteady growth occurs if singular faces can develop. The resulting special features will be explained. Sodium chloride was the model material for studying this kind of growth.

The growth process of multi-component systems is explained for miscible binary alloys. Growth in this case is governed by a combined transport of momentum, matter, and heat. So after a short description, the additional problems of alloy growth will be treated: the generation of concentration striations due to segregation and in the possible break-down of the planar interface. In the next two chapters possibilities and limits of new techniques of crystal growth in space and in a magnetic field will be discussed. What results can be expected? Kind and origin of physical and chemical defects are very important aspects. They will be treated in the next paragraph. We will see that speculation is still a governing factor in this field. Finally, results, trends, and some special efforts will be treated.

2 Aimes of Czochralski Pulling

Applying this crystal-growth method, one tries to achieve the following results. A geometric shape with the desired constant *crystal diameter* can minimize processing after growth. *Physical defects* such as dislocation lines should be below the necessary limit. Dislocations on slip bands or in sub-boundaries are in general unwanted. *Chemical defects* as impurities either should not exceed a certain limit as oxygen in silicon crystals. On the other hand doping elements in semiconductors should be distributed *homogeneously* in the crystal. *Necessary deviations from stoichiometry* (for instance for As in GaAs) should reach the desired value with a constant value in the whole crystal. − It will be shown that in many cases it is extremely difficult to achieve these goals simultaneously.

3 Important Aspects of Czochralski Pulling

The *phase transition* liquid/solid is of basic importance for Czochralski growth:

a) In the beginning the material to be crystallized is molten. Hereby boundaries and dislocations are destroyed.
b) Crystallization then is performed by moving a superimposed temperature gradient, so that a boundary solid/liquid moves through the material.
c) Crystallographic orientation is defined by using a seed for initiation the crystallization. This also avoids the problem of nucleation.

Both solidification and solidified material are depending on the *melt composition*. A melt with only one component does not exist in reality. Only dislocations and the maintenance of the purity are of interest (example: silicon, copper).

In the case of a multi-component system (miscible components) the main problem is the different solubility of the components in the liquid and in the solid state. Decomposition of the melt due to a high vapour pressure of one component (as As in GaAs) may be a big problem.

The *atomistic structure* of the interface solid/liquid but also solid/gas depends on the difference in binding energy of an atom in the solid and liquid phase at the interface. Two cases have to be distinguished: the atomically flat or singular interface and the rough interface. An interface is singular if its energy has a sharp minimum compared to those with neighbouring orientations. Its lattice sites are occupied to 100%. Growth of these interfaces, also called "facets", only occurs after building of surface nuclei which requires a supercooling of ca. 2 K.

Then fast lateral growth of the nuclei occurs. This means that growth in direction of the temperature gradient is *unsteady*. Our investigations of the growth of NaCl crystals with the {100}-singular face will be presented. {111}-faces in crystals with diamond structure as in the case of silicon are of great importance. Rough interfaces are occupied to 50% in the average. A nucleation process is not necessary for attachment of atoms from the melt. Therefore much smaller supercooling of mK lead to growth velocities of some 10 cm/h. The liquid solidifies *steadily* parallel to the temperature gradient. An example: all lattice orientations of copper. − Both interface structures can exist simultaneously on an interface during growth. This is unwanted because the incorporation of different components is different.

4 The Czochralski Method: Arrangement and Process; Steady Growth

In Fig. 2 the arrangement of the pulling system is drawn in more detail. Heating can be performed with radiofrequency coils. For big crucibles resistance heating

244

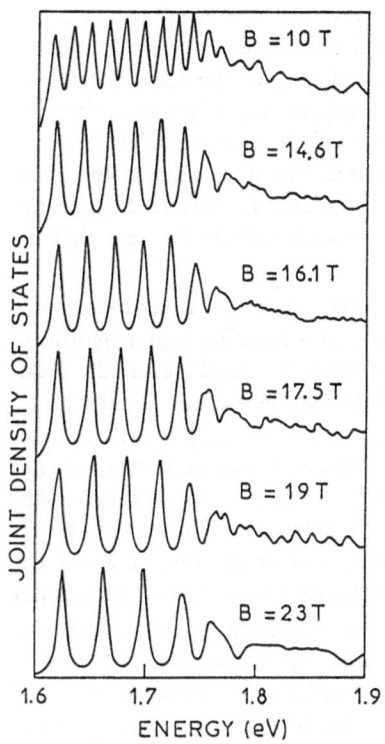

JOINT DENSITY OF STATES

ENERGY (eV)

B = 10 T

B = 14.6 T

B = 16.1 T

B = 17.5 T

B = 19 T

B = 23 T

Fig. 13

Joint density of states calculated from the
magnetic levels in the experimentally studied
superlattice (Fig. 10 and 11 bottom). The
magnetic field dependence of the maxima
are shown as the drawn lines in Fig. 4b.

The Landau levels in Fig. 10 bottom for 20 T are calculated for these parameters, and a similar calculation was performed for the holes. To compare with the experiment it is assumed that transitions take place only between the first hole and the first electron level, the second hole to the second electron, etc., whereby the cyclotron orbit center will be conserved (vertical transitions). Furthermore the matrix elements are assumed to be the same for all possible transitions. This set of assumptions are in fact the selection rules for normal Landau levels described by harmonic oscillator wavefunctions, which need not a priori to be valid in the present case. In this framework the spectra are entirely determined by the joint density of states which can be obtained from the calculated Landau levels. The result of such a calculation for a few different values of the magnetic field is shown in Fig. 14. Qualitatively, the agreement with the experimental observations is striking. More quantitatively, the magnetic field dependence of the peaks in the joint density of states is compared with the observed maxima in the excitation spectra in Fig. 4b (drawn lines). It can be seen that even quantitatively the agreement is rather good, since the correct slopes of all the transitions is reproduced by the calculations. Both experimentally and theoretically the slopes at $B_\parallel$ are somewhat shallower on average than those at $B_\perp$ even when the same mass is used for the barriers and the wells.

161

Therefore the slightly higher "parallel effective mass" is attributed to the super-lattice bandstructure. That at higher energy no peaks can be distinguished anymore in the calculated spectra of course reflects the broadening of the Landau levels in the superlattice minigap. Roughly speaking the disappearance of peaks in the joint density of states takes place when the broadening of the Landau levels becomes equal to the average energy separation between them, and this occurs at the upper edge of the superlattice miniband. However, the calculated transitions disappear at too low energies compared with the experiment, which reflects the fact that the calculated width is too small.

The main remaining discrepancy is therefore that the experimental subband width is too large compared with the theory. The difference is in effect rather significant because as can be seen from Fig. 14 it implies that, for instance, at 22 T one Landau level more is observed than is calculated. It may be noted that the rather sloppy way in which the hole states have been treated is probably not responsible for this fact, because, since the electron states are more important anyway (due to their much lighter mass), the number of observable levels is at most equal to the number of flat electron Landau levels, or in the language of Chapt. 3, to the number of closed orbits for the electrons. Treating the holes more carefully can therefore only reduce the number of calculated transitions marking the discrepancy worse. An alternative explanation might be that the sample parameters are slightly different from those used in the calculation and this explanation cannot really be excluded. However, thicknesses cannot be varied aribitrarily, because only full lattice periods can be added, and within the exploration of the parameter space for different thicknesses, for mass barrier heights discussed before no satisfactorily agreement could be found. Many more fundamental arguments may be advanced to explain this discrepancy, however, as will be shown they usually tend to make the subband width less rather than larger. For instance, it might be argued that the GaAs CB non-parabolicity has been treated too simply. However, magneto optical

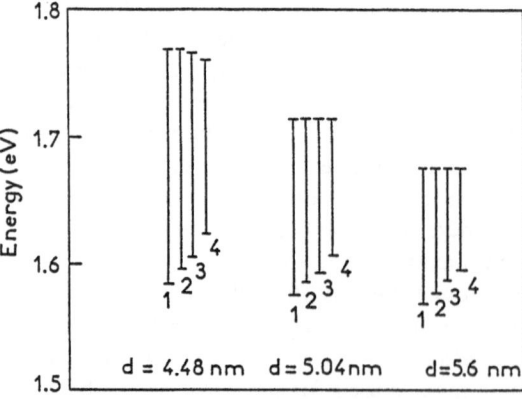

Fig. 14

The position and the width of the electron subband for a GaAs/GaAlAs superlattice with 1.12 nm barrier width and periodicity indicated in the figure. 1 and 2 are calculated using the 60/40 and 3 and 4 using the 85/15 rule for the distributing of the bandgap difference between the conduction and the valence band. 1 and 3 are calculated using the same mass in the barriers and in the wells and 2 and 4 with a 30 % higher mass in the barriers.

experiments on quantum wells [53, 55], which equally measure effective masses at energies high in the band, show that the two band model used here underestimates the non-parabolicity, and a heavier mass for GaAs would lead to an even smaller subband. Furthermore one can consider the effect of thickness variations. However, as it was argued before, the lack of long-range order in the superlattice affects the higher Landau levels first, because they are due to the interaction of wells which are far apart, contrary to experiments. For the same reason the finite number of layers of the sample (40 in this case) cannot be responsible for the discrepancy as was demonstrated in the preceding chapter. In addition, it should be noted that the Landau-level peaks disappear at the same energy independent of the magnetic field. If the finite size of the sample was responsible for the disappearance the cut-off energy should be field dependent because at lower fields the orbits are larger and they would be affected by the sample dimensions at lower energy already. In this context it is interesting to come back on the remark at the end of Chapt. 4 that there is a one to one correspondence between the number of Landau levels and the number of interacting wells. Concretely, in this experiment typically eight well resolved peaks are observed which therefore means that at least eight wells are interacting in coherent way. In the theory of the de Haas-van Alphen-effect a well-known phenomenon is magnetic breakdown. This effect is that electrons instead of following open orbit in k-space, for instance the outer orbit in Fig. 6, make a jump from some $k_{\parallel}$ to $-k_{\parallel}$ at $k_z = \pm \pi/d$ in order to follow a closed path. This effect would indeed explain the observations, namely that at energies slightly beyond the subband edge still adsorption peaks are observed. In the quasi-classical description this phenomenon is well established, although the equivalent phenomenon in the more exact description of Chapt. 4 is not directly apparent. This question too is best left for further theoretical and experimental studies.

There could be eventually a very interesting although also very speculative explanation for the results. The entire theoretical analysis of this paper is based on the validity of the envelope function approximation (effective mass theory) in this case. Concretely, a layer of two lattice constants thickness of GaAlAs (x = 0.44) is treated as some bulk material A entirely described by bulk properties, which is matched to a layer B of GaAs of seven lattice constants, equally treated as a bulk layer. First of all it may be asked whether such a very thin layer of GaAlAs can be treated in the virtual crystal approximation. Secondly, the basis of effective mass theory is the assumption that the dynamical properties of a bulk material can be described by envelope wavefunctions which are assumed to be slowly varying on the scale of the lattice periodicity, and this condition is not fullfilled here. Both theoretically and experimentally the subject alloy versus superlattice behaviour is still under discussion [56 ... 58]. For the sake of illustration in Fig. 15 schematically a superlattice with thicknesses and Al content comparable to the actual sample is drawn. Each dot represents a unit cell of GaAs or AlAs, and in the GaAlAs layers 40 % of the cells are AlAs. Looking at this figure the question

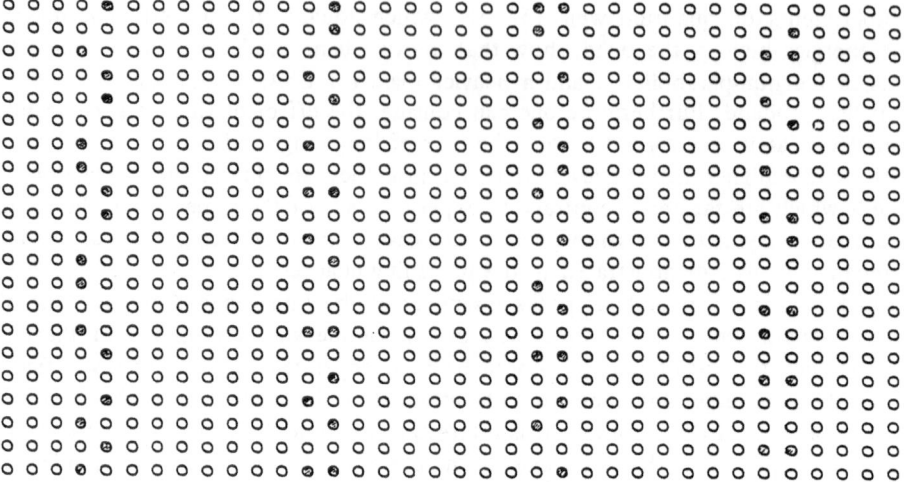

Fig. 15 Schematic representation of a superlattice with two lattice parameters of GaAlAs (x = 0.4) and 7 of GaAs. Each open circle represents one unit cell of GaAs and each dashed circle one of AlAs.

whether such a structure behaves like a superlattice or like an GaAlAs alloy with some average Al content (in this case 0.088) comes to the mind. Concretely, for the present sample it can naively be calculated that such an alloy would show a gap of ≈ 1.61 eV, have a mass which would be some 10 % higher than GaAs, would have an exciton binding energy of about 5 meV, would show Landau levels both with perpendicular and parallel magnetic field, in fact would behave almost exactly like what is seen experimentally. The important qualitative difference between the alloy and the superlattice is of course that a superlattice has minigaps due to the superperiodicity and that an alloy has not. Once more the observation of the disappearing Landau levels rules out such an explanation. Nevertheless it is very well possible that the theoretical description of a system with such thin layers, using the envelope-function approximation, is not accurate anymore. The residual discrepancy between experiment and theory might eventually be explained in such a way.

It was mentioned in the introduction that the application of a magnetic field parallel to the layers can be seen as a technique to measure "vertical transport", and the preceding sections can be seen as an illustration of this statement. To be more concrete, it is possible to estimate a relaxation time or a mobility for transport through the barriers from the data in a simple manner. The linewidth of the peaks is typically 5 meV, corresponding to a scattering time of 10^{-13} s which leads to a "mobility for vertical transport" of roughly 5000 cm/Vs. Furthermore it is interesting to note that the width of the peaks is only slightly wider in the $B_{\parallel}$ case compared to the $B_{\perp}$, showing that the scattering time is almost isotropic, and

164

this despite the fact that carriers cross several barriers in one direction. In fact, the mobility obtained is high compared to that found in dc transport measurements through the layers as reported by Palmier [27]. This is probably due to the fact that in this case no intentional doping is needed to study transport through the barriers.

Acknowledgement

It is a pleasure to acknowledge the pleasant collaboration with G. Belle during the experimental part of this work. Furthermore I would like to thank M. Altarelli for many very illuminating discussions on the theory of the subject. Finally I wish to express my gratitude to G. Weimann for the growth of the excellent superlattice samples which were indispensable for the success of the experiments.

References

[1] *L. Esaki* and *R. Tsu*, IBM J. Res. Dev. **14**, 61 (1970)

[2] *L. L. Chang, L. Esaki,* and *R. Tsu*, Appl. Phys. Lett. **24**, 593 (1974)

[3] *L. Esaki* and *L. L. Chang*, Phys. Rev. Lett. **33**, 495 (1974)

[4] *R. Dingle, W. Wiesmann,* and *C. H. Henry*, Phys. Rev. Lett. **33**, 827 (1974)

[5] *R. Dingle, A. C. Gossard,* and *W. Wiegmann*, Phys. Rev. Lett. **34**, 1327 (1975)

[6] *R. Dingle*, in: Festkörperproblem/Advances in Solid State Physics, ed. by *H. J. Queisser* (Vieweg, Braunschweig 1975) Vol. XV, p. 21

[7] *N. Holonyak, R. M. Kolbas, W. D. Laidig, B. A. Vojak, R. D. Dupuis,* and *P. D. Dapkus*, Appl. Phys. Lett. **33**, 737 (1978)

[8] *R. Dingle, H. L. Störmer, A. C. Gossard,* and *W. Wiegmann*, Appl. Phys. Lett. **33**, 665 (1978)

[9] *K. von Klitzing, G. Dorda,* and *M. Pepper*, Phys. Rev. Lett. **45**, 494 (1980)

[10] *D. C. Tsui, H. L. Störmer,* and *A. C. Gossard*, Phys. Rev. Lett. **48**, 1562 (1982)

[11] a) *K. von Klitzing*, in: Festkörperprobleme/Advances in Solid State Physics, ed. by *J. Treusch* (Vieweg, Braunschweig 1981) Vol. XXI, p. 1
   b) *G. Döhler*, in: Festkörperprobleme/Advances in Solid State Physcis, ed. by *P. Grosse* (Vieweg, Braunschweig 1983) Vol. XXIII, p. 207
   c) *N. T. Linh*, in: Festkörperprobleme/Advances in Solid State Physics, ed. by *P. Grosse* (Vieweg, Braunschweig 1983) Vol. XXIII, p. 227
   d) *H. L. Störmer*, in: Festkörperprobleme/Advances in Solid State Physics, ed. by *P. Grosse* (Vieweg, Braunschweig 1984) Vol. XXIV, p. 25
   e) *G. Abstreiter*, in: Festkörperprobleme/Advances in Solid State Physics, ed. by *P. Grosse* (Vieweg, Braunschweig 1984) Vol. XXIV, p. 291
   f) *G. Wiemann*, in: Festkörperprobleme/Advances in Solid State Physics, ed. by *P. Grosse* (Vieweg, Braunschweig 1986) Vol. 26, p. 231

[12] *L. Esaki*, in: Heterojunctions and Semiconductor Superlattices, ed. by *G. Allal, G. Bastard, N. Boccara, M. Lannoo,* and *M. Voos* (Springer, Heidelberg 1986)

[13] *L. L. Chang, H. Sakaki, C. A. Chang,* and *L. Esaki*, Phys. Rev. Lett. **38**, 1489 (1977)

[14] *L. L. Chang, E. E. Mendez, N. J. Kawai,* and *L. Esaki*, Surf. Sci. **113**, 306 (1982)

[15] *T. Ando*, J. Phys. Soc. Japan **50**, 2978 (1981)

[16] *J. C. Maan, Y. Guldner, J. P. Vieren, P. Voisin, M. Voos, L. L. Chang,* and *L. Esaki*, Solid State Commun. **39**, 683 (1981)

[17] *T. C. L. G. Sollner, W. D. Goodhue, P. E. Tannenwald, C. D. Parker,* and *D. D. Peck,* Appl. Phys. Lett. **43,** 588 (1983)

[18] *T. Nakagawa, H. Imamoto, T. Kojima,* and *K. Ohta,* Appl. Phys. Lett. **49,** 73 (1986) and also reference cited therein

[19] *A. Chomette, B. Deveaud, J. Y. Emery, A. Regreny,* and *B. Lambert,* Solid State Commun. **54,** 75 (1985)

[20] *B. Deveaud, A. Chomette, B. Lambert, A. Regreny, R. Romestain,* and *P. Edel,* Solid State Commun. **57,** 885 (1986)

[21] *D. Calecki, J. F. Palmier,* and *A. Chomette,* J. Phys. C (Solid State Physics) **17,** 5017 (1984)

[22] *J. F. Palmier, M. Leperson, C. Minot, A. Chomette, A. Regreny,* and *D. Calecki,* Superlattices and Microstructures **1,** 76 (1985)

[23] *J. Yoshino, H. Sakaki,* and *T. Furuta,* p. 519, 49, Proc. of the 17th Int. Conf. on the Physics of Semiconductors, San Francisco 1984 (Springer, New York 1985)

[24] *R. A. Davies, M. J. Kelly,* and *T. M. Kerr,* Phys. Rev. Lett. **55,** 1114 (1985)

[25] *F. Capasso, K. Mohammed, A. Y. Cho, R. Hull,* and *A. L. Hutchinson,* Phys. Rev. Lett. **55,** 1152 (1985)

[26] a) *G. Belle, J. C. Maan,* and *G. Weimann,* Solid State Commun. **56,** 65 (1985)
    b) *G. Belle, J. C. Maan,* and *G. Weimann,* Surface Sci. **170,** 611 (1986)

[27] *T. Duffield, R. Bhat, M. Koza, D. M. Hwang, P. Grabbe,* and *S. J. Allen Jr.,* Phys. Rev. Lett. **56,** 2724 (1986)

[28] *J. C. Maan,* Springer Series in Solid State Sciences, Vol. 53, ed. by *G. Bauer, F. Kuchar,* and *H. Heinrich* (Springer, Berlin 1984), p. 183

[29] *W. Beinvogl, A. Kamgar,* and *J. F. Koch,* Phys. Rev. **B14,** 4274 (1986)

[30] *U. Merkt,* Phys. Rev. **B32,** 6699 (1985)

[31] *C. Kittel,* Quantum theory of Solids (John Wiley & Sons, New York 1963)

[32] *L. D. Landau* and *E. M. Lifshitz,* Course of theoretical Physics, Vol. 9, part 2, Statistical Physics, ed. by *E. M. Lifshitz* and *L. P. Pitaevskii* (Pergamon Press, Oxford 1980)

[33] *L. D. Landau, E. M. Lifshitz,* Course of theoretical Physics, Vol. 3, Quantum Mechanics (Pergamon Press, Oxford 1977)

[34] a) *A. B. Pippard,* Phil. Trans. Roy. Soc. **A256,** 317 (1964)
    b) *A. B. Pippard,* in: The theory of metals, ed. by *J. Ziman* (Cambrdige University Press, Cambridge 1969), Ch. 3

[35] *P. G. Harper,* Proc. Phys. Soc. **68,** 879 (1955)

[36] *R. Tsui* and *J. Janak,* Phys. Rev. **B9,** 404 (1974)

[37] a) *G. E. Zil'berman,* Soviet Phys. JETP **5,** 208 (1957)
    b) *G. E. Zil'berman,* Soviet Phys. JETP **6,** 299 (1958)

[38] *G. Bastard,* Phys. Rev. **B24,** 5693 (1981)

[39] a) *M. Altarelli,* Phys. Rev. **B28,** 842 (1983)
    b) *M. Altarelli,* in: Applications of High Magnetic Field in Semiconductor Physics, ed. by *G. Landwehr* (Springer, Berlin 1983)

[40] The Kronig-Penney-bandstructure is discussed thoroughly in: *R. A. Smith,* Wave Mechanics of Crystalline Solids (John Wiley & Sons, New York 1961)

[41] a) *R. E. Doezema* and *J. F. Koch,* Phys. Rev. **B5,** 3866 (1972)
    b) *R. E. Doezema* and *J. F. Koch,* Phys. Rev. **B6,** 2071 (1972)

[42] *M. Wanner, R. E. Doezema,* and *U. Strom,* Phys. Rev. **B12,** 2883 (1975) and references cited therein

[43] *T. Ando*, J. Phys. Soc. Japan **39**, 411 (1975)

[44] *K. F. Bergren* and *D. J. Newson*, Semicond. Sci. Technol. **1**, 327 (1986)

[45] *J. C. Maan, G. Belle, A. Fasolino, M. Altarelli,* and *K. Ploog*, Phys. Rev. **B30**, 2253 (1984)

[46] *R. C. Miller, D. A. Kleinman, W. T. Tsang,* and *A. C. Gossard*, Phys. Rev. **B24**, 1134 (1981)

[47] *G. Bastard, E. E. Mendez, L. L. Chang,* and *L. Esaki*, Phys. Rev. **B26**, 1974 (1982)

[48] *P. Dawson, K. J. Moore, G. Duggan, H. I. Ralph,* and *C. T. B. Foxon*, Phys. Rev. **B34**, 6007 (1986)

[49] *A. Fasolino* and *M. Altarelli*, in: Springer Series in Solid State Sciences, Vol. 53, ed. by *G. Bauer, F. Kuchar,* and *H. Heinrich* (Springer, Berlin 1984), p. 176

[50] *J. N. Schulman* and *Y. C. Chang*, Phys. Rev. **B31**, 2056 (1985)

[51] *R. Sooryakumar, D. S. Chemla, A. Pinczuk, A. C. Gossard,* and *W. Wiegmann*, Proc. of the 17th Int. Conf. on the Physics of Semiconductors, San Francisco 1984 (Springer, New York 1985), p. 523

[52] *Y. C. Chang* and *G. D. Sanders*, Phys. Rev. **B32**, 5521 (1985)

[53] *F. Ancilotto, A. Fasolino,* and *J. C. Maan*, Proc. of the 2nd Int. Conf. on Superlattices, Gotenborg, 1986, J. of Superlattices and Microstructures, to be published

[54] *M. Kriechbaum*, Proc. of the 2nd Int. Conf. on Superlattices, Gotenborg, 1986, J. of Superlattices and Microstructures, to be published

[55] *D. C. Rogers, J. Singleton, R. J. Nicholas, C. T. Foxon,* and *K. Woodbridge*, Phys. Rev. **B34**, 4002 (1986)

[56] *W. Andreoni* and *R. Car*, Phys. Rev. **B21**, 3334 (1980)

[57] *E. Carruthers* and *P. J. Lin-Chung*, Phys. Rev. **B17**, 2705 (1978)

[58] *A. Ishibashi, Y. Mori, M. Itabashi,* and *N. Watanabe*, J. Appl. Phys. **58**, 2691 (1985)

167

Festkörperprobleme 27 (1987)

Structure and Reactivity of Solid Surfaces

Gerhard Ertl

Fritz-Haber-Institut der Max-Planck-Gesellschaft, D-1000 Berlin (West) 33, Germany

Summary: The geometric configuration of the atoms in the surface of a solid is correlated with their valence electronic properties and thereby also with their chemical reactivity towards molecules interacting from the gas phase. Several aspects of these interactions are briefly reviewed: The atomic structure of clean single crystal surfaces and their changes by bond formation (chemisorption), the formation of chemisorbed phases with long-range order and associated phase transitions, the dynamics of the gas-surface interaction processes, as well as temporal oscillations in a catalytic reaction coupled to periodic structural transformations of a surface.

1 Introduction

The ideal crystal with infinite three-dimensional periodicity will never be realized. Apart from bulk defects a crystal will always exhibit a strong distortion of this periodicity at its termination: the surface. The ratio of atoms in the surface to those in the bulk increases with decreasing particle size or film thickness. The overall properties of such systems may then be essentially determined by those of the surface.

The surface atoms are missing part of their nearest neighbors which causes variations of the valence electronic properties as well as of the equilibrium positions of the nuclei. The former effect is reflected in the chemical reactivity of the surface by which 'dangling bonds' may become saturated, while the latter manifests itself in structural parameters deviating from those of the bulk. The present contribution intents to illustrate by means of a few selected examples our present knowledge of these phenomena and their mutual interplay.

2 The Structure of Clean Surfaces

The geometric location of the atoms in the outermost layer may differ from those of a corresponding bulk plane in two respects, namely, alterations of the interlayer spacings (relaxation) and lateral displacements connected with changes of the unit cell within the surface layer (reconstruction) [1].

As a first example Fig. 1 shows the structure of the clean Ni (110) surface which exhibits the same lateral periodicity as the bulk, but where the spacing between the first and second layer is contracted by 8.5 %, while that between the second and third is expanded by 3.5 % [2]. These findings are quite general and have now

Ni(110) - clean

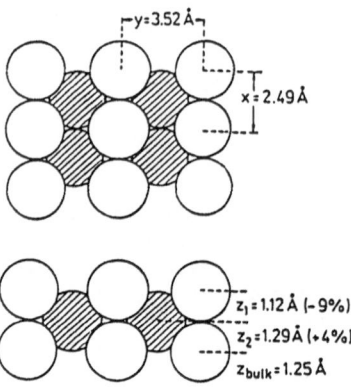

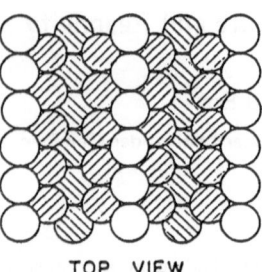

TOP VIEW

SIDE VIEW

Fig. 1 Structure model of the (110) surface of fcc metals without reconstruction, such as Ni(110).

Fig. 2 'Missing row' structure of the Pt(110) surface.

been established for a large series of metal surfaces [1]. The effects are more pronounced with the more open planes for which Δd_{12} of up 15 % were reported, while the atoms in the most densely packed planes exhibit usually only very minor deviations from their regular bulk positions.

The surface atoms will generally have the tendency to minimize their free energy by surrounding themselves by as many nearest neighbors as possible, i.e. by forming a most densely packed plane. This tendency is counterbalanced by the resulting mismatch between surface and bulk planes. This qualitative argument indicates why surfaces may undergo reconstruction: While the Ni(110) surface is unreconstructed, the Pt(110) surface is reconstructed (Fig. 2) [3]: Every second row in [$1\bar{1}0$]-direction is missing (leading to a 1×2 superstructure), and the surface may now be considered as existing of microfacets of the most densely packed (111) plane.

Similarly, the (100) planes of Ir, Pt, and Au are reconstructed in a way as illustrated by Fig. 3 [3]. The atoms of the topmost layer exhibit a quasi-hexagonal ('hex') configuration similar to that of the (111) plane placed on the square lattice of the (100) plane forming the second and deeper layer. The mismatch between first and second layer is in this case reflected by the fact that the topmost atoms do not form a perfectly flat plane, but are slightly 'buckled'. This buckling may be nicely made visible by the scanning tunneling microscope (STM). The STM image in Fig. 4 from a clean Pt(100) surface shows two domain orientations of the 'hex' surface with its corrugation periodicity of 13.5 Å, whereby the step in the center serves as domain boundary [4].

While most of the clean metal surfaces are not reconstructed, the situation is opposite with semiconductor surfaces where reconstruction is the rule. The 7 × 7-

170

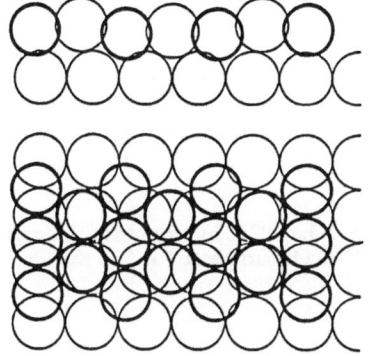

Fig. 3

Reconstructed Ir (100) surface where the atoms of the topmost layer form a hexagonal configuration yielding a 5 × 1-superstructure. Similar structures are found with clean Pt (100) and Au (100) surfaces.

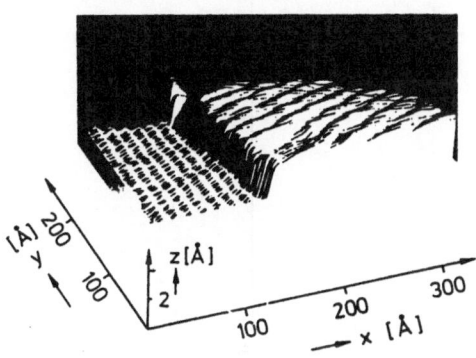

Fig. 4

Scanning tunneling microscope (STM) image of a clean Pt (100) surface with a step [4].

structure of the annealed Si (111) surface is probably the most famous example which is now — after many years of intense research — considered to be solved [5].

3 Structure of Adsorbate Covered Surfaces

The energy of surfaces may be lowered by the formation of chemical bonds with suitable particles arriving from the gas phase (chemisorption). We will restrict our discussion to cases in which this reactivity does not extend into deeper layers, eventually leading to the formation of new bulk compounds such as oxides etc. However, also those processes are initiated by chemisorption steps at the outermost atomic layer.

Fig. 5 shows the configuration of H atoms formed on a Ni (111) surface at $T \leqslant 200$ K and at a coverage $\theta = 0.5$ *) [6]. The H atoms prefer three-fold coordinated sites

---

*) The coverage θ is defined as the ratio of the density of adsorbed particles over that of the substrate atoms in the topmost layer.

171

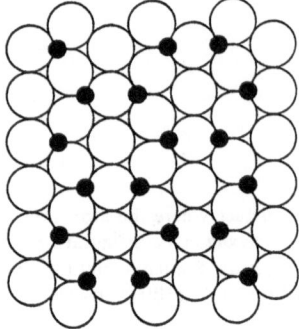

Fig. 5

Structure of the ordered 2 × 2-overlayer of H atoms adsorbed on a Ni (111) surface at T ≤ 200 K with a coverage $\theta = 0.5$.

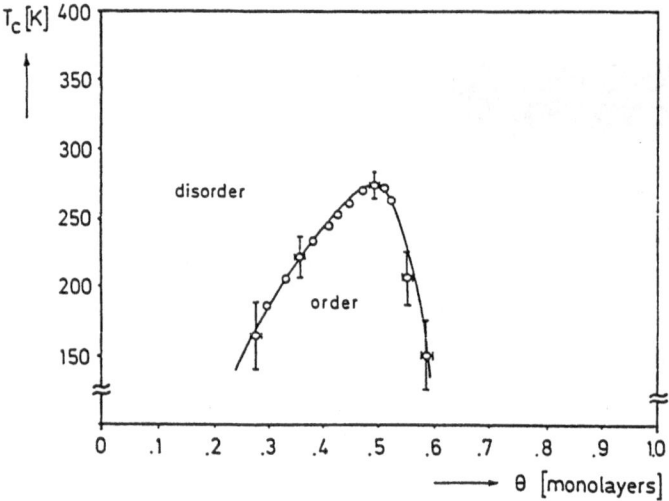

Fig. 6 Phase diagram of the 2 × 2-structure of the H/Ni (111) system.

(i.e. as many nearest neighbors as possible) to which they are attached by an energy of about 2.5 eV. The mutual configuration is under these conditions characterized by pronounced long-range order, giving rise to 'extra' beams in low energy electron diffraction (LEED). This long-range order is obviously due to the operation of interactions between the adsorbed particles, which in the present case are of the order ≲ 0.1 eV and are of the 'indirect' type, i.e. mediated through the valence electrons of the substrate metal. Increasing the temperature leads to continuous order-disorder transitions as can be followed through the variation of the respective LEED intensities. Determination of the transition temperatures at varying coverages enables to establish a phase diagram as reproduced in Fig. 6. Such two-dimensional phase diagrams have now been determined for different systems for which in turn theoretical simulation yielded good agreement if the interaction parameters were properly adjusted [7].

172

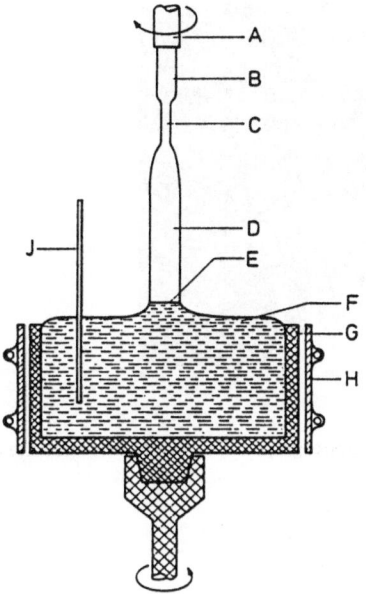

Fig. 2

Schematic diagram of a modern Czochralski crystal growth system. A. pulling rod with seed holder, B. seed crystal, C. crystal neck, D. crystal, E. interface, F. melt, G. crucible, H. RF-coil, J. thermocouple.

is preferred. The problems introduced by the rotation of the crystal and of the crucible for achieving rotational temperature symmetry will be discussed below.

The seed is fixed at the lower end of the pulling rod. In the sketched stage of the pulling process, the crystal neck, a region with a strongly decreased diameter, is already pulled, followed by the shoulder region: the transition to the final diameter.

Pulling of a crystal neck, as invented by Dash [4], is a very important part of the process. In this thin region the dislocation density, grown in from the seed or generated by the immersion of the seed in the melt, may be reduced to zero.

The growing crystal is connected with the pool of melt by the meniscus. Opaque materials allow observation of the triple-phase-line crystal/melt/gas only. Its distance down to the melt surface for nearly flat interfaces is equal to the *interface height* h.

In general growth occurs below the upper edge of the crucible. One has to watch the growing crystal with a steep viewing angle (Fig. 3a). The adhering meniscus appears as a bright ring due to light reflections of the free crubicle part and the upper meniscus part, both reflected from the lower part of the meniscus. Pulling from a filled crucible (Fig. 3b) is ideal for measuring growth determining parameters.

The actual crystal diameter, that means the *shaping process*, is controlled by the shape of the adhering meniscus and the height of the triple-phase-line. The mechanism can be described quantitatively for the growth of fcc metal crystals. Here the growth conditions are rather simple: 1) the crystal is wetted totally by the melt and

245

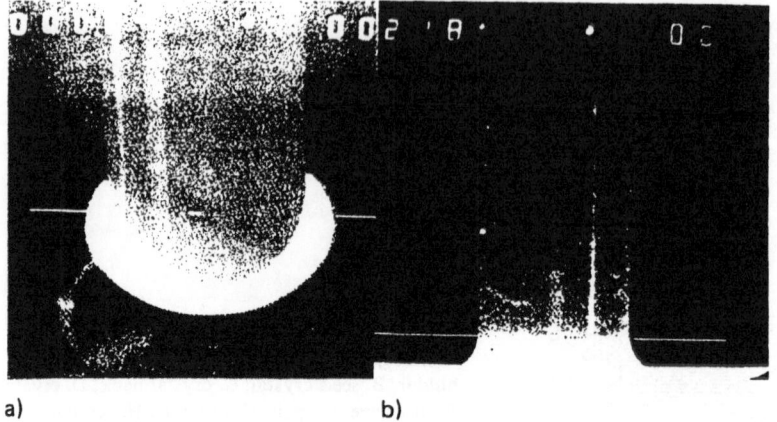

a)   b)

Fig. 3 Growing copper crystals (10 mm ϕ). TV-monitor photographs. The white line indicates the position of electronical diameter measurement.

a) The diameter of the bright ring is used for diamter control.
b) Real diameter measurement is performed at the triple-phase-line.

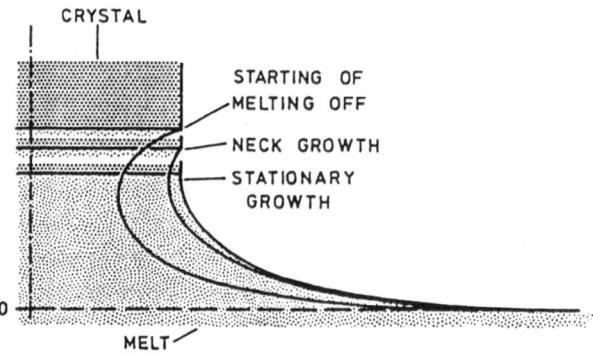

CRYSTAL

STARTING OF
MELTING OFF

NECK GROWTH

STATIONARY
GROWTH

0

MELT

Fig. 4

Meniscus shapes of different growth stages.

wetting is isotropic [5], 2) as already mentioned, the interface hight is approximately equal to the height of the triple-phase-line, 3) the interface height is proportional to the bulk melt-temperature [6, 7]. For these conditions *radial growth* occurs in the direction of the *meniscus angle* α_{men} determined by the tangent in the triple-phase-line at the meniscus and the vertical direction. Since the height of the triple-phase-line can be shifted by the bulk melt-temperature, α_{men}, and thus the geometry of the shape-building menisci [8] can be adjusted (Fig. 4), and the important stages of crystal growth as shown in Fig. 5 can be obtained.

The crystals of other materials are not wetted totally: the *contact angle* α_{con} is not zero. For example for silicon α_{con} is ca. $11°$. Singular faces (facets) of NaCl are only wetted under an angle of $\alpha_{con} \approx 30°$. In these general cases one has to

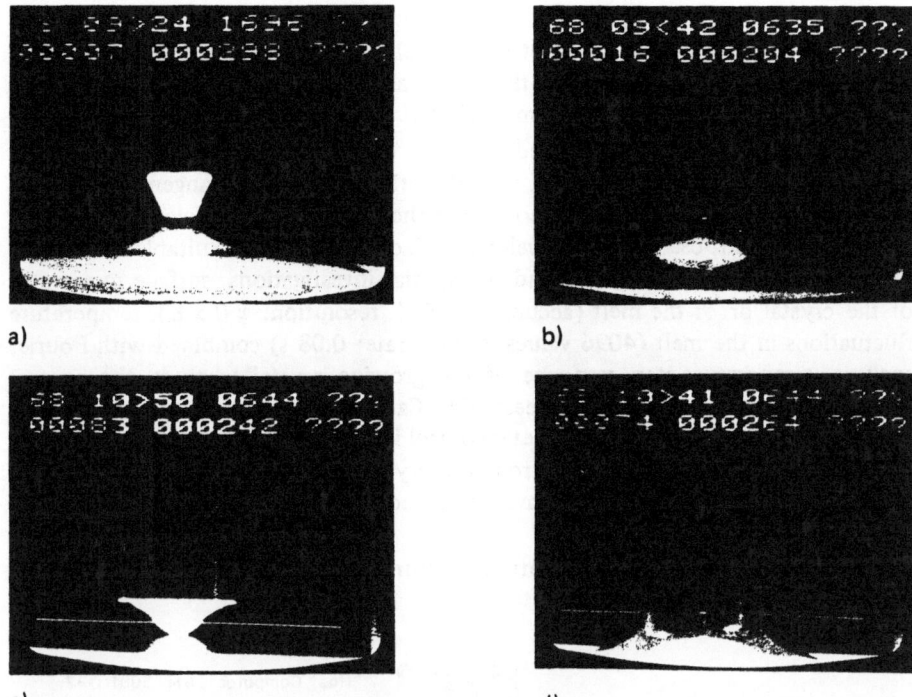

Fig. 5 Growth of a copper crystal. Except from the stage "Diameter increase" the corresponding meniscus shapes are shown in Fig. 4. a) Diameter decrease (necking process), b) diameter increase ("shoulder"), c) stationary growth (wanted diameter is reached), d) melting-off ("the end").

distinguish between the meniscus angle α_{men} as defined above and the contact angle α_{con}. Crystal growth occurs in the direction of the *growth angle* α_{gr}:

$$\alpha_{gr} = \alpha_{men} - \alpha_{con}. \tag{1}$$

Growth and crystal shape are determined by the vertical and the radial growth velocity:

$$\dot{z} = V - \dot{h}, \tag{2}$$

$$\dot{r} = (V - \dot{h}) \cdot tg\alpha_{gr}. \tag{3}$$

The growth velocity is only equal to the pulling speed V for $\dot{h} = 0$.

The *meniscus-shape* is a solution of the non-linear Gauß-Laplace-equation and is dependent on the boundary conditions at the crystal but also at the crucible wall. Therefore, vibrations of the crucible and also a hysteresis of the contact angle at the crucible wall must be avoided. When pulling crystals of dissociating melts as GaAs one has to avoid decomposition either by a layer of a non-active fluid as

B_2O_3 above the melt or by pulling in an appropriate gas atmosphere. In the case of GaAs this requires the use of a hot crystal puller for avoiding precipitation of As on the walls. The pulling speed of the Czochralski process is in the range of some cm/h (pure materials) down to 2 mm/h (alloys).

If one can use a completely filled crucible, one can watch the growth process during all stages with a horizontal view direction. With this arrangement we have measured growth parameters and compared them with a model calculation in real time. For this purpose we have developed a Czochralski system suitable for routine measurements of growth values and for special investigations: surface pyrometry of the crystal or of the melt (accuracy: -4 K, resolution: ± 0.5 K), temperature fluctuations in the melt (4096 values, sample rate: 0.08 s) combined with Fourier analysis, temperature-step response of the growing crystal to study the control behaviour of the Czochralski process. The Czochralski system with periphery is shown in Fig. 6. An automatic diameter control is necessary for obtaining a constant diameter: The energy dissipated from the crystal increases in the course of the growth; temperature deviations have to be controlled. The control accuracy of our system is $\pm 0.150\,\mu$m.

For comparison in Fig. 7 temperature measurements in a copper melt and in a nickel melt are plotted.

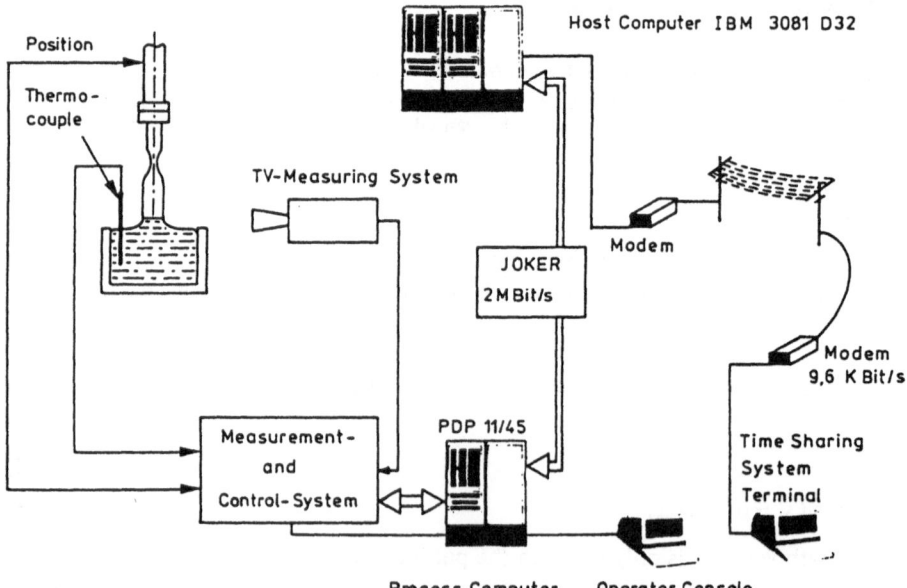

Fig. 6 Czochralski system with periphery. Directly measured quantities: Pulling rod position ($\pm$ 10 μm), by thermocouple bulk melt-temperature ($\pm$ 0.1 K), by television picture processing: diameter ($\pm$ 20 μm, range: 2.5 cm ϕ), interface position ($\pm$ 0.15 μm), local signal amplitude (pyrometry). CPU-intensive computations (for example Fourier analysis) are performed on the IBM using the fast JOKER-connection.

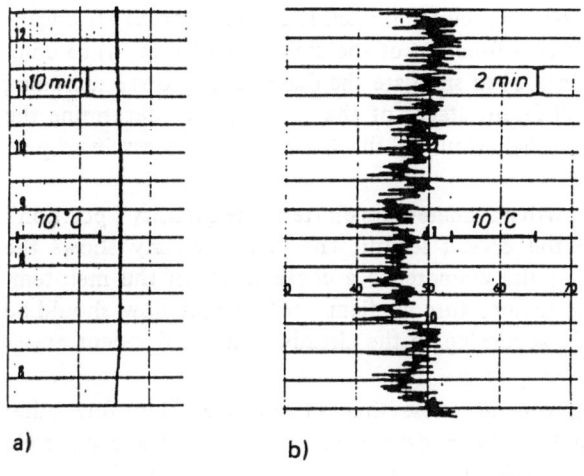

a)

b)

Fig. 7

Temperature measurements

a) In a copper melt (accuracy of control: ± 0.15 K); b) in a nickel melt. Due to the larger temperature fluctuations (higher melting point causes higher temperature gradients) control accuracy of the melt temperature is worse.

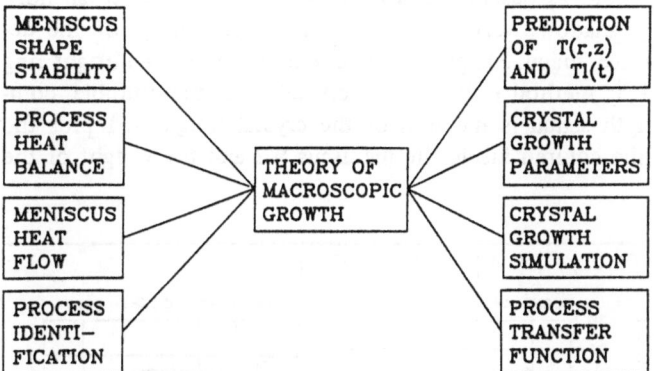

Fig. 8 Model of macroscopic Czochralski growth. At the left hand side the problems which had to be solved, at the right hand side the possibilities offered by such a model.

A model describing the shape generation during growth ("Macroscopic Growth") was developed by studying the single problems as shown in Fig. 8. We had to solve the Gauß-Laplace-equation in respect of *shape* and *stability* of the relevant Czochralski menisci [8]. Originally, the temperature of the bulk melt was assumed to be constant in respect to location and time. The temperature distribution in the crystal during all steps of the process was computed assuming high thermal conductivity and a limited diameter. Therefore we were able to apply a slice model [9]. Shape of the crystal neck and additional heat dissipation of a present atmosphere were taken into account quantitatively. The heat flow in the meniscus part is somewhat more complicated as assumed in the beginning [6]. But the recently measured temperature distribution in the meniscus [10] can be introduced in this model.

249

This results were brought together to a closed model. It allows according to Fig. 8: a) to compute the crystal shape as a function of the melt temperature, b) to compute some parameters in real time, c) to simulate the Czochralski growth including the thermal stresses acting in different stages of the growth, d) to determine the transfere function describing the behaviour of the Czochralski process in respect to control theory.

In Fig. 9 the most important growth parameters are plotted. The quantity "position" approximately corresponds to the crystal length. The picture clearly shows the necking process: the decrease of the diameter is a consequence of the melt-temperature increase. Values of the temperature gradient and the heat flow should be multiplied with a factor of 1/4 according to the already mentioned recent measurements.

The viewing conditions are in general worse than in our case. Therefore other methods for diameter control have been developed in the past. Their principle features are shown in Fig. 10. The following methods are used nowadays [11]:

1) the horizontal TV-method, as described, 2) TV-method for measuring an horizontally generated X-ray shadow, very expensiv method, possibility of the determination of the interface shape, 3) light-beam reflexion method, 4) bright ring measurement, 5) the weight method either of the crystal or of the crucible, complicated because the weigth signal is the sum of the crystal weight W1 plus the weight of the liquid in the column inside the meniscus W2 and the weight of the

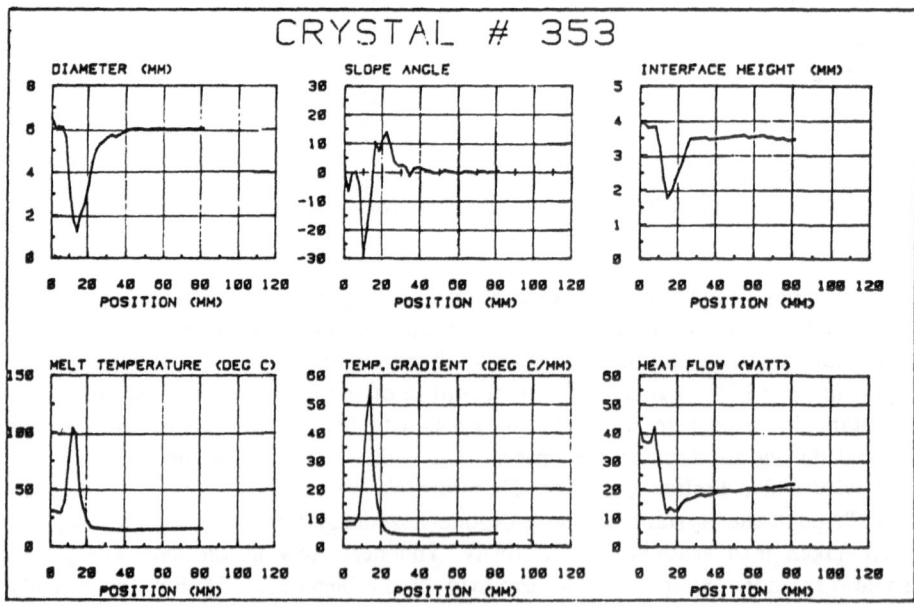

Fig. 9 Online plotted growth parameters of a copper crystal.

250

DIAMETER CONTROL METHODS

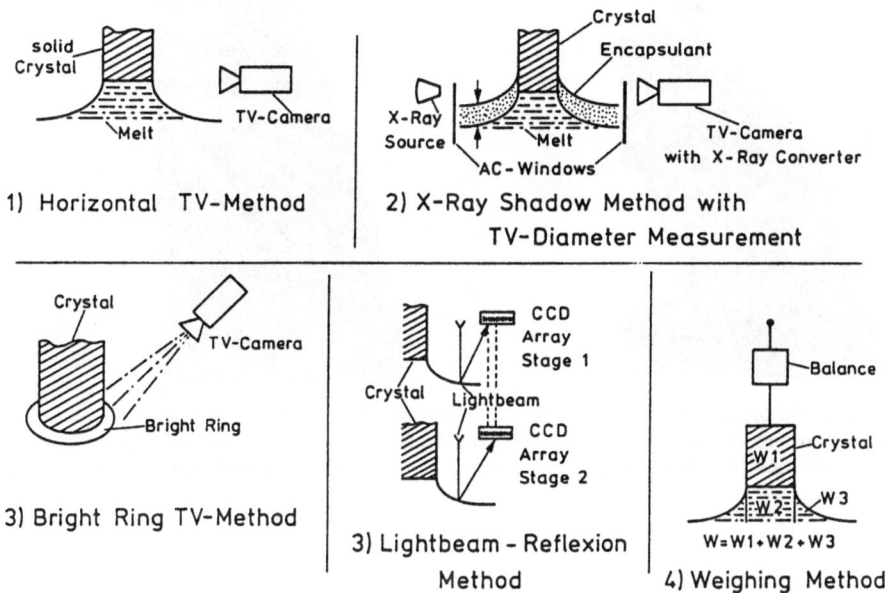

1) Horizontal TV-Method

2) X-Ray Shadow Method with
TV-Diameter Measurement

3) Bright Ring TV-Method

3) Lightbeam – Reflexion
Method

4) Weighing Method

Fig. 10 Diameter control methods.

meniscus part W3 itself. The determination of diameter changes, therefore, has to take into account possible changes of these quantities.

5 Discontinuous Facetting Controlled Growth

Quite an other mechanism of radial growth is exhibited from materials which developed singular faces (facets). For example, by lowering the bulk melt-temperature, radial growth occurs in macroscopic steps. This is demonstrated in Fig. 11 for the case of {100} -facets of sodium chloride [45]. The pulled crystal has developed these vertical facets but is rounded at the <110>-directed corners of the otherwise square cross-section. Radial growth consists a) in filling the {100} -facets by lateral growth, b) a subsequent fast non-crystallographic oriented growth below the solid/liquid interface. Therefore the contours of the [010]-shadowgraph are unsteady (Fig. 11a) whereas the contours of the [011]-edge shows a smooth radial increase (Fig. 11b) of the crystal. Local temperature measurements showed that the fast radial growth occurred only after a supercooling of 2 K was gained in front of the faces. This mechanism of discontinuous growth is controlled by a surface nucleation mechanism. The Fig. 12 demonstrates that also an impurity particle can start radial growth. Stranski [12] was the first who described this kind of discontinuous growth. He named the resulting morphology "Vergröberungen".

Fig. 11 Shoulder region of a NaCl crystal. [100]-growth direction (vertical). Height of the pictures corresponds ca. 1.0 cm crystal length. Shadowgraphs of; a) the [010]-view direction; the picture shows the step-like, unsteady growth, b) [011]-view direction; steady growth.

These singular faces, but (111)-oriented, may also develop in materials with a diamond lattice structure as in silicon. The normally curved interface in this case is partly singular and partly rough. This very unfavourable growth condition leads to crystals with "cores": the radial concentration distribution of a second component (dopant) is strongly dependent on the structure of the interface (different segregation). Facetting of the interface is avoided by achieving an interface geometry which is slightly curved towards the growing crystal.

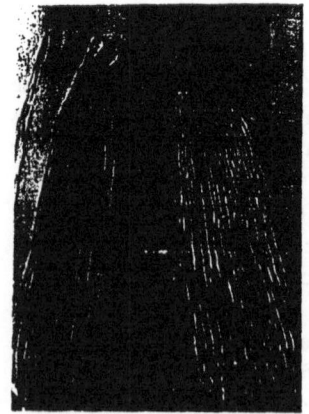

Fig. 12
An impurity particle on an (100)-facet starts lateral layer growth. Picture height corresponds ca. 0.2 cm.

252

6 Non-Rotational Symmetric Heating

A continuous growth of the crystal can only be obtained if the isotherms in the melt are symmetric in respect to the geometrical axis. This condition is never exactly fulfilled. We have to distinguish three origins of asymmetric heating:

1. At the beginning of the one turn coil, energy input into the graphite is higher than at the opposite site. In order to obtain an averaged symmetric temperature field, the *crucible has to be rotated*.
2. A slight non-symmetric energy input was observed in the meniscus region. In Fig. 13 a non-rotating crystal is sketched. The crucible is rotated. Nevertheless the interface is shifted slightly upwards in the region where the coil is beginning. Therefore, during growth the crystal at the right grows thinner than on the left hand side. By measuring the melt temperature and the inclination of the interface, we estimated an input of additional energy of 4.5% directly induced in the region of the meniscus. Thus only the *rotation of the crystal* can generate a circular cross-section by *a periodical local remelting and regrowth*.
3. Due to the orientation dependent properties, especially of the electrical resistivity and of the radiation absorption-coefficient of the graphite used as susceptor or crucible material, the reception of energy even of a rotated crucible is not rotationally symmetrical. By a slow rotation of the crucible this kind of asymmetry can be determined using a thermocouple for temperature measurement. The result of such a measurement is shown in Fig. 14. The asymmetric temperature field rotates with the angular velocity of the crucible.

Growth of a crystal for these reasons normally occurs periodically with a faster and a lower speed (changing of $\dot{h}$). In worst cases the lower speed may even convert to a periodic back melting of the crystal. These effects are not very important for

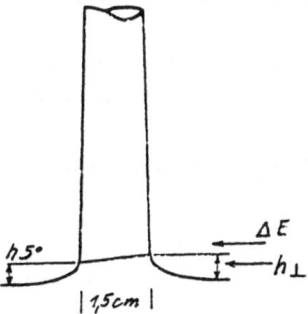

Fig. 13 Inclined solid/liquid interface of a non-rotating crystal. The left interface height h 5° corresponds a growth angle of − 5° whereas the right height $h_\perp$ leads to stationary growth.

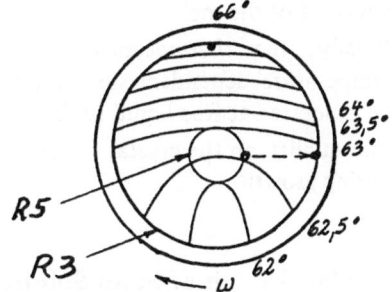

Fig. 14 Radial temperature field in a rotating crucible (without crystal). Isotherms constructed. Estimated horizontal temperature gradients on circle R5: grad $T_{l,max}$ = 0.70 deg/cm, on R3: grad $T_{l,max}$ = 0.93 deg/cm.

one-component crystal-growth but extremely important for the growth of multi-components systems. It will be shown in the following that in these cases the chemical composition in general is influenced strongly by these growth-rate fluctuations.

7 Transport Processes, Multi-Component Melts

The usual aim of crystal growth of multi-component materials is a crystal with constant *macroscopic* and *microscopic* concentrations of the components. Ideally, two conditions should be fulfilled for this purpose. We need a concentration distribution in the melt which is either homogeneous or at least independent on time. One can try to understand the flow phenomena and to achieve rules for the optimization of the rotation rates by numerical simulation. For this purpose one has to solve simultaneously the differential equations describing the following quantities:

in the melt: momentum (Navier-Stokes-equation), mass (diffusion and convection), concentrations, heat;
at the interface: boundary layer of momentum, heat and concentrations;
in the crystal: heat (dissipation by conduction and radiation).

A simulation of the whole Czochralski process at present seems to be impossible. However, simulation, also a three-dimensional one, of separated regions (crystal, melt) is performed in some laboratories (see for example [13, 9]).

The behaviour of the melt flow is determined by the ratio momentum transfer/heat transfer which may be characterized by the ratio kinematic viscosity/heat diffusivity. This value increases from metals, semiconductors to oxides. The following flow phenomena are important for crystal growth: the *free gravity driven convection* and the *forced convection* generated by rotation which is important for a mixing of the melt.

Finally, the *thermocapillary convection* (Marangoni-convection) driven by a temperature dependent surface tension $d\sigma/dT$ or a concentration field dc/dT may be generated at free liquid surfaces.

The quality on the crystal extremely depends on the kind of *convection* in the melt during growth.

8 The Growth of Solid Solutions

Now the problems that may arise during growth of solutions may be understood easily. The additional problems are the following:
1) Corresponding to the phase diagram for a given concentration C_{Liq} of a second component in the liquid, the concentration C_{Sol} in the solid at the interface is lower in our example (Fig. 15a). The ratio C_{Sol}/C_{Liq} is called the equilibrium

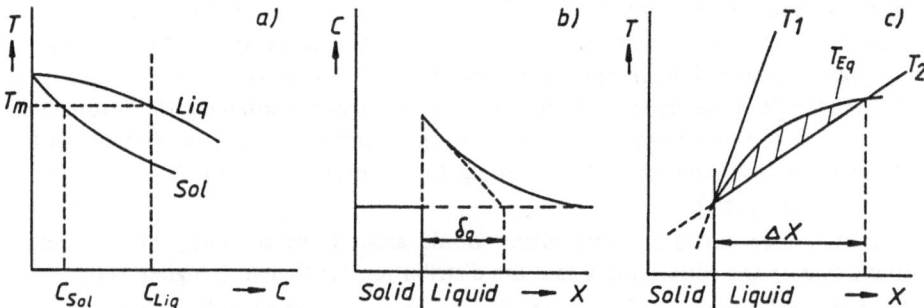

Fig. 15 Two-component melt.
a) Phase diagram, b) concentration distribution close to the interface, c) constitutional supercooling, T1: temperature gradient high enough to avoid supercooling, T2: gradient so low to generate supercooling.

Fig. 16 Striations in a gallium doped germanium crystal. Longitudinal section. Crystal diameter ca. 1.5 cm. From [14].

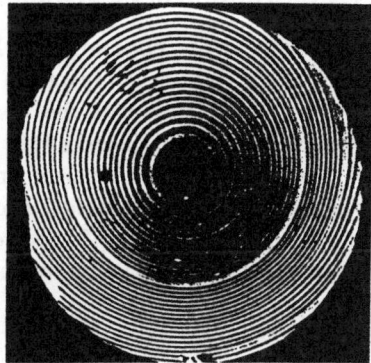

Fig. 17 As Fig. 16 but radial section. From [14].

distribution coefficient K. 2) Backward diffusion of the material is influenced by convection: a boundary layer with the thickness δ_c is generated in front of the interface (Fig. 15b). 3) So the second component is incorporated during growth with an effective distribution coefficient K_{eff}. 4) K_{eff} is a function of K and of δ_c. This quantity is dependent on the flow conditions, expressed by the rotation rate ω and the microscopic growth rate v, strongly dependent on the local melt temperature. δ_c should be kept constant. The growth is purely diffusion controlled if δ_c is larger than the characteristic diffusion layer thickness D/v. If growth is controlled by diffusion and rotation generated convection, δ_c is approximately $\sqrt{D/\omega}$. 5) Since K_{eff} on facets is different from that for rough interfaces, mixture

255

of rough and facetted growth should be avoided. δ_c is disturbed by rotation of the crystal in a non-axisymmetric heated melt or by temperature fluctuations caused by gravity convection. In this case *concentration striations* are generated. Fig. 16 shows a longitudinal section with striations of a Ge-crystal containing 10^{20} atoms/cm^3 of gallium (Lang-topography). Fig. 17 is the corresponding cross-section with a single spiral-like striation [14] according to the origin (rotation in a non-axisymmetric heated melt).

The same concentration distribution would arise if the solubility in the melt is lower than in the solid. In this case the distribution coefficient is higher than 1.

Second kind striations of quite another origin may arise according to Bauser [15] by the lateral growth of facets. They can be distinguished by their intersection with the first kind striations discussed above. Artificial striations can be generated and may be used to study the interface geometry by applying of short and regular current pulses [16, 17].

9 Constitutional Supercooling

According to the concentration layer in front of the interface the corresponding equilibrium freezing temperature T_{Eq} may be higher than the experimental temperature T_2. In this case the dashed region of Fig. 15c is constitutionally supercooled. Mullins and Sekerka [18] showed by perturbation analysis that in the case of a low temperature gradient and/or a high growth velocity the interface in the beginning becomes sinusoidal and finally breaks down. This constitutional supercooling can be avoided if the growth rate v at the interface meets the following relationship (for convection-free melts):

$$\frac{G}{v} > \frac{m\,C_s\,(1-K)}{K\,D}, \tag{4}$$

where G is the temperature gradient in the melt, m is the slope of the liquidus and C_s is the solute concentration.

Romero [19] in 1981 found out that the single crystal after break-down splits up into new grains by nucleation in front of the general interface. Fig. 18 demonstrates the transition to polycrystalline growth of Cu_3Au [20]. According to the increasing pulling speed the γ-ray reflection peaks are measured out over a large range of angles. As a consequence we may state following rules: a) for growing concentrated alloys the danger of constitutional supercooling is very high, b) pull slowly, c) avoid uncontrolled fast growth during seeding and necking, by vibrations, and by temperature oscillations.

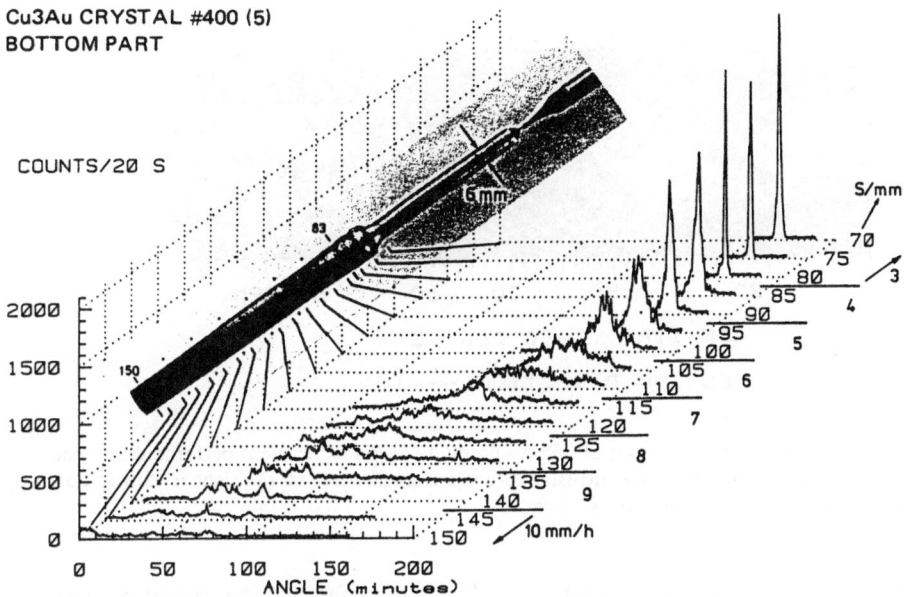

COUNTS/20 S

Fig. 18 Total break-down of the lattice by constitutional supercooling. γ-diffractometer measurements of a Cu_3-Au-crystal. The rocking curves become worse with the increasing supercooling caused by the increasing pulling speed (values at the right of the rocking-curves) and finally vanish totally.

10 Melt Growth in Space

For many years it was believed that purely diffusion controlled growth could be performed in the absence of gravity. So it was expected to obtain homogeneously doped crystals by growth under microgravity even on an industrial scale. We can now evaluate these hopes based on the results obtained by the various experiments from skylab to the D1 mission.

Different experimental choices of meniscus controlled growth can be applied under microgravity: (a) crucible-free solidifications of a sphere, (b) a Czochralski-kind method by shaping the meniscus with a mask, and (c) the crucible-free zone-melting. Only the last method finally was tested [21]. However, silicon specimens remelted by this method contained striations similar to those observed by ground based experiments. The reason is clearly the overwhelming influence of thermocapillary convection due to the high temperature gradient.

Free surfaces were avoided in Bridgman-like experiments of Witt et al. [22]. Those parts of InSb and gallium doped Ge crystals remelted and resolidified under microgravity were free of striations. So indeed in this arrangement only diffusion controlled growth occurred. However, a careful investigation of the dopant distribution

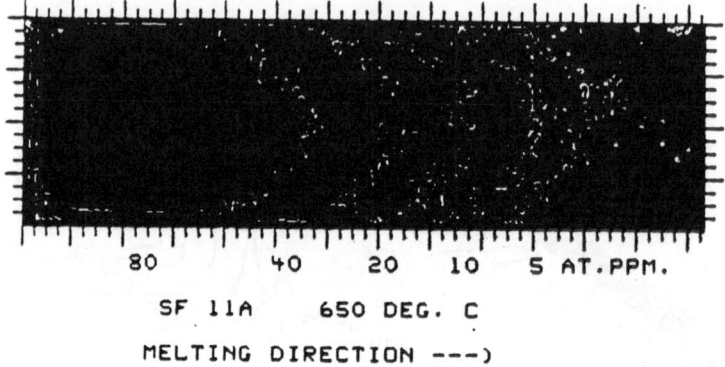

Fig. 19 Isoconcentration curves of a PbAu liquid diffusion experiment performed in space. The picture demonstrates that the diffusion process from the left to the right is slowed down by the graphite walls of the crucible (1.0 cm ϕ).

showed a radial dopant concentration [23]. This effect can be caused by a macroscopic non-planar interface influencing the lateral diffusion. This problem was investigated in a recent theoretical paper [24]. It was shown that lateral diffusion can be neglected for fast solidification rates. However, quite another principle wall effect sometimes described but not well understood may control the diffusion in liquids. This effect may be the origin of the macroscopic inhomogeneity in Bridgman crystals grown in space. An example gives Fig. 19. The picture shows the distribution of the diffused gold of a liquid PbAu diffusion experiment performed aboard the Apollo-Soyuz flight in 1975 [25]. The specimen was quenched after the diffusion soak time. The isoconcentration lines, strongly curved, demonstrate convincingly the wall effect influencing the diffusion. In a recent paper Carlberg [26] again described this phenomena.

So one may ask what is the future of crystal growth in space? In addition to the limited heating power there seem to exist fundamental problems as the impossibility to suppress Marangoni convection but also the influence of the g-jitter [27]. They have to be overcome before we can expect the growth of highly perfect crystals in space on an industrial scale.

11 Magnetic Damping of Free Convection

A strong reduction of temperature oscillations caused by free convection can be achieved by magnetic damping.

Since the fundamental work of magneto-hydrodynamics [28] it was well-known that a metallic fluid flow in a magnetic field was influenced by the induced eddy currents: the viscosity of the melt increases in the field when it is directed vertically to the flow. Hurle [29] was the first to investigate the damping of temperature oscillations in a crystal growth arrangement (horizontal boat filled with gallium).

Meanwhile during the last years crystal growth of silicon and GaAs has been performed in some institutes by applying magnetic fields. The results show that striations caused by convection are reduced strongly [30, 31, 32]. The development of crystal pullers for industry-sized semiconductor crystals, (some 40 cm diameter) equipped with superconduction magnets is in progress (Japan, Germany, United States). So indeed this method seems very promising. However, striations caused by thermal asymmetry (variable growth speed) cannot be avoided by this improvement. An excellent thermal symmetry is necessary.

12 Lattice Defects and their Mutual Origin

The mechanisms of defect generation are sketched in Fig. 20. The upper fields show the defect structures which may be observed in the grown crystal. The original microscopic defects are vacancies, interstitials, and impurities.

A one-component single crystal grown without special care will exhibit a dislocation network with small angle boundaries caused by polygonization. Alternatively, the dislocations may be arranged on slip planes with a density of 10^5 to 10^6 cm^{-2}. In this case slip with multiplication caused by thermal stress is the final defect forming mechanism during cooling the crystals after growth. Twin boundaries may be generated in the case of low stacking fault energy which is material dependent [33].

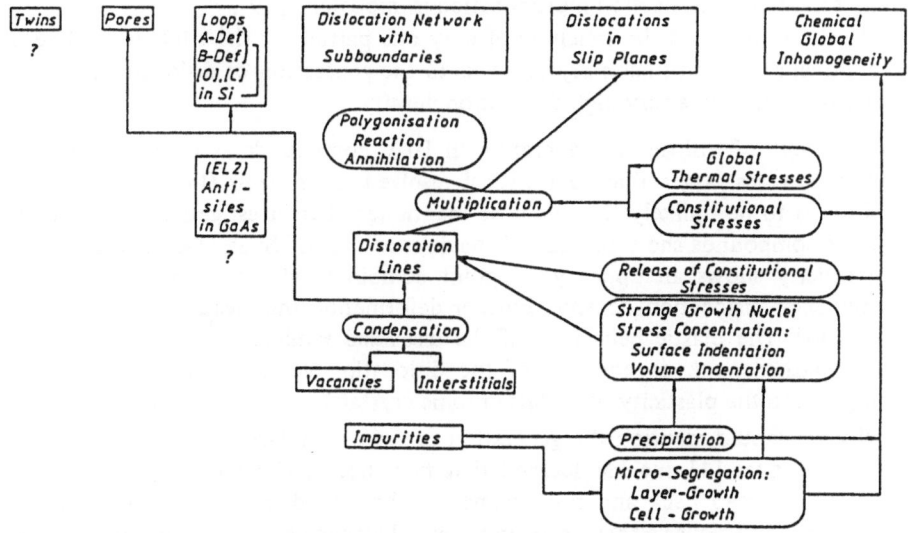

Fig. 20 Lattice defects (rectangular fields) and possible mechanisms of their generation (rounded fields). Generation of EL2 defects and antisites in GaAs is still under discussion.

Pores may be present in the crystal either by the nucleation of vacancies or by freezing-in of gas bubbles.

In binary solid solutions long-range concentration changes caused by segregation and short-range inhomogeneities (temperature oscillation introduced striations) can exist. In addition one may find impurities in the crystal not present in the starting material as oxygen in silicon crystals caused by the decomposition of the silica crucibles [34]. The impurities may precipitate, leading to local stress concentrations (volume indentation) followed by prismatic punching of dislocations [35]. A surface indentation is also a very effective dislocation source. In the case of constitutional supercooling growth nuclei may be built [19] leading to a polycrystalline sample with lineage structure. The release of (constitutional) stresses or even of long range thermal stresses due to the Tiller mechanism [36] can only generate dislocations if the theoretical shear stress, which is remarkably high (ca. 1/4 × shear modulus), is overcome. So the present discussion of stress induced dislocations in GaAs [37] seems not to be relevant. The dislocations seem to be generated at the crystal surface following decomposition or at the interface of the III-V compounds, perhaps by the contact with the encapsulating liquid B_2O_3 due to a still unknown process [38]. There are some hints supporting this assumption:

1) Dislocation glide has occurred only in thicker GaAs crystals ($> 2''$). This is understandable because thermal stresses increase with increasing diameter.
2) Even when dust particles are grown into the surface of metal crystals of smaller diameters (≈ 1 inch) one can obtain crystals with rather small dislocation densities of 10^2 cm^{-2}. This can be understood by assuming that due to the small crystal diameter and the high thermal conductivity the thermal stresses are so low that dislocation multiplication is impossible [39].
3) On the other hand the touching of only one particle at the triple-phase-line of a growing 10 inch silicon crystal leads to the generation of dislocations and by multiplication to a very high dislocation density.

The concept of dislocation generation in III-V compounds as GaAs and InP by thermal stresses larger than the critical resolved shear stress (lower yield point) [37] seems to be unlikely. It should be mentioned that investigations of plasticity of III-V compounds show the same behaviour as that of Si and Ge: The beginning of plasticity shows an upper yield point combined with the appearance of an already high dislocation density. Further deformation then leads to a lower yield point [40]. The plastic behaviour of the discussed semiconductors is determined by the rather weak dependence of the dislocation velocity on the shear stress (compared to the plasticity of metals or ionic crystals).

Finally we should mention the generation of twin boundaries. Since the investigations of Bonner [4] one has learned that twin boundaries often are generated in the stage between neck and final diameter, when the diameter increase occurs so steep that (111)-planes either on or in the shoulder can develop. Therefore, applying a sufficient slow increase rate can remove the danger of twinning.

13 Results, Trends, and Special Efforts

Applying the ingenious necking technique of Dash [4], we may grow very soft and sensitive fcc metal crystals either dislocation free (Cu) or nearly perfect with a dislocation density of less than 10^2 cm^{-2}. These crystals are mainly for research. For obtaining only very small thermal stresses the diameter is limited to 1″ [42]. For the same reason, ionic crystals such as NaCl can only be grown up to 3 mm in diameter with a low dislocation density (Schönherr [43]). Recent attempts to grow such single crystals with diameters larger than 3 mm were, however, not successful. Quite to the contrary, silicon crystals with 10 fold dimensions (10″) can be grown dislocation free. One problem is the inhomogeneous distribution of oxygen. By applying magnetic damping one tries to avoid striations and to achieve homogeneous doping. Till now homogeneous doping of silicon with P is only possible by neutron activation [34]. With this method a dopant level of some 10^{14} atoms/cm^3 of phosphorus can be obtained. This material is only used for power devices.

In the course of the last five years the effort to grow dislocation free III-V compounds such as GaAs and InP has been high. The driving force is the need of high-quality material for optoelectronic devices and for ICs. The best result, the growth of a 2″ GaAs dislocation free crystal, was published recently by a Japanese team [44]. They pulled strongly indium doped crystals (10^{20} Atoms/cm^{-3}). Indium is isoelectric and leads to a solution hardening. The crystal was surrounded by a deep layer of encapsulating B_2O_3 during the whole growth process, thereby reducing thermal stresses. During growth, convection in the melt was damped by applying a magnetic field. Another possible way may be, to avoid liquid encapsulation by adjusting the As pressure in a hot crystal puller. Indium doping for blocking the dislocation multiplication in the thermal stress fields may not be necessary in this case.

Crystal growth research is still necessary. Only one example should be given: the reason for elimination of dislocations by the Dash process is not understood very well: Climbing out of dislocations, growing out, or a smaller dislocation velocity than the applied pulling speed can be the effective mechanism to reduce the dislocations. Detailed simulations of the Czochralski process would be very instructive. But the results are only relevant if the applied boundary conditions, which in many cases cannot be measured, are realistic. Finally, new crystal-growth methods will only survive if the quality of the resulting crystals is high.

Acknowledgements

The author would like to thank A. Fattah and G. Hanke for technical assistance, Prof. H. Wenzl for valuable discussions, and the „Zentralinstitut für Elektronik" and the „Zentralinstitut für Mathematik" for the excellent technical support.

References

[1]   G. Wassermann and P. Wincierz, in: Das Metall-Laboratorium der Metallgesellschaft AG 1918–1981. Chronik und Bibliographie (Frankfurt/Main), p. 9–20

[2]   J. Czochralski, Z. Physik Chem. 92, 219 (1918)

[3]   J. C. Brice, The Growth of Crystals from the Liquids in: Selected Topics in Solid State Physics, Vol. XII, ed. by E. P. Wolfarth (North Holland, Amsterdam 1973), pp. 245

[4]   W. C. Dash, J. Appl. Phys. 30, 459 (1959)

[5]   H. Wenzl, A. Fattah, and W. Uelhoff, J. Crystal Growth 36, 319 (1976)

[6]   D. Geist and P. Grosse, Z. Angew. Phys. 14, 105 (1962)

[7]   W. Uelhoff and K. Gärtner, Rost. Kristallow. 12, 238 (1972)

[8]   K. Mika and W. Uelhoff, J. Crystal Growth 30, 9 (1975)

[9]   A. v. d. Hart and W. Uelhoff, J. Crystal Growth 51, 251 (1981)

[10] E. Quintero, Diplomarbeit, Aachen 1986

[11] W. Uelhoff, in: Dreiländer-Jahrestagung über Kristallwachstum und Kristallzüchtung, Report JÜL-Conf-18 (Jülich 1976)

[12] O. Knacke and I. N. Stranski, Ergebn. exakt. Naturw. Bd. XXVI, 383 (1952)

[13] M. Mihelcic, C. Schroeck-Pauli, K. Wingerath, H. Wenzl, W. Uelhoff, and A. v. d. Hart, J. Crystal Growth 57, 300 (1982)

[14] J. A. M. Dikhoff, Philips Techn. Rundschau 25, 441 (1963, 64)

[15] E. Bauser, in: Festkörperprobleme/Advances in Solid State Physics, Vol. XXII, ed. by P. Grosse (Vieweg, Braunschweig 1983), p. 141

[16] R. Singh. A. F. Witt, and H. C. Gatos, J. Electrochem. Soc. 115, 112 (1968)

[17] M. Lichtensteiger, A. F. Witt, and H. C. Gatos, J. Electrochem. Soc. 118, 1013 (1971)

[18] W. W. Mullins and R. F. Sekerka, J. Appl. Phys. 35, 444 (1964)

[19] J. Vidal and R. Romero, Crystal Research and Technology 16, 853 (1981)

[20] Y. K. Chang A. Fattak and W. Uelhoff, submitted to J. Crystal Growth

[21] A. Eyer H. Leiste, and R. Nitsche, J. Crystal Growth 71, 173 (1985)

[22] H. C. Gatos, A. F. Witt, M. Lichtensteiger, and C. J. Herman, in: Apollo-Soyuz Test Project, Summary Science Report, Vol. I, NASA SP-412, p. 429 (1977)

[23] H. C. Gatos, in: Materials Processing in the reduced gravity environment in Space, ed. by G. E. Rindone, (Elsevier Science, 1982), p. 355

[24] S. R. Coriell and R. F. Sekerka, J. Crystal Growth 46, 479 (1979)

[25] R. E. Reed, W. Uelhoff, and H. L. Adair, in: Apollo-Soyuz Test Project, Summary Science Report, Vol. I, NASA SP-412, p. 367 (1977)

[26] T. Carlberg, J. Crystal Growth 79, 71 (1976)

[27] H. Hamacher, R. Jilg, and U. Merbold, in: 6th European Symposium "Materials Sciences under Microgravity Conditions" (Bordeaux 1986), p. 1

[28] S. Chandrasekhar, Hydrodynamic and Hydromagnetic Stability (Clarendon Press, Oxford 1961)

[29] D. T. J. Hurle, Phil, Mag. 13, 305 (1966)

[30] K. Terashimi and T. Fukuda, J. Crystal Growth 63, 423 (1983)

[31] D. T. J. Hurle and R. W. Series, J. Crystal Growth 73, 1 (1985)

[32] T. Kimura, T. Katsumata, and T. Fukuda, J. Crystal Growth 79, 264 (1986)

[33] H. Gottschalk, G. Patzer, and H. Alexander, phys. stat. sol. (a) 45, 207 (1987)

[34] *W. Zulehner* J. Crystal Growth **65**, 189 (1983)

[35] *J. Friedel,* Dislocations (Pergamon Press, Oxford 1964)

[36] *W. A. Tiller,* J. Appl. Phys. **29**, 611 (1958)

[37] *A. S. Jordan. A. R. von Neida,* and *R Caruso,* J. Crystal Growth **76**, 243 (1981)

[38] *P. Haasen,* private communication

[39] *E. Kappler, W. Uelhoff, H. Fehmer,* and *F. Abbink,* in: Herstellung von Kupfereinkri-stallen kleiner Versetzungsdichte, Opladen 1971

[40] *H. Siethoff, J. Völkl, D. Gerthsen,* and *H. G. Brion,* submitted to Appl. Phys. Lett.

[41] *W. A. Bonner,* J. Crystal Growth **54**, 21 (1981)

[42] *H. Fehmer* and *W. Uelhoff,* J. Crystal Growth **13/14**, 257 (1972)

[43] *E. Schönherr,* in: International Conference of Crystal Growth, North Holland, Boston 1966

[44] *T. Kobayashi, J. Osaka,* and *K. Hoshikawa,* J. Crystal Growth **71**, 813 (1985)

[45] *D. Fehmer,* Diplomarbeit, Münster 1968

263

Silicon Germanium-Heterostructures on Silicon Substrates

Erich Kasper

AEG Research Center Ulm, D-7900 Ulm,
Federal Republic of Germany

Summary: Silicon based heterostructures offer the potential of monolithic integration of conventional integrated circuits with superlattices. A material concept is introduced which overcomes the problem of lattice mismatch between superlattice and silicon substrate by a homogeneous, thin, incommensurate buffer layer. Data about critical thicknesses, buffer layer design, strain adjustment, and band offsets are given for the SiGe/Si system. Room temperature mobility enhancement for modulation doped SiGe/Si heterostructures and its device application for a MODFET is described.

1 Introduction

The performance of microelectronic devices and circuits is increasing steadily with time. The forces behind this development are design cleverness, shrinkage of lateral and vertical dimensions, sophisticated fabrication methods, and advanced material systems. Research was focussed on tailoring of material properties by superlattice structure (man-made semiconductor) and on utilizing quantum effects for novel device applications. The introduction of superlattice materials and quantum effect device concepts will have a strong impact on electronics and be a challenge to manufacturing methods. The relation between the dominant producer regions may be changed dramatically by such a step in the development. Clear concepts, careful analysis of the connected problems, and co-operative research between industry and universities will help to master the step advantageously.

2 Silicon Based Heterosystem

The manufacturing of todays integrated circuits is dominated by using silicon as a semiconductor material. One can speculate which material system may compete with silicon on a broad manufacturing scale when the progress of silicon technology saturates. In the following sections arguments are given for such a future competing material system consisting of a heterostructure superlattice monolithically integrated with a conventional integrated circuit on top of a silicon substrate (Fig. 1). The superlattice regions yield the high performance core and high speed links between conventional parts of the IC.

2.1 Choice of Substrate Material

Already in existing integrated circuits the device function is confined to a thin surface layer. For bipolar circuits and with increasing tendency also for CMOS-

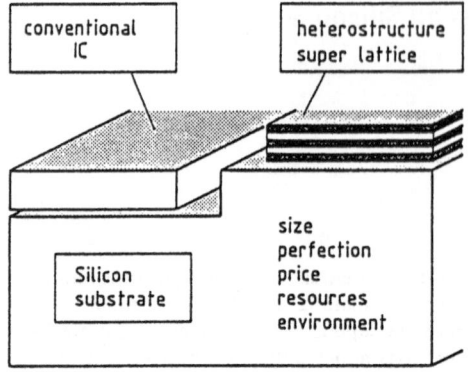

conventional IC

heterostructure super lattice

Silicon substrate

size
perfection
price
resources
environment

Fig. 1
Silicon based superlattice IC material concept.

circuits this surface layer is created by epitaxial techniques. The function of the substrate itself is confined to give
— mechanical stability,
— growth ordering information for the epitaxial layer,
— electrical insulation,
— thermal conductance for effective heat removal.

Silicon substrates offer properties superior to other semiconductor materials with regard to wafer size, crystal perfection, thermal conductivity, handling, and price. The large resources (silicon is the 2nd most frequent element of earth's crust) and the environmental harmlessness are additional factors favouring a broad usage. But the realization of the structure given in Fig. 1 requires to overcome general problems associated with mismatched heteroepitaxy. The lattice constants of a variety of group IV, III/V, and II/VI semiconductors vary between 0.54 nm and 0.65 nm. The lattice mismatch to silicon, whose lattice constant is at the lower end of the mentioned regime, amounts to up to 20%. The classical superlattice system GaAs (GaAlAs) is nearly lattice matched, but in general one may expect lattice mismatch between the two material components of an alloy superlattice. Under certain circumstances discussed later on, the layers of the superlattice accomodate the mismatch by strain relaxation. Such strained layer superlattices have gained increasing attraction not only offering a great increase of possible superlattice systems but also allowing additional tailoring of superlattice properties by strain. Growth behaviour and properties of a superlattice system are governed by strain and chemistry of the system. Silicon/silicon germanium is a model system for studying the pure influence of strain because of the close relationship in chemistry of Si and Ge.

2.2 How to Overcome the Mismatch Problem

The straightforward way is to grow the superlattice directly on top of the mismatched substrate. One has to handle at the interface both superlattice definition and mismatch accomodation. Both electronic and optoelectronic devices were successfully realized using this configuration [1]. A more general solution uses a buffer layer between substrate and superlattice. Often the rather thick buffer layer is graded to vary gradually the chemical composition from the substrate material to the mean superlattice composition.

We favour a rather thin, homogeneous buffer layer which is believed to allow better control of crystal quality and strain adjustment.

3 Concept of Strain Adjustment by a Homogeneous Buffer Layer

Consider a configuration as given in Fig. 2. A homogeneous buffer layer with Ge-content y is placed between substrate and superlattice Si/SiGe. The Ge-content of the superlattice may switch between 0 and x. The buffer layer is under compressive strain ϵ_B tetragonally distorting the cubic lattice cell of the alloy $Si_{1-y}Ge_y$. Parallel to the interface the alloy is offering a lattice constant $a_\parallel$ which is given by

$$a_\parallel = a_0 (1 + y \cdot 0.042 + \epsilon_B) \tag{1}$$

where $a_0 = 0.543$ nm is the lattice constant of silicon, $y \cdot 0.042$ is the lattice mismatch of the buffer layer (Ge is 4.2% larger than Si), ϵ_B is the built-in strain of the buffer layer (consider: ϵ_B is negative for compressive strain).

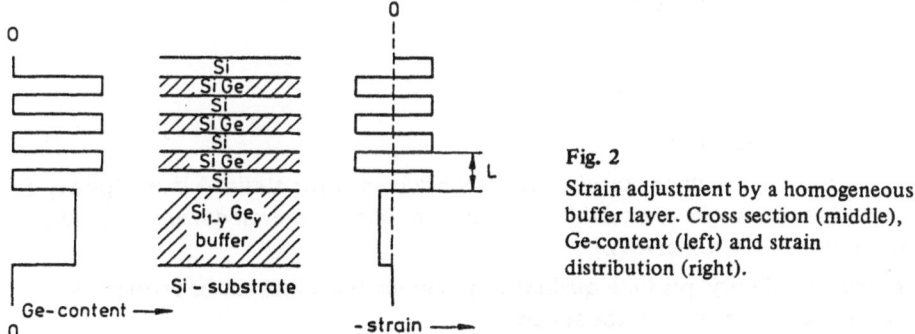

Fig. 2
Strain adjustment by a homogeneous buffer layer. Cross section (middle), Ge-content (left) and strain distribution (right).

The buffer layer acts like a virtual substrate with an effective Ge-content y^*.

$$y^* = y + 23.8 \cdot \epsilon_B . \tag{2}$$

Remember that the effective Ge-content y^* is smaller than the Ge-content y of the buffer because of the compressive (negatively valued) strain ϵ_B.

3.1 Nature's Answer to Lattice Mismatch

Nature knows two answers to accomodate lattice mismatched films (Fig. 3), namely, elastic accomodation by strain and plastic accomodation by misfit dislocation lying in the interface. Nature prefers elastic accomodation for thin films up to critical thickness t_c, whereas above the critical thickness misfit dislocations are generated relaxing the built-in strain.

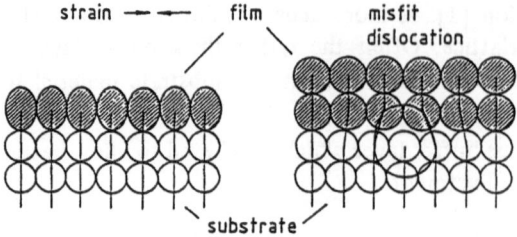

Fig. 3

Mismatch accomodation by strain (left) or misfit dislocations (right).

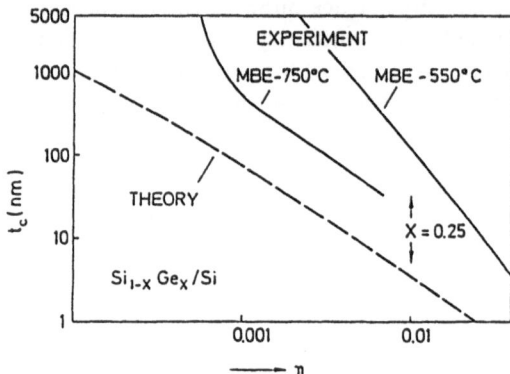

Fig. 4

Critical thickness t_c versus mismatch η_0. Comparison of theory with MBE-experiments [5].

3.2 Critical Thickness

As a guideline for understanding nature's preference for strained layer epitaxy for a thickness below the critical thickness one can consider thermodynamic equilibrium as was done by v.d.Merwe [2].

Indeed, his theory predicts qualitatively correct the observed behaviour by minimizing the total energy of the system.

However, the quantitative results of the equilibrium theory differ considerably from the results obtained by molecular beam epitaxy (MBE). This can most clearly be seen by a plot of the critical thickness t_c versus lattice mismatch η_0 (Fig. 4). Unfortunately the reader of recent literature may be confused by theory (v.d.Merwe)-curves covering a rather broad range [3...5]. This is caused by various approximations given by v.d.Merwe and also simply by misprinting. We strongly recommend the use of one data set [5] which is based on a careful analysis of the approximations valid for SiGe [6]. Most probably kinetic limitations (nucleation and glide) cause the deviations of experiments from theory.

3.3 Buffer-Layer Design

For the buffer-layer design one has not only to know the critical thickness but also the strain which partly accomodates the mismatch also above the critical thickness.

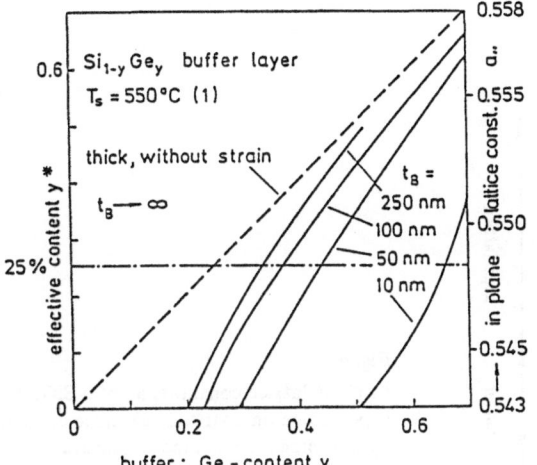

Fig. 5

Buffer-layer design. Effective Ge-content versus buffer-layer composition for various thicknesses t_B. MBE growth at 550 °C [7].

As mentioned above, v.d.Merwe-theory [6] has to be used as a qualitative guideline, for a quantitative calculation experimentally determined strain values versus thickness have to be used. As an example, Fig. 5 gives the effective Ge-content y^* of the virtual substrates as function of the buffer-layer data (Ge-content y, layer thickness t_B, growth temperature T = 550 °C) using measured strain values [7].

3.4 Strain Symmetrization

The elastic energy E_h stored in a homogeneously stressed thin layer with thickness t is given by

$$E_h = 2\,\mu\,\frac{1+\nu}{1-\nu}\,\epsilon^2\,t \tag{3}$$

where μ is shear modulus and ν is Poisson's number. For a superlattice with equally thick Si and SiGe layers the minimum energy elastically stored in the pseudomorphic superlattice is obtained for the symmetrical strain after Eq. (4). The unequally strained superlattice contains twofold energy (the elastic constants μ, ν are assumed to be equal for both layers). So, the symmetrically strained superlattice is a stable minimum energy configuration, which may be important for post epitaxial processing steps.

The plane strains ϵ_1, ϵ_2 of the Si and SiGe layers of the superlattices are then (Fig. 6)

$$\epsilon_1 = -\epsilon_2 = x \cdot 0.021 \ . \tag{4}$$

Unequally thick layers of the superlattice: Let t_1, t_2 be the thicknesses of the Si and SiGe layers of superlattice, respectively. Then, the minimum energy condition requires

$$y \cdot 0.042 + \epsilon_B = x \cdot \frac{t_2}{t_1 + t_2}\, 0.042. \tag{5}$$

269

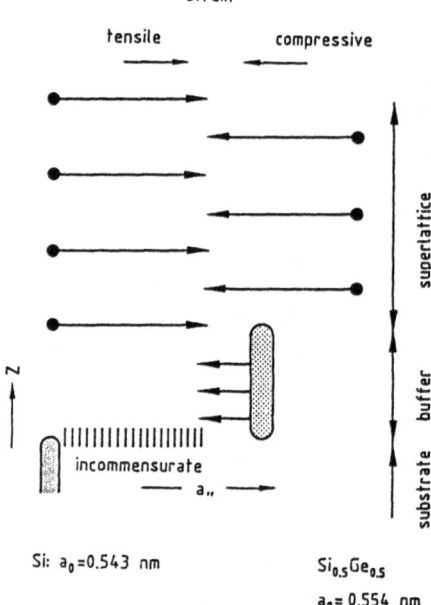

strain

tensile   compressive

superlattice

buffer

substrate

incommensurate

— $a_{\shortparallel}$ —

z

Si: $a_0 = 0.543$ nm

$Si_{0.5}Ge_{0.5}$

$a_0 = 0.554$ nm

Fig. 6

In plane lattice constants $a_{\shortparallel}$ of a SiGe/Si superlattice on a virtual substrate created by a homogeneous, incommensurate buffer layer $Si_{1-y}Ge_y$ on a Si-substrate. Arrows indicate in plane lattice constant changes caused by strain.

Then, the thinner layer is more strained than the thicker one

$$\epsilon_1 = x \cdot \frac{t_2}{t_1 + t_2} 0.042 ,$$

$$\epsilon_2 = - x \cdot \frac{t_1}{t_1 + t_2} 0.042 .$$

(6)

4 Technological Realization

Key issues for the technological realization of the concept of strain adjustment are
— low growth temperature (avoids interdiffusion),
— control of layer composition and thickness,
— crystal perfection of the buffer layer.

4.1 Silicon Molecular Beam Epitaxy (Si-MBE)

MBE is ideally suited for meeting the above requirements. For a review of Si-MBE the reader is referred to [7, 8]. The principal arrangement of our MBE apparatus which was used for growth of the SiGe superlattices is given in Fig. 7.

The following subsystems are shown; (i) material sources (Si, Ge, Sb), (ii) substrate oven, (iii) secondary implantation equipment (ionization ring, substrate voltage), (iv) in situ monitoring equipment (quartz microbalance, mass spectrometer). These subsystems are installed inside an ultrahigh vacuum chamber.

270

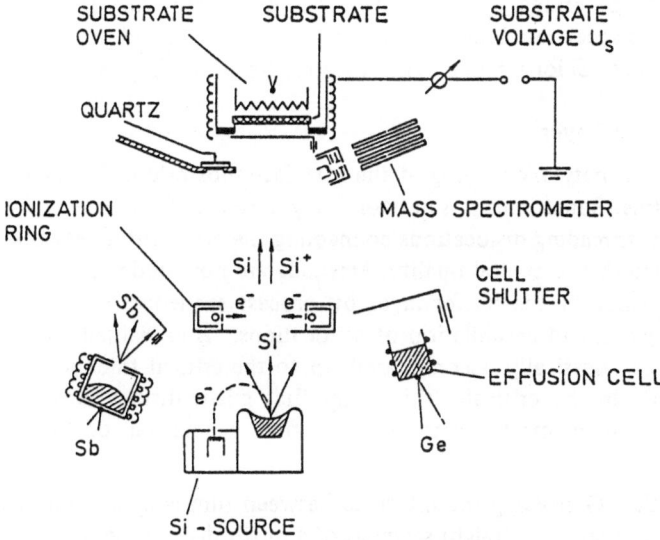

Fig. 7 Scheme of our MBE apparatus [6] used for growth of SiGe superlattices.

Three distinct features are typical for silicon molecular beam epitaxy.

(i) The industrial requirements for throughput and wafer size. We used 3 inch wafers for growth of the SiGe structures.

(ii) The existence of strong electromagnetic radiation (thermal, X-rays), electron fluxes, and ion fluxes inside the growth chamber. This radiation and charged particle fluxes which influence strongly surface physics are caused by the operation of the electron gun evaporator for silicon.

(iii) The usage of ionized matrix or dopant atoms to enhance the incorporation of dopant materials. We used the method of secondary implantation [9] for n-type doping with antimony (Fig. 8).

(DSI) DOPING by SECONDARY IMPLANTATION

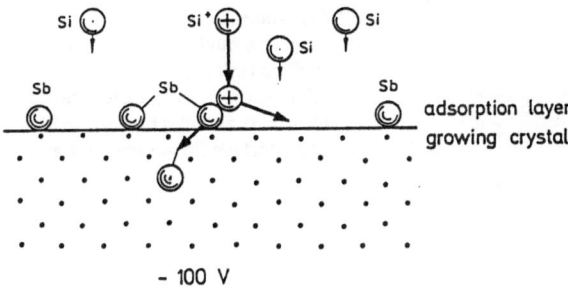

Fig. 8

Doping by secondary implantation (DSI) [9]. The adsorbed Sb atoms are implanted by recoil momentum from Si ions impinging on the growing crystal. Incorporation depth is only some atomic distances.

271

With this method Si ions generated by the electron gun evaporator and/or by an ionizer ring (see Fig. 7) are accelerated toward the substrate by an applied potential. At the substrate surface these Si ions implant the adsorbed Sb atoms (Fig. 8).

4.2 Perfection of the Buffer Layer

The ideal misfit dislocation network is lying at the interface substrate/buffer layer away from the superlattice. Sometimes this network may have beneficial effects by its internal gettering. But threading dislocations connecting the dislocation network with the surface can disturb the crystal quality. Methods are now under development to improve the quality of the buffer layer by process sequences which influence the nucleation, glide, and annihilation of dislocations. As explained above the buffer layer will grow elastically accomodated up to the critical thickness t_c (pseudomorphic growth). In the critically thick film first misfit dislocations are generated. We consider three mechanisms for this first generation of misfit dislocations (Fig. 9).

(a) A substrate dislocation G crossing the interface between film and substrate is bended by the film stress creating a straight segment of a misfit dislocation M and a

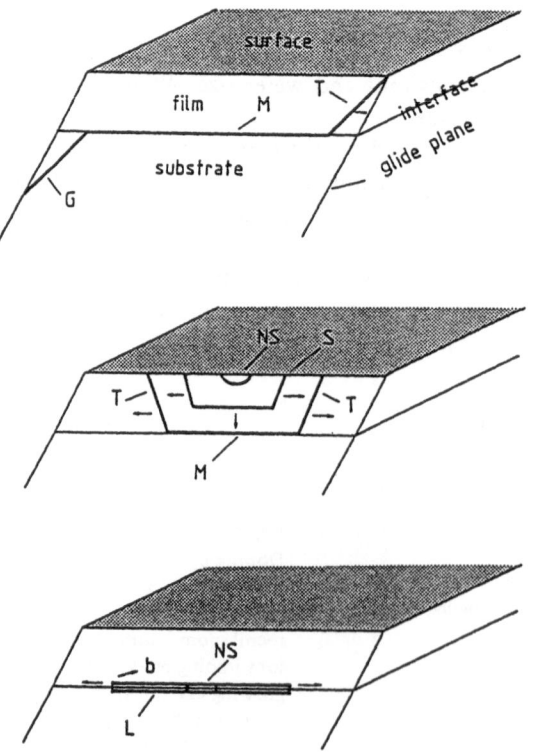

Fig. 9

Nucleation and motion of the first misfit dislocations when the film thickness is exceeding the critical thickness t_c.

(a) Bending of grown in substrate dislocation G by the film strain results in a misfit dislocation segment M and a threading dislocation segment T (upper part of figure).

(b) Nucleation of dislocation half loop at a nucleation site NS at the surface (middle).

(c) Nucleation of a dislocation loop L at a nucleation site NS at the interface (lower part of figure).

dislocation segment T threading through the film to the surface. Generally, the misfit dislocation segment will be not the ideal edge dislocation (Fig. 2) with Burger's vector b lying in the interface as was assumed by thermodynamic equilibrium theory. With nonideal Burger's vector only the edge component b_{eff} projected into the interface is effective for strain relaxation. For a parallel network of dislocations with a mean dislocation distance p the following relation between misfit η_0 and strain ϵ is valid

$$\eta_0 + \epsilon = \frac{b_{eff}}{p} \, . \tag{7}$$

The nonideal dislocations will increase the critical thickness compared to the equilibrium theory as was calculated by Matthews [10].

(b) With low dislocation density silicon substrates the foregoing mechanism will be of minor importance. But equal forces will be applied to dislocation half loops which can nucleate at surface sites (NS). Usually it is assumed that the half loop will glide to the interface creating a misfit dislocation segment M and two threading dislocation segments T. The glide process from the surface is only possible with nonideal misfit dislocations. With higher mismatch and, therefore, lower critical thicknesses ($<$ 1 nm) also nucleation and climb of ideal misfit dislocations is possible as was already demonstrated for metal systems by Cherns et al. [11].

(c) Nucleation of prismatic dislocation dipoles or loops at sites NS at the interface. Climb of the dipoles in the interface.

The kinetics of the generation of the first misfit dislocations seem to be influenced by the nucleation barrier at surface or interface sites and by the glide properties of nonideal misfit dislocations.

A rigorous investigation of the misfit dislocation generation in the material system SiGe/Si is outstanding. Preliminary results of transmission electron microscopy (TEM) of the crystal quality of the buffer layer [12] suggest that for MBE at 750 °C nucleation at surface sites and for MBE at 550 °C nucleation at interface sites is predominant.

With layer thickness exceeding the critical thickness a misfit dislocation network is created. The kinetics of network formation is complicated by additional processes as dislocation multiplication at nodes and annihilation of threading dislocations.

Performance and yield of heterostructure or superlattice devices are strongly dependent on the confinement of the misfit dislocation network to the interface substrate/buffer and on avoiding threading dislocations which connect the network with the surface. That is the technological key challenge for realization of the material concept proposed in earlier sections.

4.3 Electronic Band Ordering

The band gap of the alloy SiGe is smaller than that of silicon. The ordering of the bands at the interface determines the properties of two-dimensional carrier gases.

The most common case of band ordering between a wide band-gap and a low band-gap material is obtained with the low band-gap within the wide band-gap (type I band ordering). In this case both electrons and holes would jump from the wide gap material to the low gap material. For a staggered band ordering (type II) electrons and holes would jump in opposite directions across the interface. A rough idea about the type of ordering can be obtained by considering the electron affinities of the materials. The electron affinity measures the energetic position of the conduction-band edge against a vacuum level. This simple consideration holds if both materials are separated. For the system Si/Ge a nearly flat conduction-band ordering (between type I and type II) is proposed. Usually, interface structure, dipole moments, and strain effects contribute additionally to the band-edge position after facing both materials.

Early experiments with polycrystalline material seemed to conform the picture of a flat conduction-band ordering (with very slight type I character). Also first MBE experiments [13] of the Bell group with unsymmetrically strained Si/SiGe super-lattices demonstrating 2D-hole gas but no electron gas were in agreement with that flat band picture. In apparent contrast to that our group realized a 2D-electron gas in the wide gap material silicon [14] with a symmetrically strained Si/SiGe super-lattice. Both groups agree now that the apparent contradiction was caused by the different strain distribution within the superlattices [15]. Fig. 10 shows the calculated conduction-band offset (type II) as function of the strain distribution [15, 16] characterized by a substrate with an effective Ge-content. The band ordering is shifted toward strong type II character as the substrate is made Ge-rich.

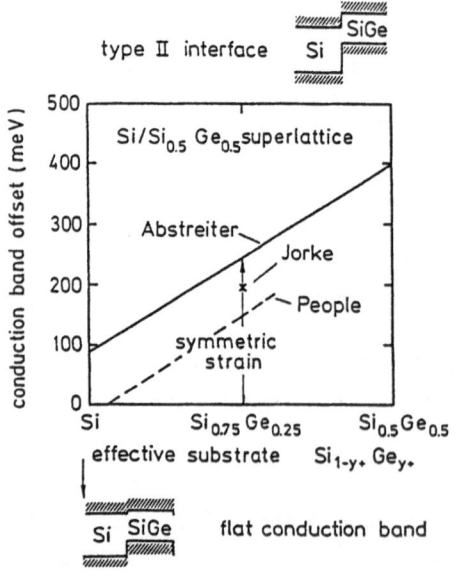

Fig. 10

Calculation of conduction-band offset (type II) for a $Si/Si_{0.5}Ge_{0.5}$ superlattice as function of the substrate which determines the strain distribution. Instead of the $Si_{1-y}Ge_y$ substrates silicon substrates with thin buffer layers of the same effective Ge-content y^* can be used [15...17].

274

The position of the symmetrically strained superlattice is marked by an arrow suggesting a conduction-band offset of 200 meV for the symmetrically strained superlattice [17].

4.4 Brillouin-Zone Folding Effects

Besides the general interest in superlattice systems there are specific technological and scientific reasons for investigation SiGe superlattices on Si substrates.

There are theoretical predictions of a band-gap conversion from indirect type to direct type by Brillouin-zone folding of the superlattice [18, 19]. One expects mini-zones in the momentum space, not observed in the host crystal.

The probability of direct optical transitions is strongly enhanced in this minizone scheme: Simple estimates give a ratio Q of the transition probability of optical direct and indirect transitions, which exceeds one hundred for photon frequencies near the band-gap energy [18, 19]. Zone-folding effects were demonstrated for the phonon spectrum by G. Abstreiter's [20] Raman scattering experiments. An experimental proof for the predicted direct optical transition in the SiGe superlattice system is announced [21].

4.5 Device Application

Generally the tailoring of material properties by superlattices will lead to improved properties of current devices and to new device classes. As an example of this material-property tailoring, Fig. 11 shows the enhancement of room-temperature mobility of an n-type $(4 \cdot 10^{18}/cm^3)$ modulation doped Si/SiGe-superlattice compared to bulk mobility [14]. The enhanced mobility was utilized in a high performance n-channel Si/SiGe-MODFET (Figs. 12, 13) using a buffer layer for strain symmetrization [22]. A main goal of the AEG-group is microwave operation of these silicon based hetero-devices. Rapid progress made for optoelectronic receivers by the Bell-group [23] lets expect a broad industrial application already in the near future.

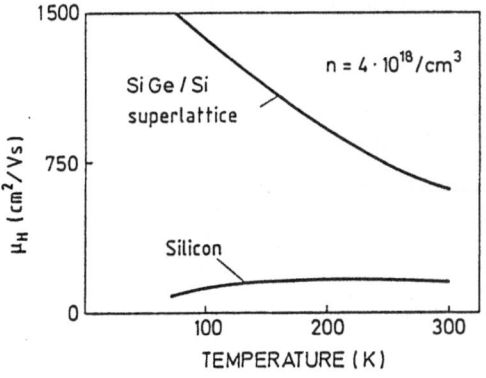

Fig. 11
Electron mobility versus temperature of a modulation doped SiGe/Si-superlattice [9].

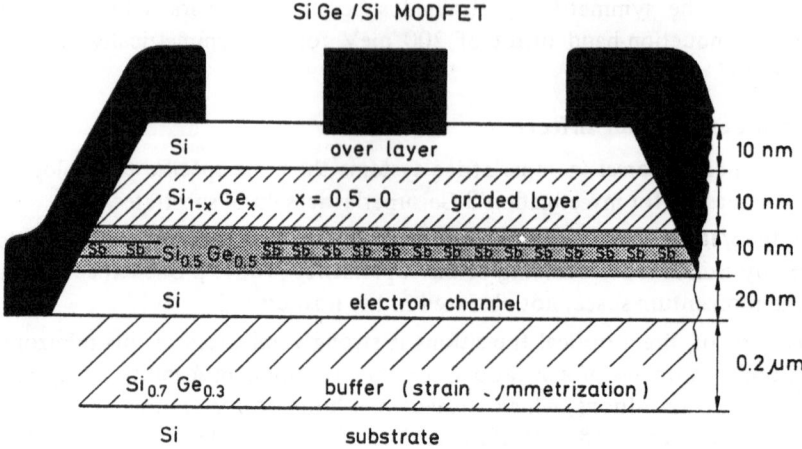

Si Ge / Si MODFET

Si   over layer   10 nm

$Si_{1-x} Ge_x$ x = 0.5 - 0   graded layer   10 nm

Sb Sb $Si_{0.5} Ge_{0.5}$ Sb Sb Sb Sb Sb Sb Sb Sb Sb Sb Sb Sb Sb   10 nm

Si   electron channel   20 nm

$Si_{0.7} Ge_{0.3}$ buffer (strain symmetrization)   0.2 μm

Si   substrate

Fig. 12 n-channel SiGe-MODFET structure with symmetrical strain distribution [22].

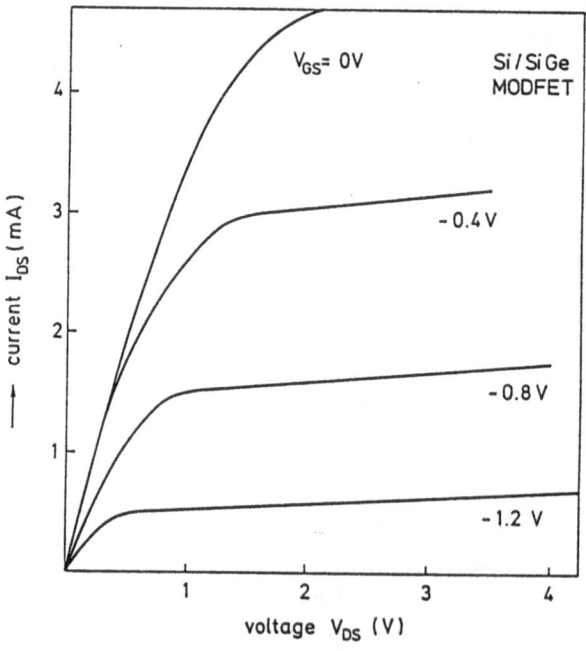

Fig. 13 Room-temperature characteristics of an n-channel SiGe-MODFET with 1.6 μm gate length. Current versus voltage for different gate voltages.

Acknowledgement

Cooperation of my collegues H. Kibbel, H.-J. Herzog, H. Jorke, H. Dämbkes, A. Casel and discussions with Th. Ricker are acknowledged. The project was partly sponsored by the German Ministry of Science and Technology.

References

[1] *J. C. Bean*, Physics Today, p. 36, Oct. 1986

[2] *J. H. v. d. Merwe*, Surf. Sci. **31**, 198 (1972)

[3] a) *R. People* and *J. Bean*, Appl. Phys. Lett. **47**, 322 (1985)
   b) *R. People* and *J. Bean*, Appl. Phys. Lett. **49**, 229 (1986)

[4] *B. W. Dodson* and *P. A. Taylor*, Appl. Phys. Lett. **49**, 642 (1986)

[5] a) *E. Kasper*, Surf. Sci. **174**, 630 (1986)
   b) *E. Kasper, H.-J. Herzog, H. Dämbkes*, and *Th. Ricker*, in: Twodimensional Systems: Physics and New Devices, ed. by *G. Bauer, F. Kuchar*, and *H. Heinrich*, Springer Series in Solid State Sciences 67 (Springer, Berlin 1986), p. 52
   c) *E. Kasper, H.-J. Herzog, H. Dämbkes*, and *G. Abstreiter*, Mat. Res. Soc. Proc. Vol. 56, 347 (1986)

[6] *E. Kasper* and *H.-J. Herzog*, Thin Solid Films **44**, 357 (1977)

[7] Silicon-Molecular Beam Epitaxy, ed. by *E. Kasper* and *J. C. Bean*, CRC Press, Boca Raton (USA), in press

[8] *Y. Shiraki*, Silicon Molecular Beam Deposition, in: The Technology and Physics of MBE, ed. by *E. H. C. Parker* (Plenum Press, New York 1986)

[9] *H. Jorke* and *H. Kibbel*, J. Electrochem. Soc. **133**, 774 (1986)

[10] *J. W. Matthews* and *A. E. Blakeslee*, J. Cryst. Growth **27**, 118 (1974)

[11] *D. Cherns* and *M. J. Stowell*, Scripta Metallurgica 489 (1973)

[12] *H.-J. Herzog*, pers. comm.

[13] *R. People, J. C. Bean, D. V. Lang, A. M. Sergent, H. L. Störmer, K. W. Wecht, R. T. Lynch*, and *K. Baldwin*, Appl. Phys. Lett. **45**, 1231 (1984)

[14] *G. Abstreiter, H. Brugger, T. Wolf, H. Jorke*, and *H.-J. Herzog*, Phys. Rev. Lett. **54**, 2441 (1985)

[15] *G. Abstreiter, H. Brugger, T. Wolf, R. Zachai*, and *Ch. Zeller*, in: Twodimensional Systems: Physics and New Devices (Springer Series in Solid-State Sciences 67), ed. by *G. Bauer, F. Kuchar*, and *H. Heinrich* (Springer, Berlin 1986), p. 130

[16] *R. People* and *J. C. Bean*, Appl. Phys. Lett. **48**, 538 (1986)

[17] *H. Jorke, H.-J. Herzog, E. Kasper*, and *H. Kibbel*, J. Cryst. Growth **81**, 440 (1987)

[18] *U. Gnutzmann* and *K. Clausecker*, Appl. Phys. **3**, 9 (1974)

[19] *S. A. Jackson* and *R. People*, Proc. MRS, Vol. 56, p. 365, Boston, Dec. 85

[20] *H. Brugger, G. Abstreiter, H. Jorke, H.-J. Herzog*, and *E. Kasper*, Phys. Rev. **B33**, 5928 (1986)

[21] *T. P. Pearsall, J. Bevk, J. C. Bean, J. M. Bonar*, and *J. P. Mannaerts*, MRS Spring Meeting, Anaheim April 1987

[22] *H. Dämbkes, H.-J. Herzog, H. Jorke, H. Kibbel*, and *E. Kasper*, IEEE Trans. **ED-33**, 633 (1986)

[23] *S. Luriy* and *F. Capasso*, in [15], p. 140

Large-Area Electronics Based on Amorphous Silicon

Karl Kempter

Siemens A.G., Corporate Research and Development, D-8000 München 83, Federal Republic of Germany

Summary: Large-area electronics is mainly concerned with the development of input/output devices such as xerographic drums, page-width image sensors, and addressing circuits for LCD panels and printer heads. The introduction of amorphous hydrogenated silicon (a-Si:H) as a large-area thin-film semiconductor with low-cost fabrication capability has given a new impetus to this field. While the excellent photoconductivity of a-Si:H was initially exploited (solar cells, xerography), the development of thin-film transistors and their application for addressing circuits has now assumed the key role. Examples of 22.8 cm large electronic devices are shown and their technological and design implications discussed.

1 Introduction

The hardware of modern information systems is based upon two rather different pieces of equipment: electronic processors and input/output devices. Whereas the improvement of processors has been associated with continuous miniaturization on crystalline silicon chips, input/output devices have to continue to be of large dimensions. Electronic displays, printers, keyboards, or document scanners demand an electronics technology which operates with large formats: large-area electronics.

This special domain of modern electronics was vitally stimulated by the introduction of a new material, amorphous hydrogenated silicon (a-Si:H), to application-related development in about 1980. This thin-film material combines the capability of large-area manufacturing with almost the full range of semiconductor virtues. During the first stage of a-Si:H development, its excellent photoconduction was primarily exploited (solar cells, drums for laser printers). Today, the development of thin-film transistors for addressing circuits on large areas (image sensors, displays, printing heads) has caught up.

Beginning with the material basis, i.e. the distinctive features of a-Si:H, this review describes the principal applications presently under development and illustrates the characteristic features of the design and fabrication of large-area electronic devices.

Solar cells have deen described in numerous review papers [1] and will be excluded here.

2 Material Basis

2.1 Choice of Material

Economic fabrication of large-area electronic devices requires a homogeneous material basis which provides the necessary semiconductor properties. Large-area semiconducting materials with sizes up to 25 cm may only be achieved at reasonable cost by thin-film technologies on low-cost substrates (excluding epitaxy). Three approaches, based on the well established element silicon, are known to date. The first of these are the successful efforts to approximate the conventional monocrystalline state by recrystallizing thin polycrystalline films of silicon [2]. But limitations in the attainable uniformity caused by remnant grain boundaries have hitherto presented fundamental problems when large homogeneous devices are to be fabricated. Improved uniformity is attained with polycrystalline material. However, the processing temperature for poly-silicon is 600°C for deposition and 1000°C for oxidation and annealing [3] which leads to severe substrate warpage problems in the photolithographic steps. The thermal expansion coefficients of silicon $(1.9 \cdot 10^{-6}K^{-1})$ and of the suitable substrate material, fused quartz $(0.54 \cdot 10^{-6}K^{-1})$ differ widely. A transparent substrate is required for imaging application. This disadvantage limits the use of poly-silicon when the substrate size increases beyond about 12 cm. Moreover, poly-silicon exhibits a very low photoconduction, thus excluding its use in image sensors. Finally, amorphous semiconductors – like monocrystals – offer optimum structural uniformity and are distinguished by moderate processing temperatures which do not exceed 250°C. This technological advantage may now be fully exploited by using amorphous hydrogenated silicon, which represents the amorphous semiconductor with the broadest spectrum of useful semiconductor properties. Reviews of this field are available, see for example the monograph [4]. Comparisons between various candidate materials for large-area devices are given in [5, 6, 7].

2.2 Physical Basis of the Amorphous Semiconductor a-Si : H

The physical description of amorphous semiconductors requires some particular completions to the physics of crystalline semiconductors. In consequence of its deviation from the periodic atomic structure, the forbidden gap between valence band and conduction band is now filled with a multitude of electronic states continuously distributed over the energy. The sharp boundaries between these localized states and the extended electronic states in the bands define the mobility gap of an amorphous semiconductor.

Whether an amorphous semiconductor may be useful for applications or not is easily identified by its most characteristic property, the density of states distribution within the gap.

The unique importance of the density of states will be illustrated with the help of Fig. 1. The fundamental property which makes a semiconductor so useful for devices is that it permits a controlled variation of the number of electrons (holes)

280

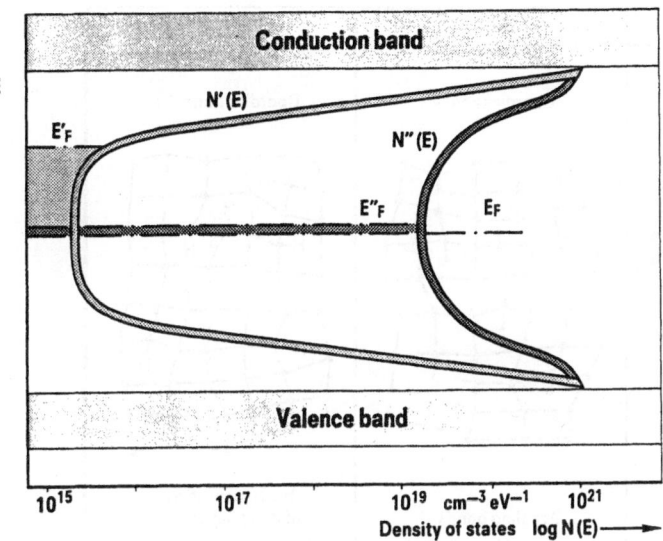

Fig. 1 Low density of states N'(E) within the bandgap of an amorphous semiconductor is required to allow a significant shift of the Fermi level: $E'_F - E_F$. High density N"(E) impedes the shift: $E''_F - E_F$. Schematic diagram.

in the conduction (valence) band, which is usually expressed as a shift of the Fermi level. A distinct shift of the Fermi level, for example under the influence of doping, will only occur if the density of states in the gap is low enough (Fig. 1, curve N'(E)). With high density of states, all additional electrons (or holes) are absorbed by localized states near E_F, and hence the Fermi level appears to be pinned. The same mechanism determines the case of photoconduction, the field effect, Schottky barriers, etc.

To reduce the density of states near the midgap of an amorphous semiconductor means to reduce the point defects in the random atomic network. The exceptional feature of a-Si:H with respect to defects becomes clear when we compare it with the other amorphous semiconductors.

The amorphous chalcogenides, up to now the key material for photoconductor drums in xerography, possess a twofold covalent atomic coordination, which favors an amorphous structure. Long atomic chains are mingled to produce a random network. However, the outer electron shells of the chalcogenides contain lone pairs of electrons, and hence they easily undergo a change in coordination to a single or threefold coordination. Since the creation energy of those defects (valence alternation pairs) is very low, a rather high defect density at midgap of about $10^{18} cm^{-3} eV^{-1}$ is found in all chalcogenide semiconductors [8], see Fig. 2a. The resulting lack of dopability and field effect renders the chalcogenides unsuitable for the full range of electronic applications.

281

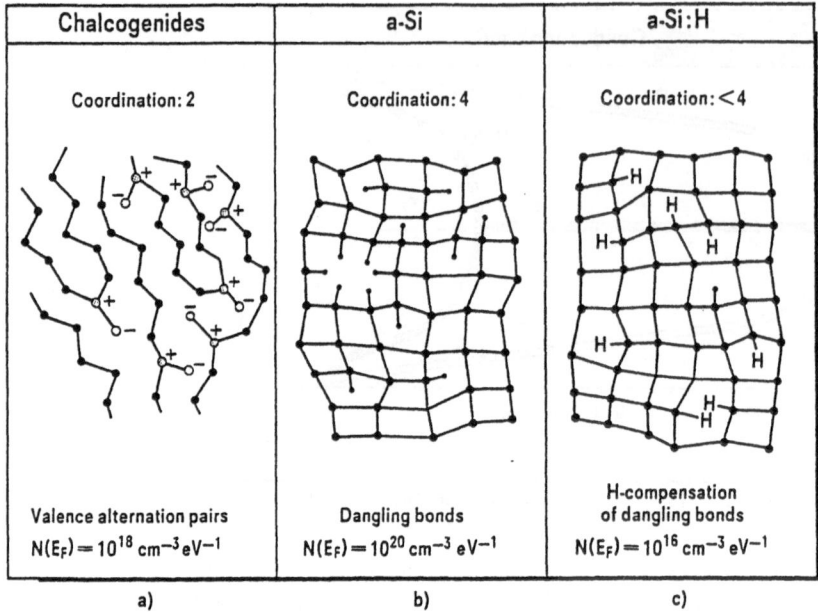

Chalcogenides	a-Si	a-Si:H
Coordination: 2	Coordination: 4	Coordination: <4
Valence alternation pairs $N(E_F) = 10^{18}$ cm^{-3}eV^{-1}	Dangling bonds $N(E_F) = 10^{20}$ cm^{-3} eV^{-1}	H-compensation of dangling bonds $N(E_F) = 10^{16}$ cm^{-3}eV^{-1}
a)	b)	c)

Fig. 2 Structural defects in different types of amorphous semiconductors. $N(E_F)$ is the density of states at the Fermi level.

The group of semiconducting materials without lone pairs are the tetrahedral bonded elements such as silicon and germanium. As shown by Phillips [9], the constraints of covalent chemical bonding, such as bonding angle and bonding length, may only be satisfied in a 3-dimensional random network if the coordination number is smaller than 2.45. In the case of the fourfold coordinated silicon we therefore find a considerable amount of strain and a multitude of unpaired bonds (dangling bonds) incorporated in the network. The energy state of an electron in a dangling bond state is about in the middle of the mobility gap. As a result, normal amorphous silicon has a defect density at midgap as high as 10^{20}cm^{-3}eV^{-1} (see Fig. 2b) and hence is completely useless for electronics applications.

The discovery of hydrogenated amorphous silicon in the early seventies [10] marked a genuine breakthrough because this material attains a density of states as low as $5 \cdot 10^{15}$ cm^{-3}eV^{-1} at midgap [11]. The special quality of this material is due to the incorporation of hydrogen, which compensates most of the inevitable dangling bonds, Fig. 2c. In this way we obtain a true semiconductor with all its useful properties, despite it being an amorphous material.

282

2.3 Properties of a-Si:H Relevant to Applications

An overview of the significant advantages and also of the problems of a-Si:H is given in Table 1.

The great interest in a-Si:H is due to its unique combination of a single fabrication process (upper part of Table 1) with an almost complete set of semiconductor properties (lower part of Table 1).

The deposition (and doping if neccessary) of thin-films of a-Si:H is performed by a single-step process at a temperature of 230°C in a plasma-CVD reactor. No subsequent treatment such as annealing or implanting is required. The process is shown schematically in Fig. 3. Gaseous SiH_4 (silane) is dissociated by an RF

Table 1

Advantages	Problems
Large-area fabrication Low temperature process Multilayer structures Compatible with microelectronics technology	Low deposition rate Little experience in mass production
Photoconduction Dopability } similar Field effect } to c-Si Interfaces blocking or ohmic Light sensitivity in the visible range Optical absorption greater than with c-Si Lateral carrier diffusion is small	Small drift mobility Schubweg smaller than $50\,\mu m$ Light degradation

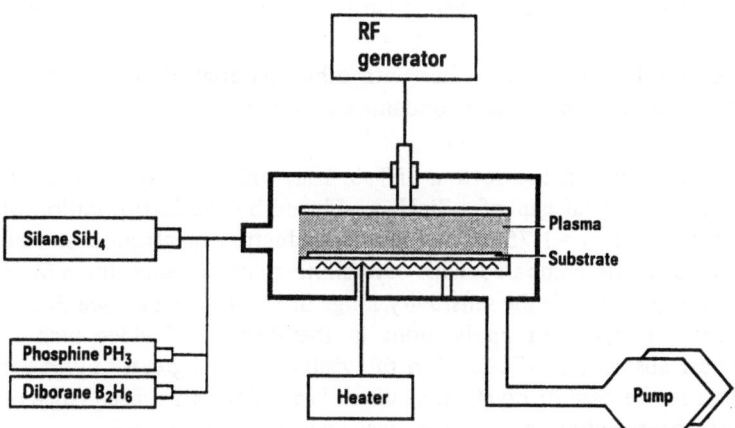

Fig. 3 Principle of deposition of a-Si:H on a substrate by plasma CVD. Phosphine is admixed for n-type doping and diborane for p-type doping.

283

plasma at low ambient temperature (in contrast to high-temperature induced CVD) thus ensuring the amorphous state of the deposited silicon. The excess of hydrogen in the reaction chamber incorporates this gas into the growing film to a proportion of about 10%. Doping is performed by adding the appropriate gas during film growth. The low processing temperature enables the use of inexpensive substrates like glass. No fundamental limitation is known for the size of the deposition area. Typical substrate sizes currently deposited are 25 cm × 25 cm. Detailed reviews of optimum deposition parameters have been published, see for example [12].

A-Si:H may easily be combined for devices in multilayer systems (most dielectrics) because its amorphous structure exhibits no lattice matching properties and because of the great number of amorphous materials which may be deposited in plasma-CVD chambers. Plasma-CVD equipment is similar to dry etching systems, which are very common in microelectronics. Compatibility is further ensured by the use of the element silicon.

On the problem side of Tab. 1 is the limited deposition rate, usually 1μm/h. Higher rates (5μm/h) have been achieved for special purpose material (xerographic drums [13]. Experience in mass production has hitherto been confined to the solar cell manufacture. The physical properties listed in the lower part of Tab. 1 once more summarize the semiconductor properties explained qualitatively in the foregoing chapter. The photogeneration of carriers for photoconduction is not field dependent (as in the case of chalcogenide semiconductors) and therefore ensures a quantum efficiency near unity provided the schubweg in the device is longer than the electrode spacing. By n-type and p-type doping the conducitivity of a-Si:H may be controlled over 10 orders of magnitude [14].

The field effect allows the sheet conductance of a a-Si:H layer to be switched over 6 orders of magnitude [15]. The fabrication of TFT thin-film transistors is facilitated by the fact that the gate dielectric layer (mostly SiN_x), the channel layer of a-Si:H, and the ohmic source and drain n^+-contacts can all be deposited in the same plasma-CVD reactor.

The band bending of a-Si:H at the interface with other materials generates ohmic or blocking contacts, which represents a fundamental requirement for an electronic device.

The next three features listed at the bottom of Tab. 1 are unique to the amorphous state of silicon. As the optical gap of a-Si:H is widened by the incorporation of about 10% hydrogen to $E_{opt} = 1.75$ eV, the photoconduction spectrum is shifted to shorter light wavelengths compared with crystalline silicon. Hence the a-Si:H spectrum coincides very well with the sensitivity range of the human eye, see Fig. 4. This is particularly important for applications in the domain of video pick-up systems. The optical absorption of a-Si:H in the visible range is greater by about a factor of 20 than in the case of crystalline silicon [16]. This is a consequence of the irrelevance of the selection rules for optical transitions in amorphous semiconductors (non-conservation of k-vector [17]).

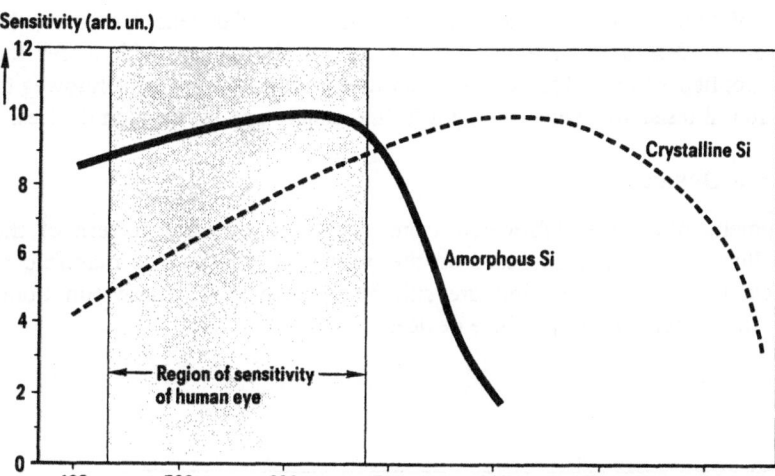

Fig. 4 Photoconductivity spectrum of amorphous and crystalline silicon.

Thus a thin film of only $1\,\mu$m thickness absorbs most of the visible light, an advantage which has primarily stimulated solar cell development. The strong light absorption in a $1\,\mu$m a-Si:H layer also significantly facilitates the fabrication of high-efficiency photosensors.

To reduce crosstalk and blooming in image sensors and to ensure high local resolution in imaging systems, it is important that the lateral spreading of electrons and holes in the photoconductor be as small as possible. The very low conductivity of undoped a-Si:H (10^{-10} to $10^{-11}\,\Omega^{-1}\mathrm{cm}^{-1}$) results in a dielectric relaxation time of 10 to 100ms, which is a safe value for imaging systems with line scan times of 1 ms.

As a result of its disordered structure, the a-Si:H shows a sharply reduced drift mobility for electrons ($1\,\mathrm{cm^2/Vs}$, [18]) and for holes ($1.5 \cdot 10^{-2}\,\mathrm{cm^2/Vs}$, [19]). This limits the speed of devices to around a Megahertz due to the limited transit time of the carriers.

The limited schubweg ($50\,\mu$m for electrons in an electric field of 10^4V/cm, [20]) of a-Si:H results in a design rule requiring the spacing of the collecting electrodes to be smaller than $50\,\mu$m. This is fullfilled in most devices very easily by directing the current along the direction of the film thickness which is of the order of $1\,\mu$m.

The degradation of the electric properties of a-Si:H under the influence of strong light irradiation (Staebler-Wronski-effect, [21]) is associated with an increase of dangling bonds. The energy for bond breaking is provided from the recombining carriers. The reduction of the $\mu\tau$-product and of the carrier diffusion length in consequence of light soaking is a serious constraint for solar cells. However, the electronic devices described here usually depend on the carrier drift under the action of an applied voltage. This situation ensures enough reserve of schubweg to prove the material resistant against the light influx usually applied to such devices.

3 Large-Area Devices

The development of devices fabricated from a-Si:H follows the pattern of the three steps illustrated in Fig. 5. Based on the properties of a-Si:H, a manifold of functional elements have been and are still being developed. These functional elements are then assembled to produce devices.

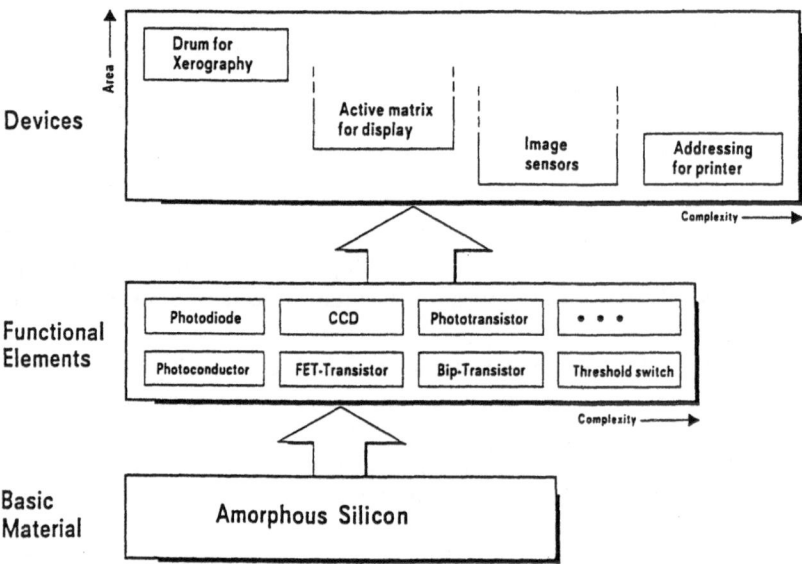

Fig. 5 Development steps of large-area electronics based on amorphous silicon.

3.1 Xerographic Drums

This application [22] of a-Si:H is based on its photoconduction and on its capability to coat drums of large size with uniform layers. Xerography is an important business and thus a very attractive application. A-Si:H offers higher abrasive strength and does not contain toxic components in comparison with the presently used chalcogenide photoconductors. The deposition on drums is effected in deposi-

286

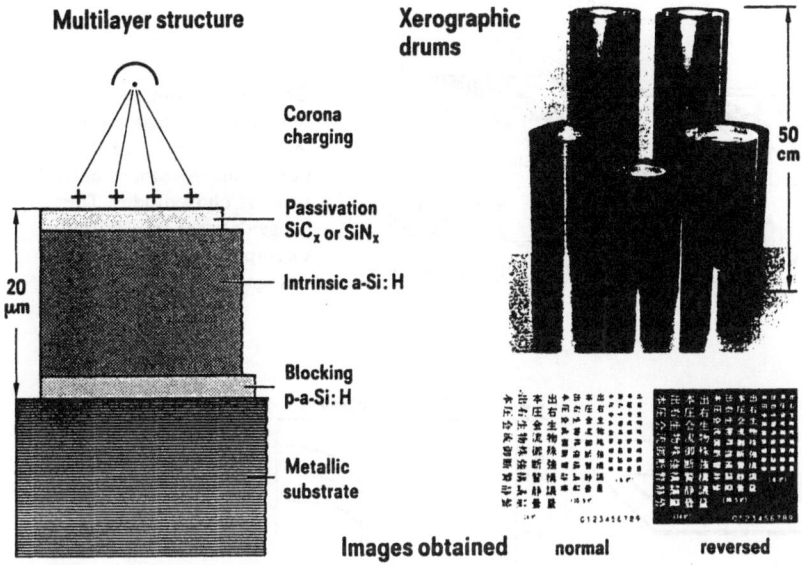

Multilayer structure

Corona charging

Passivation SiC$_x$ or SiN$_x$

Intrinsic a-Si: H

20 µm

Blocking p-a-Si: H

Metallic substrate

Xerographic drums

50 cm

Images obtained normal reversed

Fig. 6 A-Si:H photoconductor for xerographic drums. Photograph and images taken from [23].

tion chambers with cylindrical electrodes. Xerographic photoconductors are operated at several hundred volts, which demands a layer thickness of about 20 µm. With the object of economic fabrication, the deposition rate could be increased to 5 µm/h [13].

To ensure protection against corona charging, a multilayer structure for a-Si:H was developed. A thin layer of silicon carbide or silicon nitride is deposited on top of a-Si:H for protection, see Fig. 6. An additional bottom layer completes the devices, thus preventing charge injection from the metallic substrate and guaranteeing primary photocurrent. Such drums have already been marketed by Canon for one of its office copiers. Their high sensitivity at the wavelength of He-Ne laser light and the possibility of shifting the sensitivity maximum to the near IR of laser diodes by alloying with Ge opens up good opportunities for a-Si:H xerographic drums in future laser printers.

3.2 Image Sensors

Image sensors are composed of a linear (or even two-dimensional) array of photodiodes, which are scanned by an electronic circuit. Such devices [24, 25, 26] are used for reading A4 wide documents in office automation applications. The length of a linear sensor must equal the width of the document (21.6 cm) in order to scan the document in a 1:1 projection onto the sensor, see Fig. 7.

287

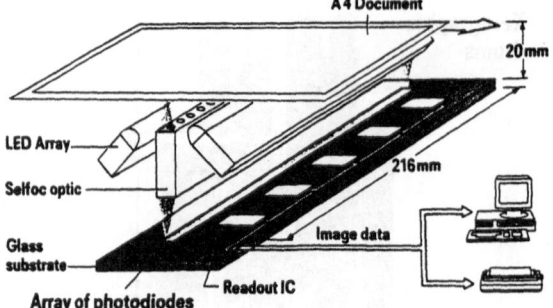

Fig. 7
Document scanning with a large-format image sensor. The selfoc optics gives a 1:1 projection of the document onto the array of photodiodes. The image data are transmitted to a computer for processing, storage, and printout.

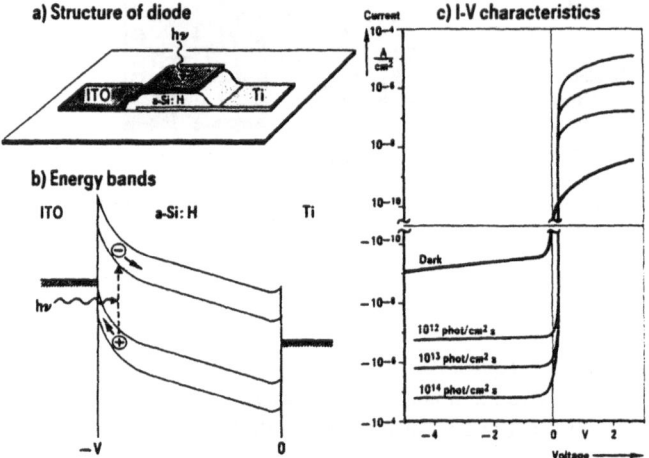

Fig. 8 A-Si:H thin film photodiode with a Schottky barrier. The voltage given in the characteristic is applied to the ITO electrode.

The basic functional elements of image sensors are thin-film photodiodes. In contrast to photoconductive elements, photodiodes operated with reversed bias offer a lower dark current, a faster response, and a photocurrent largely independent of the applied voltage.

The p-i-n-structure of the diodes mostly used for a-Si:H solar cells requires the three a-Si:H layers plus electrodes which is necessary for the high efficiency of this device, which has to generate current and voltage by itself. However, photodiodes of image sensors are powered from outside and hence may be built-up much more easily. Transparent contact electrodes were shown to act as a Schottky barrier when properly prepared on the surface of a-Si:H. The result is therefore the simple sandwich structure shown in Fig. 8a.

288

The sandwich type diode with a $1\mu m$ thick a-Si:H in between for complete light absorption is preferred because of its short drift path for the photogenerated carriers. The penetration depth of the commonly used LED light ($\lambda = 560\,nm$) is less than $0.2\mu m$ so that the sensor diode operates as a one-carrier (electrons in most cases) device. This suppresses the carrier recombination and hence the Staebler-Wronski-effect.

An effective Schottky barrier with a large upward bending of the semiconductor bands requires a contact material with a high work function. Semitransparent metal [27] and silicide layers [28] have been proposed, but ITO (indium-tin-oxide) is most commonly used and results in a barrier height of about 0.86 eV [29]. This large barrier for electrons ensures a very low dark current and a primary photocurrent when the ITO is biased negatively. A typical current voltage characteristic is shown in Fig. 8c. The metallic bottom contact must consist of a low work function material such as titanium or chromium. A bias voltage of about 2 volts is sufficient to achieve a continuous slope of the energy bands from the top electrode to the bottom electrode, see Fig. 8b. The very short rise and decay times of the photocurrent while switching the incident light on and off are shown in Fig. 9.

The assembly of an image sensor composed of these diodes is shown schematically in Fig. 10. For local resolution (typically 8 or 12 pixel/mm) only the bottom electrode is patterned; the lateral carrier drift between the single diodes is negligible. The successive readout of the individual photodiodes is effected by a crystalline silicon IC bonded onto the same substrate as the diodes. The data acquisition is performed mostly in charge collecting mode, i.e. the readout circuit registers the voltage drop of the diode capacitance at the end of the line scan time which is of the order of 1 ms [30]. Since a page-wide image sensor contains some 2000 diodes, the switching speed of the readout chips is about 2 MHz. A photograph of a fully assembled sensor array sufficient for A4 width which was fabricated in our labora-

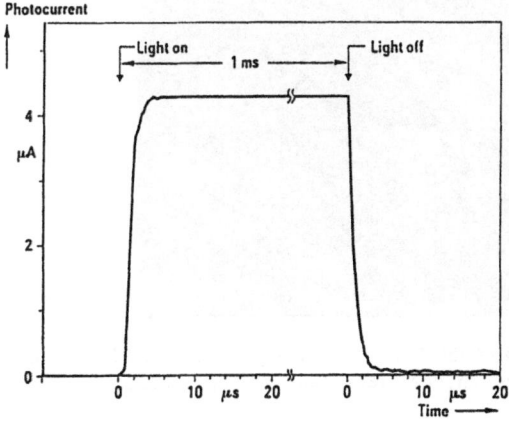

Fig. 9
Rise and decay times of a-Si:H photodiode with a Schottky barrier. Applied (reversed) voltage $-5\,V$, wavelength of light $515\,nm$.

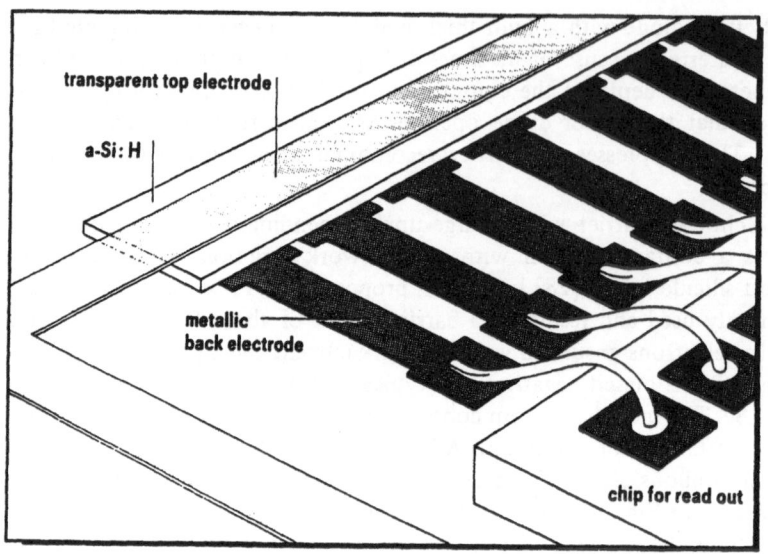

Fig. 10 Basic setup of an a-Si:H linear image sensor. Thickness of a-Si:H layer is $1\,\mu$m.

Fig. 11 Page-width a-Si:H image sensor assembled with 14 readout chips. The 22.8 cm long sensor array consists of 1792 photodiodes. The local resolution is 8 P/mm, the scanning rate per diode is 2 MHz.

290

tories is shown in Fig. 11. This sensor consists of 1792 single diodes corresponding to a local resolution of 8 pixel/mm. It should be noted that most of the device area is covered by the metallization, a typical phenomenon of large area devices. The active a-Si:H layer is merely a narrow (100μm) strip running along the upper part of the device.

This type of device combining thin film elements and monocrystalline chips wire-bonded together (hybrid sensor) represents an intermediate step towards a fully integrated thin film sensor array operated with a thin-film transistor readout circuit (see Chapt. 3.5).

3.3 Thin Film Transistors (TFT)

Field-effect transistors of a-Si:H [31] are built-up in a multilayer system as shown in Fig. 12a. The inverted staggered structure is employed because placing the dielectric on the bottom yields a lesser density of interface states than on the top [32]. The gate dielectric, mostly silicon nitride or silicon oxide, must be deposited as an extra layer because the amorphous silicon does not allow easy thermal oxidation like crystalline silicon in MOS technology. The source and drain contacts are prepared with a thin layer of n^+ doped a-Si:H for good ohmic contact. The a-Si:H

a) Layer structure

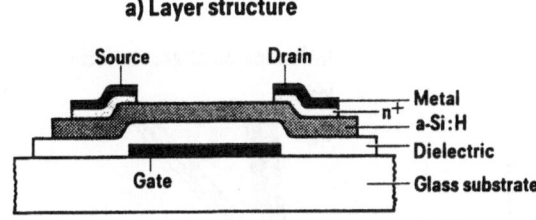

b) Transfer characteristic

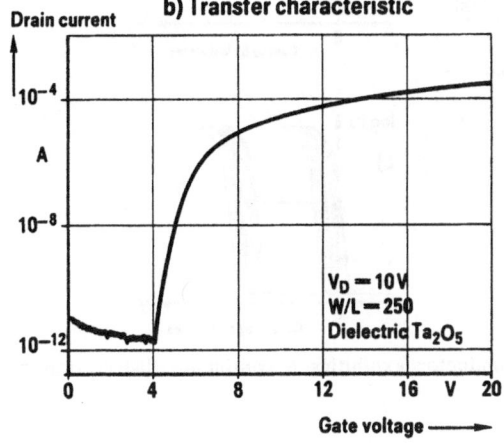

Fig. 12
a-Si:H thin-film transistor (TFT).

in the channel must be undoped because every doping step increases the density of states [23]. The transistor operates in the insolated gate accumulation mode. A transfer characteristic of an a-Si:H TFT is shown in Fig. 12b. The ratio of the ON to OFF current in this case (Ta_2O_5 as dielectric) is 10^8, the channel length was $12\,\mu m$. To understand the particulate features of an a-Si:H TFT, a Fermi shift must be considered through a continuous distribution of states as shown in Fig. 13a. The shift itself is restricted to those portions in the gap where the density of states is low, i.e. in the energy range between equilibrium Fermi level and the onset of the conduction-band tails.

The charge induced by positive gate potential initially occupies the localized states above the original Fermi level and most of it becomes trapped. Only a small proportion reaches the extended states of the conduction band to increase the conductivity. As a consequence, the steepness of the I_D-V_G characteristic is lowered by a great density of (fast rechargeable) states, see Fig. 13b. On the other hand, deep states (at the interface or within the dielectric) which are slowly charged and discharged, shift the curve to higher V_G values during the application of a positive gate voltage [35], see Fig. 13c. Recent improvements in the deposition process have reduced this shift of V_G to values lower than 0.7 V [36].

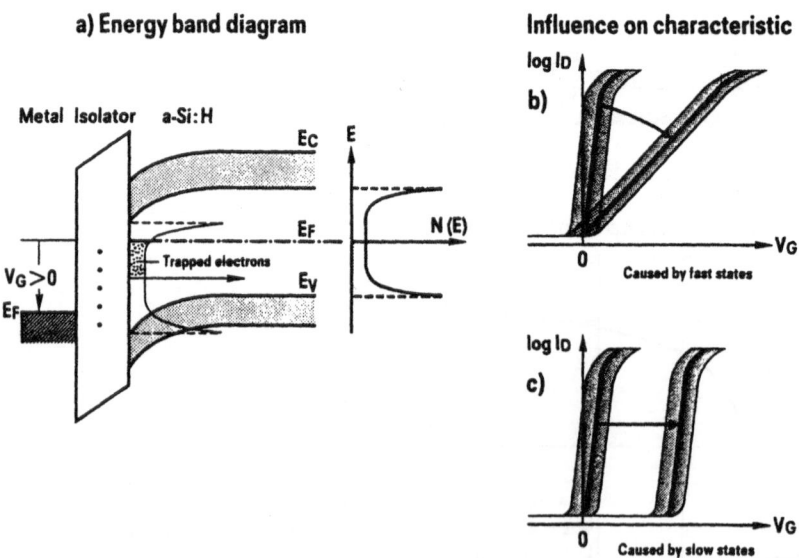

Fig. 13 Field effect in amorphous semiconductor exhibiting a continuous distribution of electron states within the gap [34].

The switching process of an a-Si:H TFT itself takes less than $1\mu s$. However, as a consequence of the restricted conductivity of a-Si:H, the sheet conductance of the channel in the ON state is limited to about $5 \cdot 10^{-8}\,\Omega^{-1}\,\square$ [15].

A practical width-to-length ratio for the channel of 100 leads to an ON resistance of $200\,k\Omega$. This means, when charging a capacitive load (say 20pF), that the operation frequency is limited (to about 50 kHz). The applications of a-Si:H TFT are therefore currently confined to medium speed devices.

3.4 TFT-Addressed LCD Panel

The low speed requirement of an LCD active matrix addressing circuit (20μs charging time [37]) coincides very well with the a-Si:H-TFT performance, and it has therefore become the first a-Si:H-TFT application. The TFT operates as an active element to switch each pixel in the display ON and OFF [38]. Fig. 14a shows the basic circuit composed of vertical and horizontal bus lines which are connected to source and gate of the individual switching transistors. A transistor powered simultaneously by both lines switches to the ON state and charges the pixel electrode of the liquid crystal cell. The OFF current of the a-Si:H TFT is low enough (see Fig. 12) to hold the pixel charge during the frame time of about 20ms. A cross section of a fully assembled active matrix display is shown in Fig. 14b.

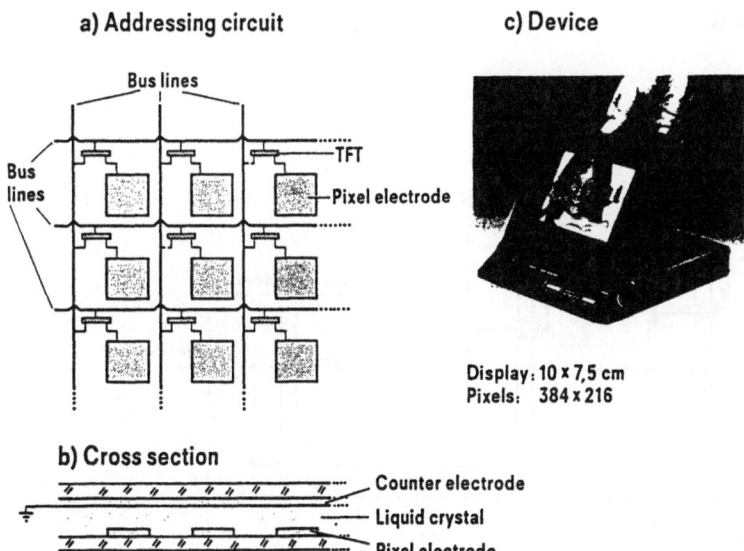

a) Addressing circuit   c) Device

Bus lines

Bus lines

TFT

Pixel electrode

Display: 10 x 7,5 cm
Pixels: 384 x 216

b) Cross section

Counter electrode

Liquid crystal

Pixel electrode

Fig. 14 a-Si:H TFTs as active elements in a LCD addressing circuit. The portable TV device was demonstrated by Sanyo Electric Co. [39]. Courtesy to Y. Kuwano who provided the photograph.

A commercially available product, a pocket TV receiver with an a-Si:H addressing circuit for the display, was recently demonstrated by Sanyo, [39] see Fig. 14c. The 5 inch TV display has full color capability. Employing a redundancy technology Matsushita succeeded in fabricating a 12.5 inch LCD containing 640 × 480 dots each addressed by two a-Si:H TFTs [40]. Since these displays operate under intense illumination it is essential that the transistors are not affected by light. Whereas the early transistors are mostly provided with light shields (at the expense of increased capacitance), van Berkel and Powell [41] recently proposed a tailoring of a-Si:H for TFT with sharply reduced photoconductivity. By selecting suitable deposition parameters for a-Si:H it seems indeed possible [39] to increase mainly the number of recombination centers, which reduce the photoconductivity without influencing the field-effect performance.

3.5 Image Sensor with TFT Readout

The switching speed required for an image sensor is of the order of MHz. This overall speed requirement may be met with the aid of TFTs by combining a number of diodes to sections which are switched simultaneously. The most common solution is a multiplex circuit shown schematically in Fig. 15. Each photodiode is connected to a TFT. The gates of a number of transistors are connected together, so that all the transistors of one section are turned on simultaneously, the charge being transferred from the diodes to the inputs of a single fast IC multiplexer. Each section is addressed sequentially. The requirement for the switching speed of the single TFT is thus lowered by a factor given roughly by the number of diodes (and TFTs) combined in one section.

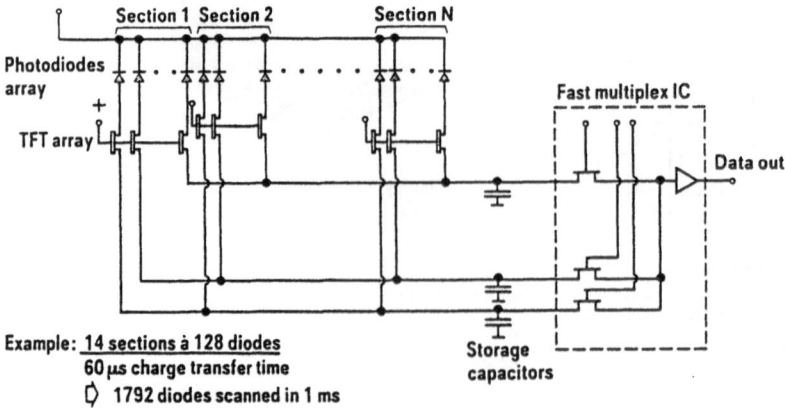

Fig. 15 Equivalent circuit of an integrated image sensor with a TFT readout system.

294

The calculated example given in the insert of Fig. 15 demonstrates the feasibility of this multiplexing system. Page-wide image sensors with integrated TFT readout (including one single IC) are under intensive development in several laboratories [42, 43, 44]. This is because significantly lower costs are expected than in the case of hybride image sensors, which need a great number of IC multiplexers. Additionally, the wire bonding associated with some yield risk is sharply reduced. The overall thin-film technology of such an integrated sensor ensures uniformity over the whole length of the large area device.

3.6 TFT-Driven Printer Heads

A recently proposed application for TFT is in the field of printing, another domain of input/output devices. Page-width printer heads need driving switches for each individual printing element spread locally over a long distance. This is an ideal situation for a-Si:H transistors. An output multiplexing circuit similar to the input-image sensor circuit is used, so that the transistor switching time or the charging time of the printer element does not have to be very short [5, 7].

The activation of a printer element usually requires a rather high voltage. H. C. Tuan from the Xerox Research Center succeeded in developing a high-voltage thin-film transistor which works at source drain voltages of up to 400 V [45]. This transistor is not based on the usual field effect but controls the carrier injection of the source contact. The carriers are then swept over to the drain by the high applied voltage. The gate is therefore situated just beneath the source electrode as illustrated in the basic set-up of Fig. 16a. Fig. 16b shows the output characteristic of this novel element.

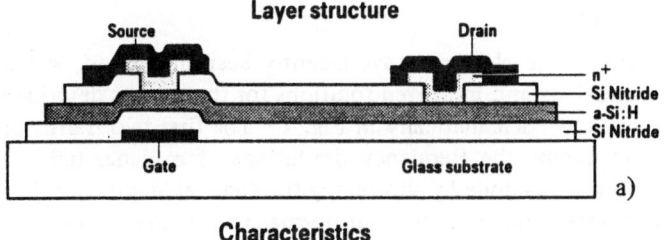

a)

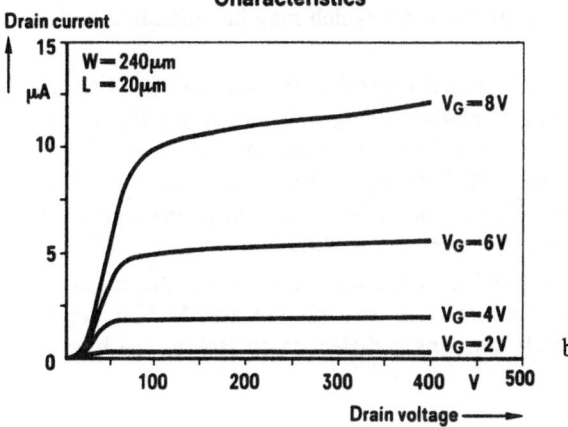

b)

Fig. 16
High voltage a-Si:H TFT, from [45].

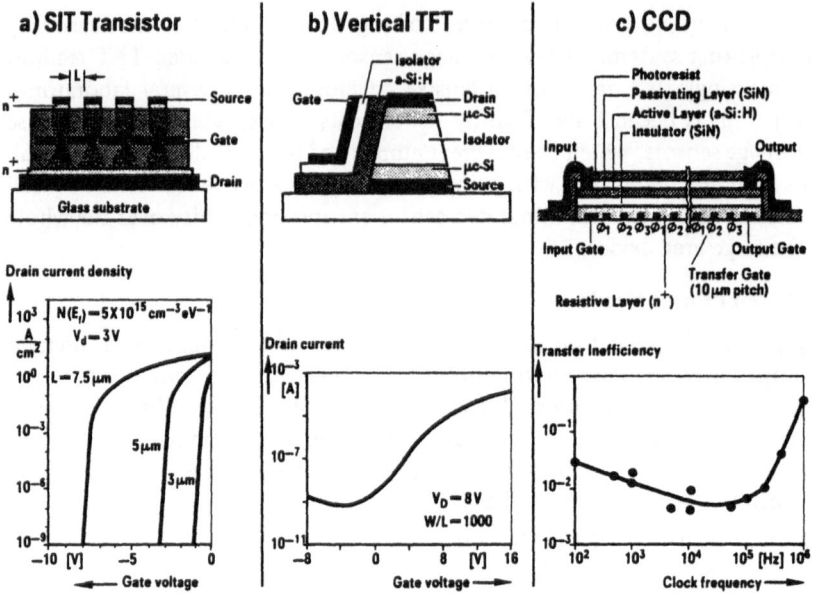

Fig. 17 Novel a-Si:H functional elements. a) from [46], b) from [47], c) from [48].

3.7 Novel Functional Elements

Another series of novel functional elements have recently been developed, which are intended for new applications and improved solutions for the devices described above. Three examples are shown schematically in Fig. 17. The first two transistor systems are intended to overcome the frequency limitations of a planar field-effect transistor. In both cases this is done by shortening the channel length and thus increasing the ON conductance. The channel is set normal to the direction of the substrate and is determined by a film thickness which may be fabricated with very thin values in a reproducible way.

In the first system, Fig. 17a, a static induction transistor was built-up with a-Si:H as the channel material and a grid of platinum fingers in between [46]. From the platinum interface, a depletion layer originates which at high reversal voltage expands from one finger to the next. This strangles the carrier flow such that the current flow (dotted lines in Fig. 17a) is controlled as shown in the characteristic below, see Fig. 17a. The calculated ON resistivity is $100 \, k\Omega$.

Fig. 17b illustrates a vertical field-effect transistor realized by Matsumura and coworkers [47]. The transistor channel runs through the a-Si:H film deposited (almost) vertically at a step which contains a dielectric to isolate the horizontal

296

source and drain contacts. The thickness of this dielectric layer determines the channel length. The relatively high ON current of $150\mu A$ (see transfer characteristic) demonstrates the feasibility of this promising type of transistor, which is specific to the thin-film technology of a-Si:H.

The third example is a thin-film CCD device, which was also realized by the Matsumura group [48], Fig. 17c. The low carrier drift mobility in a-Si:H requires a device structure which generates a uniform horizontal electric field between the transfer gate electrodes. This is accomplished by a resistive layer of n^+ doped a-Si:H inserted between the electrodes. With this structure, a transfer inefficiency of less than 1% was obtained at clock frequencies between 10 kHz and 200 kHz.

4 Process Technology for Large-Area Electronics

In technological terms, the novel field of large-area-electronics is defined by two characteristic dimensions: wafer size and minimum pattern size as shown in Fig. 18. A typical wafer (substrate) size today is 25 cm × 25 cm ($10''$ × $10''$), the minimum pattern size is not smaller than $10\mu m$. Hence this new field of electronics is situated just in the gap between printed circuit boards (thick-film technique) and microelectronics. The process technology is pure thin-film technology throughout.

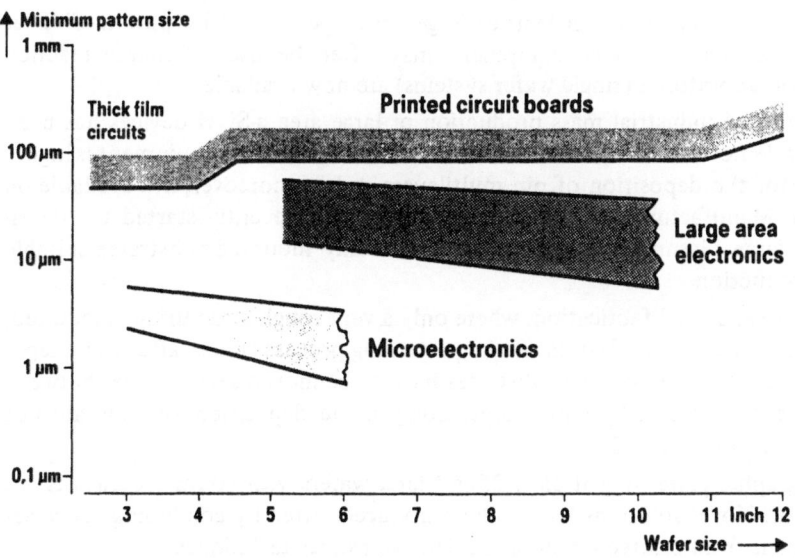

Fig. 18 Technological definition of large-area electronics.

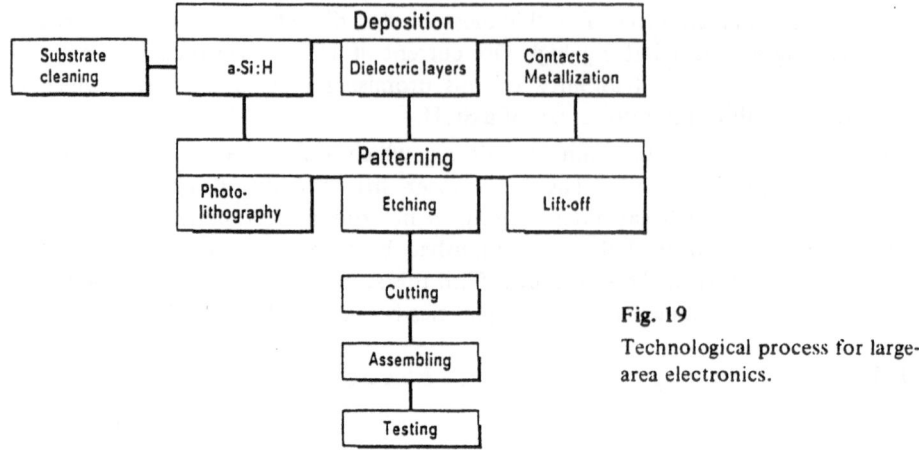

Fig. 19

Technological process for large-area electronics.

The principal steps for fabricating large-area electronic devices are summarized in Fig. 19. The deposition of a-Si:H is only one of numerous other steps. Corning 7059 glass is mostly used for substrates; it yields a low leakage current. The deposition of metallic layers on large sized substrates has been well established for some time. The deposition of the dielectrics is generally performed by plasma CVD as for a-Si:H, so that the same equipment may often be used. Suitable reaction chambers for laboratories (single wafer systems) are now available.

The feasibility of industrial mass production of large area a-Si:H devices has been successfully demonstrated by solar cell manufacturers. But their equipment is highly specialized for the deposition of pin multilayers and is, moreover, not available on the market. Manufacturers of vacuum equipment only recently started the development of large plasma CVD systems with vertically mounted substrates suitable for mass production.

In contrast to solar cell fabrication, where only a very rough structuring is required, most large-area electronic devices need a patterning process almost after each deposition step. In consequence, the substrates have to be moved several times between deposition and photolithographic steps. Long in-line deposition systems are not therefore generally required.

Photolithographic patterning of $25 \times 25\,cm^2$ large wafers with $10\,\mu m$ accuracy is not a simple task. Novel solutions have to be introduced, often by combining processes and equipment from printed circuit and microelectronics techniques.

Spinning systems are preferred for photoresist coating as well as for resist development and for wet chemical etching processes. These systems are suited for clean operation, reducing device defects due to particles [5].

298

However, some arrangements have to be applied to the wafer holder to compensate the mostly square format of the wafers with respect to the rotating motion, otherwise significant non-uniformities arise at the corners of the squares. Huge photomasks with sizes of 30cm × 30cm (12 inch × 12 inch) are needed which have to be written directly by a pattern generator. The alignment of one layer to an underlaying layer is the step which greatly limits positional accuracy. Bending and insufficient flatness of substrate and mask reduce the alignment accuracy. With a typical mask aligner for 25cm substrates and sufficient skill, ± 5μm overlap accuracy is achieved over the whole area.

Testing of the devices presents a challenge due to the large number of single elements. In the case of image sensors or display panels, not even a single element may malfunction. A single element test of 10^3 to 10^4 diodes or transistors by means of a wafer prober is extremely time consuming and costly. In general, it is preferred to test the final performance of the device and repair it, if necessary, at built-in redundancies [40].

5 Specific Issues Related to Large-Area Electronics

5.1 Yield of Device Fabrication

In the first instance, the fabrication of devices covering an area more than 100 times larger than normal silicon chips raises the question of yield. A reliable fabrication of these large area devices can only be attained by reducing two other technological requirements: a larger minimum pattern size (10μm) and a smaller number of process steps. In each case, the reduction is almost one order of magnitude. Only 4 to 6 layers have to be processed for a large-area device.

Losses in fabrication yield must be differentiated according to its origin in two categories: systematic failures and random failures. The group of systematic failures comprises mask defects and non-optimized parameters in process operation. As most of the processes in large-area device fabrication are novel, this category of failures usually prevails. It takes up some time and experience to investigate the details of these processes and eliminate systematic failures. For the remaining device failures which occur randomly a Poisson function is assumed for the local distribution of defects which generate failures. To a first approximation (supposing constant defect density across the wafer), the yield Y is calculated from the defect density D and the area A (which is susceptible to defects) with the equation [49]

$$Y = e^{-DA}. \tag{1}$$

To give an example, we examine the case of an integrated image sensor with TFT readout designed according to the circuit shown in Fig. 15. As emphasized earlier, the greatest part of large-area devices is covered by metallization: data and clock lines including a great number of crossovers. As the linewidth is of the order of 20 to 50μm, line opens do not usually belong to the category of random failures. However, shorts at the crossovers represent the principal source of random defects.

A sensor with 1792 photodiodes (8P/mm) and 1792 TFTs subdivided into 14 sections contains more than 100,000 crossovers, mainly in the data line system. In a rough estimate, we simply add together the areas of the crossovers of the data and clock lines ($625\,\mu m^2$ each), the channel area of the TFTs ($100\,\mu m^2$ each) and the area of the photodiodes ($12\,100\,\mu m^2$ each). The total multilayer area of the sensor susceptible to shorts is then $0.928\,cm^2$ for a subdivision into 14 section and $0.647\,cm^2$ for 28 sections. The yield (i.e. probability for a device with no defects) calculated according to Eq. 1 and shown in Fig. 20 demonstrates the rigorous requirements for low defect density. A yield of 50%, which is the minimum for economic manufacturing, requires a defect density smaller than $1\,cm^{-2}$. This low value (and even lower values in the case of LCD addressing circuit [50]) are characteristic for large-area devices. To meet this requirement is one of the fundamental challenges of large-area electronics. The greatest help is provided by simplicity of the structure, e.g. the crossovers of the metallization which contribute by far the greatest part to the area susceptible to defects may be isolated by a dielectric layer thick enough to guarantee the desired low defect density.

5.2 Influence of Capacitances

Many implications have to be considered which arise from the capacitances of the metallization lines and the crossovers. In consequence of the large area of these devices, extremely long lines are inevitable, and these introduce significantly large capacitances into the readout or addressing circuits. The mostly negative influences of capacitance on the performance of the device are apparent in three major fields: response speed, signal amplitude, and crosstalk.

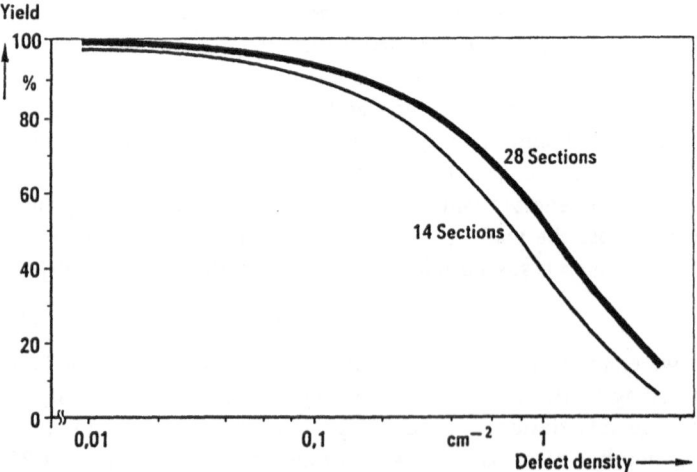

Fig. 20 Calculated yield for an integrated image sensor. Details see text.

With respect to the response speed, the line capacitance is particularly important when TFTs are the driving elements because of their large ON resistance. A required switching speed of, say, $4\mu s$ restricts the maximum load capacitance to $20\,pF$ given a typical ON resistance of $200k\Omega$ (see chapter 3.3). This confines the application of TFT to switching an adjacent individual element, whereas peripheral driving transistors for sections or columns as a whole require lower ON resistances.

The signals of image sensors are usually acquired in charge collecting mode i.e. the originally charged photodiode is discharged by illumination during the line scan time, at the end of which the diode potential is detected by any readout preamplifier at the device output. The signal thus transformed from a (photogenerated) charge to a voltage drop is proportional to the reciprocal value of all capacitances connected to the transmission line. Hence the signal amplitude and its uniformity from diode to diode is directly affected by the line capacitances. A way of eliminating this direct influence is to feed the photocurrent itself continuously to integration stages at the device output [51]. This method transfers the uniformity requirement to the integration stages.

The signal amplitude of a large-area sensor has to compete with the switching noise generated by the transistors connected to the data line. The switching noise amplitude increases with the speed capability of the TFTs due to the enlarged channel width increasing the gate-channel capacitance.

These brief considerations demonstrate the rigorous requirement for a careful system design in order to obtain a workable signal-to-noise ratio. The design should be supported by a computer simulation adapted to the particularities of a thin-film device. Suitable analytical systems are just beginning to be developed [52, 53]. The crosstalk between neighboring lines is a consequence of the coupling capacitance. The charge exchange between adjacent lines depends on whether the potentials of these are equal or not. This generates crosstalk of a sensor array whose lines run in parallel for long distances. The result of a computer simulation of this effect is illustrated in Fig. 21. The amplitudes of the voltage signals (charge collection mode) of the individual diodes decreases when the neighboring diodes have different illumination states and hence different potentials. This represents a simple example of the rather complex series of computer simulations which are of vital importance in designing a large-area device.

5.3 Uniformity

Sensors or displays transmitting optical information have to meet the high uniformity standard set by human visual perception. The large-area devices described here passes excellent preconditions for uniformity as a consequence of their thin-film technology. In this, technological processes are applied to the whole substrate area in a homogeneous manner. Fig. 22 shows as an example results taken from our sensor development. The photocurrent and the dark current of almost 1000 individually tested photodiodes exhibit a deviation from the mean value which is only a few percent. Similar results on the uniformity of TFTs have been obtained by others [7].

301

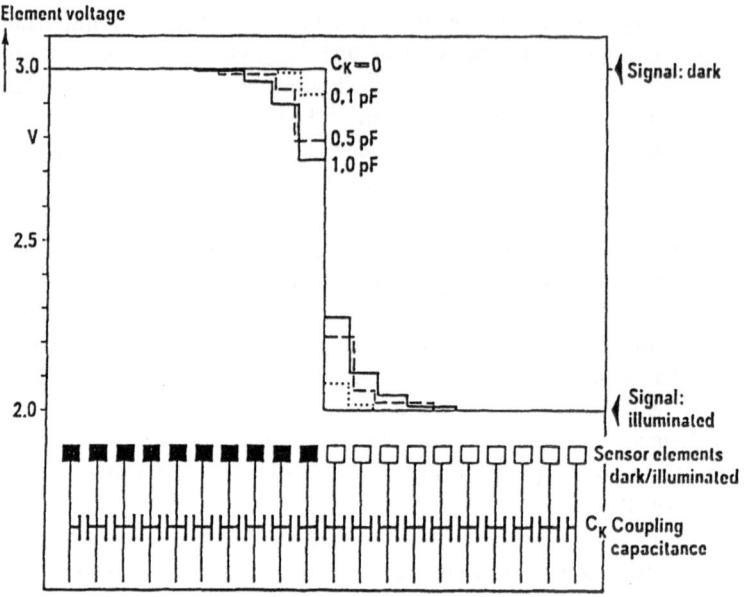

Fig. 21 Simulation of output signal for an array of image-sensor elements. The equivalent capacitances between adjacent data lines cause smearing of the signal over the boundary between dark and illuminated elements.

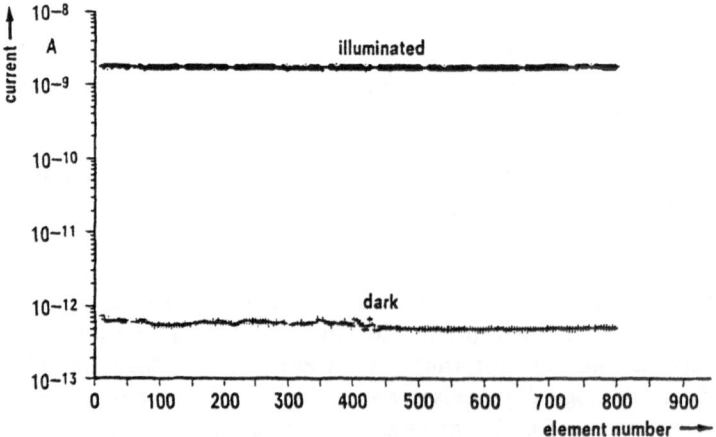

Fig. 22 Uniformity of a-Si:H image-sensor elements.

An essential task in designing large-area devices is to preserve this excellent uniformity of the a-Si:H elements from being destroyed by the geometrical variations of the data lines and/or by inhomogeneities in the properties of crystalline Si-circuits used as drivers, multiplexers, or preamplifiers of a large-area device.

6 Conclusions

Beginning with the demand for large-area input/output devices, we concluded that thin film technology alone appears to be suitable for this field of electronics. After several efforts to develop this technology on the basis of recrystallized silicon and polycristalline silicon, the recent introduction of amorphous hydrogenated silicon has strongly stimulated the sector of large-area electronics.

The principal advantages of a-Si:H are low-cost fabrication due to a low temperature process and its intrinsically perfect uniformity. Its semiconductor properties are excellent for photosensors and sufficient to cover medium speed applications driven by TFTs. Limitations in high speed switching arise from the restricted ON current of today TFTs. Since the speed of the charge carriers in a-Si:H is fast enough for a MHz response, we have to wait for specifically developed a-Si:H device structures. The next field which demands intensive development is the technology for processing large-area wafers. Existing techniques have to progress from the exploratory state of laboratory operation to reliable industrial manufacturing.

Finally, the design of optimized systems and circuits for large-area devices must be pushed forward. A careful balance between several competing measures for signal-to-noise ratio, speed, and yield is necessary to achieve the desired performance.

Considering the extreme ambitious detailed requirements for a large-area device on the one hand and the limited properties of the materials on the other hand, it would seem good policy to confine the overall specifications of these devices to the low end of the performance spectrum in the first step. This will then provide a firm basis from which to proceed to the high end in a second step.

Acknowledgements

The author is very grateful to his colleagues Dr. G. Brunst, N. Brutscher, Dr. S. Griep, Dr. M. Hoheisel, K. Rosan, and H. Wieczorek who provided experimental results prior to publication and contributed to this paper by fruitful discussions.

References

[1] *D. E. Carlson*, in: Semiconductor and Semimetals, Vol. 21, Part D, ed. by *J. I. Pankove*, 1984, p. 7

[2] *W. G. Hawkins, D. J. Drake, N. B. Goodman*, and *P. J. Hartmann*, Mat. Res. Soc. Symp. Proc. Vol. 33, 231 (1984)

[3] *W. G. Hawkins*, Mat. Res. Soc. Symp. Vol. 49, 443 (1985)

[4] Hydrogenated Amorphous Silicon, Part A–D, ed. by *J. I. Pankove*, in: Semiconductors and Semimetals, Vol. 21, ed. by *R. K. Willardson* and *A. C. Beer* (Academic Press, New York, London 1984)

[5] H. C. Tuan, Mat. Res. Soc. Symp. Proc. Vol. 33, 247 (1984)

[6] M. J. Thompson, J. Vac. Sci. Technol. B 2, 827 (1984)

[7] M. J. Thompson, Mat. Res. Soc. Symp. Vol. Proc. 70, 613 (1986)

[8] D. Adler, in [4] Part A, p. 291

[9] J. C. Phillips, J. Non-Crystalline Solids 34, 153 (1979)

[10] W. E. Spear and P. G. LeComber, J. Non-Crystalline Solids 8–10, 727 (1972)

[11] M. Vanecek, A. Abraham, O. Stika, J. Stuchlik, and J. Kocka, phys. stat. sol. (a) 83, 617 (1984)

[12] M. Hirose in [4], Part A, p. 9

[13] Y. Nakayama, T. Natsuhara, N. Nagasawa, and T. Kawamura, Jpn. J. Appl. Phys. 21, L 604 (1982)

[14] W. E. Spear, and P. G. LeComber, Solid State Comm. 17, 1193 (1975)

[15] K. D. Mackenzie, A. J. Snell, I. French, P. G. LeComber, and W. E. Spear, Appl. Phys. A 31, 87 (1983)

[16] G. Myburg, and R. Swanepoel, J. Non-Crystalline Solids 89, 13 (1987)

[17] J. Tauc, in: Amorphous and Liquid Semiconductors, ed. by J. Tauc (Plenum Press, 1974), p. 175

[18] W. E. Spear, H. L. Steemers, and H. Mannsperger, Phil. Mag. B 48, L 49 (1983)

[19] J. M. Marshall, R. A. Street, and M. J. Thompson, Phys. Rev. B 29, 2331 (1984)

[20] R. A. Street, J. Zesch, and M. J. Thompson, Appl. Phys. Lett. 43, 672 (1983)

[21] E. S. Sabisky, J. Non-Crystalline Solids 87, 43 (1986)

[22] J. Mort, J. Vac. Sci. Technol. B 2, 823 (1984)

[23] Y. Nakayama, A. Sugimura, M. Nakano, and T. Kawamura, Photogr. Sci. Eng. 26, 188 (1982)

[24] T. Hamano, H. Ito, T. Nakamura, T. Ozawa, M. Fuse, and M. Takenouchi, Jpn. J. Appl. Phys. 21, Suppl. 21–1, 245 (1982)

[25] S. Kaneko, Y. Kajiwara, F. Okumura, and T. Ohkubo, Mat. Res. Soc. Symp. Proc. Vol. 49, 423 (1985)

[26] K. Kempter, Amorphous Semiconductors for Microelectronics, ed. by D. Adler, Proc. SPIE 617, 120 (1986)

[27] M. Hoheisel, H. Wieczorek, and K. Kempter, J. Non-Crystalline Solids 77 & 78, 1413 (1985)

[28] T. Tsukada, K. Seki, Y. Tanaka, H. Yamamoto, and A. Sasano, J. Non-Crystalline Solids 77 & 78, 1027 (1985)

[29] N. Brutscher, to be published

[30] E. Holzenkämpfer, K. Rosan, R. Primig, and K. Kempter, Poly-micro-crystalline and amorphous semiconductor Proc. MRS Meeting 1984, Strasbourg, p. 575

[31] P. G. LeComber, and W. E. Spear in [4] Part D, p. 89

[32] R. A. Street, and M. J. Thompson, Appl. Phys. Lett. 45, 769 (1984)

[33] R. A. Street, J. Non-Crystalline Solids 77 & 78, 1 (1985)

[34] O. Sugiura, and M. Matsumura, J. Electron. Commun. (Denshi-Tushin Gakkai shi) J 65–C, 914 (1982) (in japanese)

[35] R. A. Street, and C. C. Tsai, Appl. Phys. Lett. 48, 1672 (1986)

[36] Y. Kaneko, A. Sasano, T. Tsukada, R. Oritsuki, and K. Suzuki, Extd. Abstracts 18th' Int. Conference Sol. St. Devices and Material, Tokyo, 1986, p. 699

[37] Z. Yaniv, V. Canella, Y. Baron, A. Lien, and J. McGill, Mat. Res. Soc. Symp. Proc. Vol. 70, 625 (1986)

[38] D. G. Ast in [4], Part D, p. 115

[39] M. Yamano, and H. Takesada, J. Non-Crystalline Solids 77 & 78, 1383 (1985)

[40] M. Takeda, S. Ogo, T. Tamura, H. Kamiura, H. Noda, T. Kawaguchi, I. Yamashita, D. Ando, and H. Kuroda, Proc. 6th Int. Display Res. Conference (Japan Display '86), Tokyo 1986, p. 204

[41] C. van Berkel, and M. J. Powell, J. Non-Crystalline Solids 77 & 78, 1393 (1985)

[42] A. J. Snell, A. Doghmane, P. G. LeComber, and W. E. Spear, Appl. Phys. A 34, 175 (1984)

[43] F. Okumura, and S. Kaneko, Mat. Res. Soc. Symp. Proc. Vol. 33, 275 (1984)

[44] H. Ito, Y. Nishihara, M. Nobue, M. Fuse, T. Nakamura, T. Ozawa, S. Tomiyama, R. Weisfield, H. Tuan, and M. Thompson, Int. Electron. Device Meeting, Washington 1985 (Techn. Digest), p. 436

[45] H. C. Tuan, Mat. Res., Soc. Symp. Proc. Vol. 70, 651 (1986)

[46] T. Tsukude, S. Akamatsu, M. Hirose, M. Ueda, and Y. Osaka, J. Non-Crystalline Solids 77 & 78, 1389 (1985)

[47] Y. Uchida, Y. Watanabe, M. Takabatake, and M. Matsumura, Jpn. J. Appl. Phys. 25, L 798 (1986)

[48] S. Kishida, Y. Naruke, Y. Uchida, and M. Matsumura, J. Non-Crystalline Solids 59 & 60, 1281 (1983)

[49] H. Murrmann, and D. Kranzer, Siemens Forsch. und Entwickl. Ber. 9, 38 (1980)

[50] A. I. Lakatos, Int. Electron. Device Meeting, Washington 1985 (Techn. Digest), p. 428

[51] K. Rosan, and G. Brunst, Mat. Res. Soc. Symp. Proc. Vol. 70, 683 (1986)

[52] T. Leroux, Solid-State Electronics 29, 47 (1986)

[53] M. Shur, C. Hyun, and M. Hack, J. Appl. Phys. 59, 2488 (1986)

Contents of volumes I ... 27

Author index

308

310

313

314

315

316

320

322